Roswell in the 21st Century

The Evidence as it Exists Today

Books by Kevin D. Randle

WINGS OVER NAM

Chopper Pilot
The Wild Weasels
Linebacker
Carrier War
Bird Dog
Eagle Eve

SEALS

Ambush!
Blackbird
Rescue!
Target!
Breakout!
Desert Raid
Recon
Infiltrate!
Assault
Sniper
Attack!
Stronghold
Crisis
Treasure

Coming Soon!

Case MJ-12

Roswell in the 21st Century

The Evidence as it Exists Today

Kevin D. Randle
Lt. Col. USAR (Ret)

SPEAKING VOLUMES, LLC
NAPLES, FLORIDA
2016

Roswell in the 21st Century

Alien symbols by Spyros Melaris

ISBN 978-1-62815-513-6

Introduction

It has been said that the Roswell UFO crash case has been overexposed by all the books, magazine articles, news reports and documentaries created about it. It has been said that it is a confusing mass of contradictory information written by people with an agenda in mind that is not always related to the truth of the case. It has been said that there is simply too much information and all the work done, rather than clarify, has merely obscured. It has been said that we know no more about it today than we did in 1947 when a crash was first reported by a Public Affairs officer whose story was not as comprehensive as it could have been.

On the one hand it is said that the Roswell crash is the best of the UFO cases because it provides everything that science could want for a paradigm shifting event. There is eyewitness testimony from dozens of participants, from those who saw the object in the air, to those who saw wreckage on the ground, to those who assisted in the recovery and those who saw the bodies of alien creatures proving that this was no terrestrial craft. There is documentation that suggests a cover up that does not lead to a secret balloon project but to something much more important. There is evidence of the government hiding the truth, successfully, for more than 69 years…

And when the Air Force weather balloon cover-up began to crumble in the 1990s, they initiated another, more public investigation that came up with the same basic weather balloon answer. They just dressed it in the new clothes of something called Project Mogul, gave it a proper classification of top secret and waited. This was enough to convince the press, which would have asked hundreds of questions had the subject not been UFOs to dismiss the evidence. Reporters are too sophisticated to believe that there is alien visitation; therefore, there is no alien visitation and therefore the Air Force must be believed on this because there can't possibly be an alien spacecraft crash.

On the other hand, it does seem that much of that testimony has been discredited. Some of the witnesses simply lied about their involvement, others exaggerated what they had seen and some weren't even in Roswell in 1947.

What looked like a solid and intriguing case that would prove there had been alien visitation eroded to a point where there was little of interest to intelligent, rational people and certainly nothing persuasive about this case.

Given what some believe is this sorry state of the Roswell investigation, why begin another?

It is something that I had been thinking about for a very long time. That someone needed to shift through all the information, the eyewitness reports, the documentation and the various solutions in an attempt to put it all in order that might lead to a proper conclusion. Someone needed to make a dispassionate investigation. There needed to be an ultimate Roswell book.

Tom Carey had been thinking about the ultimate Roswell book for a long time as well. He believed that much of the best information had been overlooked or obscured by various points of view and personal agendas. While we both were in Roswell in 2011,[1] he approached me with his idea about the book. He wondered if I was interested in such a project. I said that I was.

We made some tentative plans but Tom said that he needed to finish a book with Don Schmitt about Wright-Patterson Air Force Base. At this point it wasn't much of a hurdle since we had nothing on paper and we had only just started the process with this somewhat brief conversation. The important points here were that we were going to work on a book about Roswell and it would be just the two of us.

Sometime later, after we all had left Roswell, Tom emailed me that he had told Don about the project because he felt some loyalty toward Don and Don had lots of media contacts. Since he had told Don about the book, there was nothing much that I could do at that point.

In July 2012 we were all, that is, Tom, Don and I, again in Roswell for the festival, and we met late at night in the hotel's breakfast area to talk about the

---

[1] This was during one of the regular Roswell Festivals which have become an annual affair since the late 1990s. Rather than being a research symposium about the case, it has evolved into a series of lectures on a wide variety of UFO related topics with excursions into the paranormal and science fiction.

book project. I mentioned that for this to work that we would need some additional information, some new witnesses, something that would set this book apart from the others. Tom and Don nodded their agreement but said nothing else about that.

There was one other thing that I mentioned then and it was the nun's diary because that was becoming an important point.[2] The story had been told to Don and me by an ex-Special Forces (Green Beret) officer and a woman who might have been a nun or who was still a nun and who verified the existence of the diary and what was written in it.[3] Because these were Catholic nuns and I didn't know the customs and courtesies of the Catholic Church, this was an avenue for Don to explore which he did. I mentioned at the book meeting that this was something we should corroborate because it was a document created at the time that would provide some very important evidence. No one was going to dispute the provenance because it was held by the church and a photograph of the page would provide the evidence of the diary and the information and the information contained in it.

I was confident I had it but I now wanted to be sure. Then I learned that the Special Forces officer had been a PFC when in the Army, wasn't a Green Beret, wasn't an officer and hadn't been in Vietnam which shot his credibility all to hell.[4] And I learned that the diaries had been traced to Oklahoma City and while I had believed they had been seen and verified, such was not the case.[5] Once I learned where they were and arrangements could be made, I'd

---

[2] The real point of contention for some was the way it had been documented back in 1991. This was something that could be resolved in the future.

[3] The information was shared with Don Schmitt and me after a presentation in Alamogordo, New Mexico. The sources seemed to be reliable.

[4] This is based on the documentation recovered from the Army Records Center in St. Louis, MO, which houses the records of everyone who had served in the Army. A fire in 1973 destroyed some of those records, but those needed for this investigation had survived the fire and were available.

[5] After we learned about the diaries, kept by Catholic nuns, Don Schmitt investigated this. He apparently traced them to Oklahoma City. Don told me about digging through

go look at them myself. I pushed and learned the diaries had been moved to Wisconsin, which was good, but that they only went back to 1960 which was bad. I asked questions, was told many things, but in the end, I don't know where the diaries are, I don't know what they say, and I don't know if they even exist. All that can be said is that we had the names of two nuns who had been in Roswell in 1947.

There were other things that I didn't believe important at the beginning, that is 1989, simply because I had no reason to doubt the reliability of some of the research being conducted. For example as we wrapped up the final draft of the first book, Don added a footnote about a Lieutenant Colonel Albert Lovejoy Duran. It suggested that Duran had been interacted with Frank Kaufmann during the recovery. We later determined that Kaufmann's testimony was unreliable.[6]

Karl Pflock attempted to chase down Duran but was unable to do so. He wrote, about Duran, "... as Randle told me in 1999, both of these men [Jay West, a stringer for the UPI in 1960[7] and Duran] and anything even hinting at confirmation of their alleged accounts remain more than a little elusive."[8]

Others would suggest that Duran didn't exist, but according to Don, Carrie Wallace of Alamogordo introduced us to Juanita "Theresa" Valenzuela who is

---

dusty boxes of old church records to find the names of the nuns. I believed that this meant he had seen the diaries. He later told Tom Carey that I had investigated this, but Schmitt was the Catholic and understood the customs of the church. I do not. In 2012, he said that he had not seen the diaries.

[6] It was Mark Rodeghier, Mark Chesney and Don Schmitt who found the documentation that Kaufmann was unreliable. See Randle, Kevin D. and Mark Rodeghier. "Frank Kaufmann Reconsidered." *International UFO Reporter* 27,3 (Fall 2002, published January 2003): 8, 17 – 19; Rodeghier, Mark. "Frank Kaufmann Exposed." *International UFO Reporter* 27,3 (Fall 2002, published January 2003): 9 – 11, 26.

[7] I do have in my possession an interview with Jay West taped in St. Petersburg, Florida.

[8] Pflock, Karl. *Roswell: Inconvenient Facts and the Will to Believe*. Amherst, NY: Prometheus Books, 2001: 181.

the daughter of Duran. She said that her father was something of an alcoholic and when drunk he would sometimes talk of unit from White Sands that was dispatched out to the desert north of Roswell. He would sometimes mention bodies found at one of those sites.[9] So Lieutenant Colonel Duran does exist. He is mentioned in an official document that I was able to locate after Karl's book was published and one that Karl didn't access. That doesn't prove that what he told his daughter was accurate, or the assumption that he had been part of the team working with Kaufmann was correct. At the time, there were plans to talk with Duran but the circumstances complicated the issue.

There is a somewhat similar problem that developed with another witness. We had published the fact the First Lieutenant Jesse Johnson,[10] who was in Roswell in 1947, had been a pathologist. Don said that he had consulted *The ABMS Compendium of Medical Specialists* to confirm the fact. It was true that Johnson had trained as a pathologist, but according to the book, he had trained at the University of Texas Medical Branch in Galveston from 1948 to 1949. Don had apparently missed the dates of his training. He wouldn't have been a pathologist in 1947, though he certainly could have had an interest in it in 1947.[11]

But the partnership that had been suggested in 2011 broke down under the pressure from the Roswell Slides.[12] While we discussed what the book would

---

[9] This is obviously second-hand testimony from a source that is described as "alcoholic." While it now seems to be unreliable, at the time it seemed to provide corroboration for another source. We traced Duran from White Sands to Utah and eventually to Colorado but he was suffering from Cirrhosis of the Liver which complicated the attempted interviews which was why he was only mentioned in a footnote.

[10] His photograph appears in the Yearbook, which, of course, proves that he was in assigned to the Roswell Army Air Field in the right time frame.

[11] Randle, Kevin D. *The Roswell Encyclopedia*. New York: Quill (A HarperResource Book), 2000: 202 – 203.

[12] Both Carey and Schmitt have said that the name "Roswell Slides" was invented by the skeptics. They would claim that they had not tied them to Roswell, but in various interviews, they made exactly that claim. In August 2015, on Paranormal Podcast 393 hosted by Jim Harold, Carey said, "To me it was one of the Roswell aliens."

need to push it beyond all the other books and information out there, neither Tom nor Don mentioned that they might have just the sort of thing we needed. They said nothing to me and while they had signed a nondisclosure agreement, they could have suggested that they had something. To make it worse, they had shown an inkjet scan of part of one of the slides to David Rudiak at that same Roswell Festival.[13]

That didn't mean that the research had been completed. I thought of this as a cold case and was reviewing my notes, interviews including the audio and video tapes and documents assembled over more than two decades. At the same time I was conducting new interviews with soldiers who had been assigned to Roswell in 1947.[14] Most of the telephone numbers came from information supplied by Don. None of these interviews provided any new or exciting information.

As I worked through all that material I was beginning to see something that was disturbing. We had lost some of those we thought of as first-hand witnesses and who had provided some of the best of the eyewitness testimony.[15] They turned out to having invented their tales but as they disappeared, others surfaced. Unfortunately we were dealing with an event that more than 65 years old, and someone who was thirty in 1947 was now 95 if they had survived. Interviewing a man who was that old is a difficult task and you have to wonder how reliable their memories might be, especially after more than two decades of books, articles, documentaries, rumors, half-truths and outright

---

[13] David Rudiak in an email to me mentioned that he had seen the scan in 2012 while in Roswell for their annual festival.

[14] As a single example on January 18, 2012, I interviewed Stanley Muelling who was 87 at the time. He was there in July 1947, but said, "That was all highly classified at the time." Asked if he had seen anything, he said, "If I did I already forgot about it. I'm supposed to forget."

[15] These included Frank Kaufmann, Glenn Dennis and Gerald Anderson. We were finding that many others were expanding their roles for a variety of reasons.

lies.[16] Now we were left with the family of those who had been there and you have to wonder how much was shared and how much of those tales were accurately remembered. In other words, we had reached the point of diminishing returns.

There were some interesting discoveries that had gone almost unnoticed and supplied some interesting data. Lydia Sleppy had been interviewed in 1974 when her son mentioned a UFO sighting.[17] In two paragraphs buried in a larger article, she mentioned that her transmission of a UFO story over a news wire had been interrupted by an incoming message ordering her to cease that transmission.

Frank Edwards did mention the Roswell case in a book published in 1966.[18] He got almost all the details wrong, but he did report, accurately, that something had fallen. His mention didn't provide anything other than the location and there wasn't much of a way to follow up on his claims. All it showed is that the story was out there, somewhere.

So, the new investigation, which now was in my hands while Tom and Don chased the Roswell Slides, was progressing which lead, indirectly to a severing of our participation in the book project. This then is the result of the research that I conducted and I fear this book will annoy just about everyone. For those who wish to argue endlessly, there is something for them, regardless of the side of the debate they find themselves on. For those who wish some sort of definitive conclusion, you won't find it here. For those who believe that Mogul was responsible, you'll learn that all is not as you have been led to believe.

This, then, is Roswell boiled down to its essence in the 21 century.

---

[16] Muelling, who was 87, seemed to be quite alert, quick, and had no trouble in talking about what he had seen Roswell in 1947.

[17] Slate, B. Ann and Stanton T. Friedman. "UFO Battles the Air Force Couldn't Cover Up," *Saga's UFO Report*, 2,2 (Winter 1974), p. 60.

[18] Edwards, Frank. *Flying Saucers – Serious Business*. New York: Bantam Books, 1966. pp. 41 – 42.

Chapter 1:

Tuesday, July 8, 1947

Although it was on Tuesday, July 8, 1947, that the world learned that there had been a "flying saucer" recovered near Roswell, New Mexico, that wasn't the first that those at the higher headquarters in Fort Worth knew about it according to the eyewitness testimony of one high-ranking officer. Colonel Thomas J. DuBose, the Eighth Air Force Chief of Staff in July, 1947 said that "two or three days earlier"[19] he received a telephone call from Major General Clements McMullen[20] in Washington, D.C. and was told, "...there was talk of some elements had been found on the ground outside Roswell, New Mexico... The debris or elements were to be placed in a suitable container and [Colonel William] Blanchard was to see they were delivered [to Fort Worth]. [Colonel] Al Clark, the base commander[21] at Fort Worth would pick them up and deliver

---

[19] Thomas J. DuBose, personal interview by Don Schmitt, March 31, 1990; see also Randle, Kevin D. and Donald R. Schmitt. *UFO Crash at Roswell*. New York: Avon Books, 1991. p. 74 – 75; Carey, Thomas J. and Donald R. Schmitt. *Witness to Roswell*. Pompton Plains, NJ: New Page Books, 2009. p. 125; Klass, Philip. *The Real Roswell Crashed-Sauce Coverup*. Amherst, NY: *Prometheus Books*, 1997. pp.89 – 94; Friedman, Stan and Don Berliner. *Crash at Corona*. New York: Paragon House, 1992. p. 111. Other interviews with DuBose by Don Schmitt and Stanton Friedman August 10, 1991.

[20] From the Air Force Fact Sheet about Clements McMullen: In November 1946, he took over the 8th Air Force at Fort Worth, Texas, and the following January was designated Deputy Commander of the Strategic Air Command, with headquarters at Andrews Field, Maryland. In March he took over the additional duty as Chief of Staff of SAC, retaining his position as Deputy Commander.

[21] The base commander position is analogous to the mayor of a city. He is responsible for the base facilities, security and the like. In the modern Army it is referred to as the "Mayor's Cell." Position verified by the FWAAF base telephone directory and by the 8th Air Force Unit History, July 1947.

them to McMullen in Washington. Nobody, and I must stress this, no one was to discuss this with their wives, me with Ramey, with anyone. The matter as far as we're concerned was closed."[22]

DuBose then called Colonel Blanchard [509th Bomb Group commanding officer] and relayed the instructions to him. Blanchard then ordered the samples that W. W. Brazel had taken to the Chaves County Sheriff, George Wilcox, brought to the base. DuBose said, "… [I] told him that the material his S2 [Major Jesse A. Marcel, Sr.] had found… to be put in a suitable container by this major, and you are to see that it is sealed, put in your little command aircraft, and flown by a proper courier,[23] flown to Carswell and delivered to Al Clark, who will deliver it to McMullen."[24]

Because it was a hot day, DuBose waited in his office until he was told that the aircraft from Roswell was in the traffic pattern. Once he learned it was about to land, he drove out to the ramp area on the airfield proper and waited for the aircraft. In 1991, DuBose couldn't remember the exact type of aircraft that came in but did remember that it wasn't a B-29. McMullen wouldn't have approved of using one of the big bombers as a transport when smaller aircraft were available.[25]

Once the aircraft was on the ground, the courier transferred the material to Clark. DuBose said, "Clark took the package… to the B-26 [the FWAAF aircraft standing by] … He handed it to somebody and climbed in there. And that's the last I saw of it. In a couple, three hours, it was delivered to McMullen, and that's the last I heard of it."[26]

---

[22] Thomas J. DuBose, videotaped interview by Don Schmitt and Stanton Friedman, August 10, 1991.

[23] Here meaning that it was carried by an officer or NCO detailed as a classified courier and is authorized to carry classified material, whatever form it might take.

[24] The assumption here is that if debris was delivered to Fort Worth, then it must have been some of that Brazel gathered and took to the sheriff's office on Sunday, July 6.

[25] DuBose, personal interview by Schmitt and Friedman, August 10, 1991.

[26] Ibid.

There were no guards on the flight from Roswell to Fort Worth nor were there any on the flight to Andrews Army Air Field in the Washington, D.C. area, according to the best information available. DuBose also made it clear that he hadn't seen any of the debris at that time. He said, "I only saw the container and the container was a plastic bag that I would say weighed fifteen to twenty pounds. It was sealed [with] a lead seal around the top... The only way to get into it was cut it.[27]

That, according to what DuBose told various UFO investigators in several interviews, was the only package on that aircraft. He said that the debris in the bag was not the same as that displayed in Ramey's office on July 8 after Marcel had arrived.[28] The only flight that DuBose knew about was the one made on Sunday, July 6. There would be other flights but because the destination wasn't FWAAF there was no reason for him, their operations or the airfield to have

---

[27] Ibid. See also, Randle, Kevin and Donald Schmitt. *The Truth about UFO Crash at Roswell.* New York: M. Evans and Company, 1994: pp. 44 – 45. He repeated that statement in several interviews with various investigators.

[28] There would be controversy about exactly what DuBose said about this. Jaime Shandera would insist that the material in Ramey's office was the real debris picked up near Roswell. He would insist that DuBose had not said the debris on the floor was only a weather balloon, but Shandera had made no recording of his conversations with DuBose and his description of events was different than what was heard on the tapes made of various DuBose interviews by several others. See Shandera, Jaime. "New Revelations about Roswell Wreckage: A General Speaks Up." *The MUFON UFO Journal* 273 (January 1991):

been directly involved.[29] DuBose said, "McMullen told me you are not to discuss this, and this is a point which this is more than top secret, beyond that… That is the highest priority, and you will say nothing.[30]

These last statements are wrapped in controversy because, if as DuBose claimed, debris was transported to Fort Worth and then onto Washington, D.C. on Sunday, July 6, then Marcel's revelation on July 8 would be secondary and certainly not surprising to the military officers. They would have known on Sunday afternoon that they were dealing with something unusual though they might not have been thinking in terms of an extraterrestrial craft, and some of the things that happened in the days to follow would have been radically altered. Rather than dealing with something ordinary, they would have been more circumspect with what they would say publicly because of the extraordinary nature of the debris and the press release ordered by Blanchard would not have been issued.

The Events in Roswell on Tuesday, July 8

However, and according to other information developed during the investigations of the Roswell event, the story can be said to have actually begun early on July 8 after Major Jesse Marcel had returned from the debris field bringing more of the strange, metallic material with him. The regularly scheduled 9:00 staff meeting was held, instead, at 0730 hours (7:30 a. m.) on the orders of

---

[29] It must be noted here that this testimony raises a problem. If this material had been sent through Fort Worth on Sunday, and if it was as strange as DuBose suggested, then the press release of July 8, 1947, makes no sense. It is revealing something that was classified two days earlier.

[30] DuBose, personal interview by Schmitt and Friedman, August 10, 1991; See also, Randle and Schmitt. *Truth about UFO Crash.* pp. 44 – 45; Carey and Schmitt, *Witness to Roswell*, pp. 124 – 125.

Colonel Blanchard.[31] The change in time might have been prompted by Marcel, possibly with Cavitt, briefing Blanchard on what they had seen on the debris field found by Brazel.[32] It would apparently be the first hint they had of the nature of the crash.

Besides the normal staff officers there, including Marcel and Cavitt, were Brigadier General Roger Ramey and his Chief of Staff, Colonel Thomas J. DuBose, both who arrived early from Fort Worth.[33] Samples of the material were passed around so that everyone there had a chance to feel them and no one was able to identify the debris.[34] It was at this conference that Ramey proposed the plan to announce the recovery, but to limit it to the debris field that was closer to the town as a way to curb interest in what had been found.[35]

There is another problem with all this which calls Haut's memories into question. According to a newspaper report, "Marcel brought back the discovery to Roswell Army Air Field early Tuesday morning and at 8 a.m. reported

---

[31] Affidavit signed by Walter Haut on December 26, 2002; Carey and Schmitt, *Witness to Roswell*, p. 252. Haut had said that the regular staff meeting had been moved to 7:30, indicating that the time had been changed. See also Pflock, Karl, *Roswell: Inconvenient Facts and the Will to Believe*, Amherst, NY: Prometheus Books, 2001: pp. 26 -27. Lieutenant Colonel Joe Briley, in a telephone interview with Kevin Randle on October 20, 1989, however, said regular staff meeting was on Monday and that Blanchard was in touch with his officers daily. According to the *Fort Worth Star-Telegram*, Marcel said that he met with Blanchard at 0800. See Chapter 8, "The True Timeline."

[32] Pflock, Karl, *Inconvenient Facts*, pp. 26 - 27; Information confirmed in a personal interview of Lewis Rickett by Mark Rodeghier, January, 1990.

[33] Walter Haut, signed affidavit, December 26, 2002.

[34] There has been no corroboration for this claim by Haut. The records available do not indicate that either Ramey or DuBose had traveled to Roswell for this briefing. These include the unit histories for both the 509th Bomb Group and the 8th Air Force, the newspapers of the day, and the communication records that have survived.

[35] Walter Haut, signed affidavit, December 26, 2002. This idea, however, was first proposed by Frank Kaufmann during interviews in Roswell, New Mexico, in the early 1990s.

to his commanding officer, Col. William H. Blanchard… Blanchard, in turn, reported to General Ramey, who ordered the find flown to Fort Worth."[36]

At about 9:30 that same morning, once everyone had returned to his duties, Blanchard called Haut in his office and dictated the press release to him.[37] He was told to deliver the press release to the four media outlets in Roswell, that is, the two radio stations and the two newspapers. In one of many discrepancies, Haut would say that he drove the release into town and later say that he used the telephone and dictated it to them.[38] Either way, the press release went out to the media, and then was put on the news wire both by Frank Joyce[39] or George Walsh.[40]

Walsh remembered that Haut had telephoned the press release to him "about mid-day."[41] He copied the press release exactly as Haut read it to him

---

[36] "New Mexico Rancher's 'Flying Disk' Proves to Be Weather Balloon-Kite." *Fort Worth Star-Telegram*, July 9, 1947, p. 1.

[37] In other interviews, Haut would say that he didn't remember if Blanchard had called him and given him the facts of the story or if he had dictated the finished release to him. He said didn't remember if he had gone to Blanchard's office, or received a telephone call about it. See Carey and Schmitt, *Witness to Roswell*, pp. 252 – 253; Pflock, *Inconvenient Facts*, p. 26; Philip Klass, *Real Roswell Cover-up*, pp. 31 -32; Randle and Schmitt, *UFO Crash*, pp. 68 -76; Walter Haut personal interview by Randle, April 1, 1990.

[38] In the documentation provided by Frank Joyce, there had been a query from Denver, to Santa Fe, asking for the text of the announcement. Santa Fe responded, "Army gave verbal ann[oun]c[e]ment. No text."

[39] Joyce would suggest that after he received the press release, he called Haut and warned him that the wording of it was incorrect. It had to do with who was issuing the release and what it said. Joyce claimed he warned Haut that he would get into trouble based on that wording, that is, the release was by the authority of Marcel rather than Blanchard.

[40] Walsh worked at one of the two Roswell radio station, KSWS.

[41] George Walsh, affidavit signed September 13, 1993.

over the phone. Walsh, in turn called it into the Associated Press in Albuquerque. From there the release was put on the AP wire and that story was published in a number of newspapers.[42]

Art McQuiddy, who was the editor of the *Roswell Morning Dispatch* said, "I can remember quite a bit about what happened that day. It was about noon and Walter brought in a press release. He'd already been to one of the radio stations, and I raised hell with him about playing favorites."[43]

Unfortunately for McQuiddy, the *Dispatch* was a morning newspaper, so there wasn't much for him to do with the story. He said, "By the time Haut got to me, it hadn't been ten minutes[44] and the phone started ringing. I didn't get off the phone until late afternoon… The story died, literally, as fast as it started."[45]

It was just after 1300 hours (one o'clock p.m.) mountain time or 1400 hours central time on July 8, 1947 that a special flight took off for Fort Worth.[46] This is the flight that took Jesse Marcel and several wrapped packages from Roswell to Eighth Air Force headquarters. It was originally planned to continue onto Wright Field with just an hour or so in Fort Worth. This was also about the time that the story of the UFO crash was being transmitted over various news wires.

---

[42] Pflock, *Inconvenient Facts*, p. 62.

[43] Art McQuiddy, personal interview by Kevin Randle, January 19, 1990.

[44] Given the sequence of events, based on time lines published in other newspapers and interviews conducted with other witnesses, the telephone calls to McQuiddy probably didn't start for two hours. He, like others in Roswell said that he received telephone calls from all over the world, mentioning London, Rome, Paris and Hong Kong.

[45] McQuiddy, personal interview by Randle, January 19, 1990.

[46] Carey and Schmitt, Witness to Roswell, p. 123. There are newspaper articles that suggest the flight was ordered at 10:00, but it unclear if that meant the flight took-off at that time or if that was when the order was issued. See *Fort Worth Star-Telegram*, July 9, 1947

There is some controversy about this as well. According to Jesse Marcel, he said, "So my CO, early the next morning, sent me to Carswell to stop over and talk to General Ramey… [I took] all the stuff in a B-29. My CO told me to go ahead and fly it to Wright-Patterson air field in Ohio, but when I got to Carswell, General Ramey wasn't there but they had a lot of news reporters and a slew of microphones that wanted to talk to me, but I couldn't say anything. I couldn't say anything until I talked to the General. I had to go under his orders. And he [Ramey] said, "Well, just don't say anything."[47]

Major Ralph Taylor, the 830th Squadron Operations officer ordered the flight. The pilot was Lieutenant Colonel Payne Jennings with Lieutenant Colonel Robert Barrowclough[48] riding in what would have been the bombardier's seat. Marcel, of course was on board, holding a package on his lap.[49]

At 2:30 p.m. (MST), Blanchard announced that he was going on leave. He would not be available to take telephone calls about the flying saucer and would be out of touch for four days.[50] Lieutenant Colonel Joe Briley, a squad-

---

[47] The news release wasn't made until the afternoon on July 8 which means that no one would have been at the airfield to interview Marcel when he arrived because no one would have known about the crash. There is no evidence that the aircraft was met by reporters. The evidence suggests that Marcel saw a single reporter which was J. Bond Johnson, though both DuBose and Marcel talked of several reporters in Ramey's office. Marcel didn't arrive at the airfield until sometime after 1600 hours.

[48] LTC Robert Barrowclough would later communicate with Kent Jeffrey with a handwritten note on June 15, 1997, that said, "Thank you for the copy of the UFO Journal on the Roswell myth. Maybe some of those crack pots will quit calling me up and say I'm covering a deep gov't secret. You pretty well covered the subject." See Randle, Kevin D. *The Roswell Encyclopedia*. New York: Quill Books, 2000: pp.51, 166 – 194; Kent, Jeffrey. "Roswell – Anatomy of a Myth." *The MUFON UFO Journal* 350 (June 1997). 3 – 17.

[49] Robert Porter, personal interview with Don Schmitt, April 8, 1990. See also Randle and Schmitt. *UFO Crash at Roswel*l. pp. 81 – 82, Carey and Schmitt. *Witness to Roswell.* pp. 124 – 126.

[50] Donald Schmitt, *UFO Crash at Roswell II*, The Author, 1997: p. 60

ron commander until the middle of July 1947 when he became the 509th Operations Officer, confirmed that the leave was a blind and that Blanchard had gone out to the crash site. He said, "I'm sure of it."[51]

The Timeline as Published in 1947

There is a document, created in 1947, that provides the exact times for some of this. According *The Daily Illini,* the first of the stories on the Associated Press wire appeared at 4:26 p.m. on the east coast.[52] That would mean that the stories went out from Albuquerque, sometime prior to 2:26 p.m.[53]

The Associated Press version, as it appeared in a number of west coast newspapers said:

> The many rumors regarding the flying disc became a reality yesterday when the intelligence office of the 509th Bomb Group of the Eighth Air Force, Roswell Army Air Field, was fortunate enough to gain possession of a disc through the cooperation of one of the local ranchers and the sheriff's office of Chavez County.
>
> The flying object landed on a ranch near Roswell sometime last week. Not having phone facilities, the rancher stored the disc until such time as he was able to contact the sheriff's office, who in turn notified Major Jesse A. Marcel of the 509th Bomb Group Intelligence Office.

---

[51] Joe Briley, telephone interview with Kevin Randle, October 20, 1989.

[52] We have no information how long the story was in Albuquerque, how it was vetted, or how long it took to type it into the news wire.

[53] "AP Wires Burn With 'Captured Disk' Story," July 9, 1947, p. 5. Although the *Daily Illini* story gives all times in relation to the Associated Press, the line quoted at the beginning for the article is actually from the United Press International.

> Action was immediately taken and the disc was picked up at the rancher's home. It was inspected at the Roswell Army Air Field and subsequently loaned by Major Marcel to higher headquarters.[54]

At 4:30 p.m. (EST), there is the first "add" to the AP story, which mentioned "Lt. Warren Haught [Walter Haut]," who was described as the public information officer at Roswell Field. This new information suggested that the object had been found "last week" and that the object had been sent onto "higher headquarters."

The original United Press bulletin, which went out fifteen minutes later, at 4:41 p.m. (EST), according to newspaper sources, said:

Roswell, N.M. – The army air forces here today announced a flying disc had been found on a ranch near Roswell and is in army possession.

The Intelligence office reports that it gained possession of the 'Dis:' [sic] through the cooperation of a Roswell rancher and Sheriff George Wilson [sic] of Roswell.

> The disc landed on a ranch near Roswell sometime last week. Not having phone facilities, the rancher, whose name has not yet been obtained, stored the disc until such time as he was able to contact the Roswell sheriff's office.
>
> The sheriff's office notified a major of the 509th Intelligence Office.
>
> Action was taken immediately and the disc was picked up at the rancher's home and taken to the Roswell Air Base. Following examination, the disc was flown by intelligence officers in a superfortress (B-29) to an undisclosed "Higher Headquarters."

---

[54] "Disc Solution Collapses," *San Francisco Chronicle*, July 9, 1947, p. 1

> The air base has refused to give details of construction of the disc or its appearance.
>
> Residents near the ranch on which the disc was found reported seeing a strange blue light several days ago about three o'clock in the morning.[55]

When the aircraft carrying Marcel and three packages apparently containing wreckage landed in Fort Worth, the enlisted soldiers had been ordered to remain with the plane until a guard was posted and Marcel got off, with the package he had been carrying himself. According to what Robert Porter, one of the NCOs on the flight said, once the guard was posted, they went to the mess hall. The debris that had been loaded onto that aircraft was transferred to a B-25 that was going on to Wright Field in Ohio. When they got back to the aircraft, they were told the debris was nothing more exotic than a weather balloon.[56]

The Story Moves to Fort Worth

Marcel, carrying a small sample of the debris, went to Ramey's office, put the box on Ramey's desk and then followed Ramey to what was the map room to show him where the debris had been found.[57] While he was in the map room

---

[55] "AP Wires Burn With 'Captured Disk' Story," July 9, 1947, p. 5, which only published the first line of the story, and had attributed it to the Associated Press. Frank Joyce retained the teletype copy which he had sent to the United Press. He provided copies of these documents to various UFO researchers including Randle, Schmitt, Carey, and Pflock. See also, Pflock, *Inconvenient Facts*, pp. 244 – 248; Randle and Schmitt, *Truth About*, pp.46 - 50.

[56] Randle and Schmitt. *UFO Crash at Roswel*l. pp. 81 – 82, based on interviews with Porter.

[57] Ibid.

with Ramey, according to what he would say later, the debris he had brought was removed from Ramey's office and a torn up weather balloon was substituted.[58]

In Fort Worth, Texas (3:30 p.m. CST, 4:30 p.m. EST) Cullen Greene, an editor at the *Fort Worth Star Telegram* would have read the story as it came over the wire. J. Bond Johnson who worked at the newspaper in July 1947 said, "I don't know the mechanics. We'd get those alerts. The bells would ring and it would be an attention thing. It would be an editor thing."[59]

At 4:55 p.m. (EST) on the east coast 2:55 p.m. (MST) in New Mexico, the location of the discovery, that is New Mexico, is given. This bulletin, described as a "95" which is just below bulletin in importance, was repeated at 5:08 (EST) and a minute later, at 5:09,[60] there was another repeat of the story that said the information came from a radio reporter, but the identity of the reporter was not given.[61]

---

[58] DuBose, personal interview by Schmitt and Friedman, August 10, 1991; See also, Randle and Schmitt. *Truth about UFO Crash*. pp. 44 – 45; Carey and Schmitt, *Witness to Roswell*, pp. 124 – 125.

[59] J. Bond Johnson, audio taped telephone interview by Kevin Randle, February 27, 1989, March 24, 1990; Johnson in interviews conducted by Bill Moore and Jaime Shandera would later change much of his story. For these alternative views, see, Shandera and Moore, "Three Hours that Shook the Press," *Focus*, 1990, 3 – 7; Shandera and Moore, "Three Hours that Shook the Press," *MUFON UFO Journal* 269, September 1990: 3 - 10; Schmitt and Randle, "What Happened in Ramey's Office*?" MUFON UFO Journal* 276, April 1991: 3 – 9.

[60] "AP Wires Burn With 'Captured Disk' Story," *Daily Illini,* July 9, 1947, p. 5.

[61] Years later, both George Walsh and Frank Joyce would claim credit for breaking this story. It would seem that Walsh put the story on the AP wire, which did get it out nationally first. Joyce, a stringer for the United Press had to take the story to the Western Union office for transmittal to Albuquerque. Documentation suggests that the AP story hit the wires about fifteen minutes before the UP story, which gives the nod to Walsh.

Johnson, who described himself as a reporter in July 1947,[62] said, "...late in the afternoon, I returned to the office... My city editor... ran over and said, 'Bond, have you got your camera?' I said, 'Yes, I had it out in my car.' He said to get out to General Ramey's office and... he said they've got something there... [and] get a picture... He said something crashed out there or whatever and they're... we just got an alert on the AP wire."[63]

At 5:10 p.m. (EST) or 3:10 p. m. (MST), there was a message that was addressed to the newspaper editors to let them know that the Associated Press had now gone to work on the story.[64] This fact becomes important when the information from Jason Kellahin is added. Kellahin had, at one time, been the editor of the *Roswell Morning Dispatch*, but in July 1947, worked for the Associated Press in Albuquerque.[65]

According to the *Daily Illini*, "One minute later, at 5:11 [p. m. EST], the third add [additional information] to the bulletin announced, 'The war department in Washington had nothing to say immediately about the reported find.' That meant the AP was on the job investigating."[66]

It was about 4:30 p.m. (CST, 5:30 EST) that Johnson arrived at the Fort Worth Army Air Field. He told Bill Moore and Jaime Shandera that it was about a twenty minute trip from the newspaper office out to the airfield.[67] He said that he routinely covered activities at the airfield, so when he reached the

---

[62] Johnson, audio taped interview by Kevin Randle, March 24, 1989.

[63] Ibid. For additional information see Schmitt and Randle, "Fort Worth, July 8, 1947: The Cover up Begins," *International UFO Reporter*, 15,2, March/April 1990; 21 – 23; Schmitt and Randle, "The Fort Worth Press Conference: The J. Bond Johnson Connection," *International UFO Reporter*, 15,6, November/December 1990: 5 – 16.

[64] "AP Wires Burn With 'Captured Disk' Story," *Daily Illini,* July 9, 1947, p. 5.

[65] George Walsh, affidavit signed, September 13, 1993.

[66] "AP Wires Burn With 'Captured Disk' Story," *Daily Illini,* July 9, 1947, p. 5.

[67] "Three Hours that Shook the Press," *MUFON UFO Journal* 269, September 1990: 3 - 10.

gate, he showed his press pass. He also had a Civil Air Patrol sticker on his car, which would have made it easier for him to enter the airfield. He had been told to go to Ramey's office, though he normally would visit the Public Information Officer.[68]

Johnson said that when he entered the office, he was met by Ramey and DuBose. According to him, in one interview, there was no one else in the office.[69] The pictures taken at that time would suggest that Marcel, if not there, was brought in later and there was no sign of Ramey's aide, Captain Roy Showalter.[70] He should have been close to Ramey.

Johnson said, "I posed General Ramey with this debris piled in the middle of his rather large and plush office. It seemed incongruous to have this smelly garbage piled up on the floor… spread out on the floor of this rather plush, big office… I posed General Ramey with this debris. At that time, I was briefed on the idea that it was not a flying saucer but in fact was a weather balloon that had crashed."[71]

Johnson would say that he had only taken two photographs; one of Ramey and one of Ramey and DuBose. Later he would amend this, saying that he'd take two of Ramey and two of Ramey and DuBose and later still, it would become six pictures, now including two of Jesse Marcel.[72] Finally he would say that he had not taken those of Marcel.

---

[68] Johnson, audio taped interview by Kevin Randle, March 24, 1989.

[69] "Three Hours that Shook the Press," *MUFON UFO Journal* 269, September 1990: 3 - 10.

[70] The name of Ramey's aide is based on the Fort Worth Army Air Field telephone directory and on the Eighth Air Force Unit history for July 1947.

[71] Johnson, audio taped interview by Kevin Randle, March 24, 1989.

[72] Johnson said that he had only brought enough film for the two pictures. When it became clear from an examination of the photographs that he had taken others, he changed his story saying he had taken four, and then six. Eventually he would say that he had taken four but had no idea who had taken the two pictures of Marcel. From the look of them, and the fact that one of the Marcel negatives was part of the Fort Worth Special Collection, it seems only logical to believe that Johnson had taken all six.

Johnson, according to what he said, didn't stay long in Ramey's office because generals were busy. He said, "As I remember, I probably wasn't there more than twenty minutes, which was not unusual." He took the photographs, gathered some information and left.[73]

He said, "It was entirely possible that I was briefed by the PIO."[74]

This last quote could be important. The story that Johnson wrote to accompany the pictures, contained no direct quotes from Ramey, DuBose or Marcel, but did quote Irving Newton, the weather officer called in to identify the debris.[75] But the timing seems to suggest that Johnson had arrived before Newton got to Ramey's office so it is puzzling. How is it that Newton is quoted in the first story directly, but none of the others are?

At 5:53 p.m. (EST), 4:53 p.m. (CST), there was another bulletin which had a Washington dateline but was a statement by Ramey, which had to originate in Fort Worth, which said the disk had been sent to Wright Field.[76] What is critical here is the use of the past tense. The story didn't say it would be forwarded, but that it had already been sent.

At 6:02 p.m. (EST, 5:02 p.m. CST), the AP put together the whole story and started the transmission of the "First Lead Disk." This story, datelined Albuquerque said, "The army air forces has gained possession of a flying disk, Lt. Warren Haught [Walter Haut], public information officer at Roswell army

---

[73] Johnson, interview by Randle, March 24, 1989.

[74] Ibid. The problem is that the PIO would have been in the office. In other interviews, Irving Newton suggested he was there when all the pictures were taken, which would mean that Johnson saw him as well. Johnson's claim that only Ramey and DuBose were there does not hold up when other information is considered.

[75] J. Bond Johnson, "'Disk-overy' Near Roswell Identified As Weather Balloon by FWAAF Officer," *Fort Worth Star-Telegram*, July 9, 1947: pp. 1 – 2

[76] "AP Wires Burn With 'Captured Disk' Story," *Daily Illini,* July 9, 1947, p. 5.

airfield announced today." That new lead was to be integrated into the stories that had already been transmitted.[77]

The FBI entered the picture at 5:17 p.m. (CST, 6:17 EST).[78] The FBI office in Dallas sent a message to the FBI office in Cincinnati, about the story. The message was sent to the director, J. Edgar Hoover, and to the SAC (Special Agent in Charge) and was titled, "Flying Disc, Information Concerning." The text said:

> Major Curtan [sic, Edwin Kirton], Headquarters Eighth Air Force, telephonically advised this office that an object purporting to be a flying disc was recovered near Roswell, New Mexico, this date [July 8, 1947]. The disc was hexagonal in shape and was suspended from a balloon by cable, which baloon [sic] was approximately twenty feet in diameter. Major Curtan further advised that the object found resembles a high altitude weather balloon with a radar reflector, but that telephonic conversation between their office and Wright Field had not borne out this belief. Disc and balloon being transported to Wright Field by special plane for examin [sic]. Information provided this office because of national interest in case and fact that National Broadcasting Company, Associated Press, and others attempting to break the story of location of the disc today. Major Curtan advised would request Wright

[77] Ibid

[78] The time on the document said, "6:17" but it turns out that all FBI documents used Eastern Time for the convenience of the Director, J. Edgar Hoover. Therefore, rather than a transmittal time of 6:17 Eastern Time, it was made at 5:17 Central Time. There is one other complication to this. Daylight Savings Time was not universal in the United States, so there were pockets where standard time was still in effect. The point is that the time on the document was based on the time in Washington, D.C. and not in Fort Worth.

> Field to advise Cincinnati office results of examination. No further investigation being conducted.[79]

Dallas Morning News reporters called out to the Fort Worth Army Air Field, according to them, at 5:30 p.m. (CST, 6:30 EST) and interviewed Major E. M. Kirton, an intelligence officer at Eighth Air Force Headquarters. He told reporters that "there is nothing to it... It was a rawin high altitude sounding device." Kirton said that the identification was final and there was no reason to send it on to Wright Field for confirmation. He added that the material had been flown to Fort Worth on a B-29.[80]

Warrant Officer Irving Newton was a weather officer at the base in 1947 and said that he was alone in the weather office when he received a call ordering him to General Ramey's office. Newton said that he was the only one there and couldn't leave. General Ramey then called and told him to "get your ass over here now. Use a car and if you have to, take the first one with the keys in it."[81]

Newton said, "I was met at the General's office by a Lt. Col. or Col. who told me that some one [sic] had found a flying saucer in New Mexico and they had it in the General's office... but the General suspicioned that it might be meteorological equipment or something of that nature and wanted it examined by qualified meteorological personnel."[82]

---

[79] FBI Telex, July 8, 1947.

[80] "Suspected 'Disk' Only Flying Weather Vane," *Dallas Morning News*, July 9, 1947.

[81] Irving Newton, telephone Interviews with Kevin Randle, October 20, 1989, March 24, 1990 and January 1991. See also, Weaver and McAndrew, *The Roswell Report*, Statement Thirty (Irving Newton). A similar account is found in Berlitz and Moore, *The Roswell Incident*, pp. 31 – 37.

[82] Weaver and McAndrew, *Roswell Report*. (Statement 30). Also Newton said much the same thing in his interview with Randle, though he said a colonel as the man who briefed him.

Newton said that when he entered the office there were several others there including reporters. He said, "...when I went in... [there were] a couple of press people, a Major, I learned to be Major Marcel and some other folks. Someone introduced Major Marcel as the person who found this material."[83]

Newton added something new to his interview when he spoke to the Air Force in 1994. He told the Air Force investigator, "While I was examining the debris, Major Marcel was picking up pieces of the target sticks and trying to convince me that some notations on the sticks were alien writing. There were figures on the sticks lavender or pink in color, appeared to be weather faded markings with no rhyme or reason."[84]

The problem is that Newton's testimony in the mid-1990s does not agree with what he had said in the past. In his interview with Bill Moore, he was asked, "But wouldn't the people at Roswell have been able to identify a balloon on their own?"

Newton said, "They certain should have. It was a regular Rawin sonde. They must have seen thousands of them."[85]

Later, he would tell Air Force investigators that "We did not use them at Fort Worth... These were used mostly on special projects and overseas."[86]

Johnson, based on the timing of the events, probably left Ramey's office around 5:00 p.m. He said, "I probably wasn't there more than twenty minutes

---

[83] Weaver and McAndrew, *Roswell Report*. (Statement 30).

[84] Weaver and McAndrew, *Roswell Report*. (Statement 30). This is the first time that Newton has mentioned this aspect of what happened in Ramey's office. In earlier interviews, first with Bill Moore and later with Kevin Randle, he mentioned nothing about figures on the sticks. He did say that the major was picking up bit and pieces of the debris and asking if it would be on a balloon.

[85] Berlitz and Moore. *The Roswell Incident*. p. 34. The Roswell Daily Record of July 9, 1947, suggested that the balloon could have come from the Albuquerque weather bureau because they used rawins in their weather gathering observations, which suggests that Newton's statement then not being used is in error.

[86] Weaver and McAndrew, *Roswell Report*. (Statement 30).

which was not unusual. Generals are pretty busy. You get in and I didn't have a whole lot to question him on."[87]

He did not see Newton and did not photograph Newton.[88] That picture was taken by an unidentified photographer and given the lack of any sort of credit line, suggests it was taken by an Army photographer for distribution to the newspapers.

The Turmoil in General Ramey's Office

There is a problem developing here, and it is based on the seven photographs that were taken in Ramey's office. Six seemed to belong together and these six were apparently taken by Johnson. The seventh is of Irving Newton and is clearly from a different camera, using a different film stock. More to the point is that in the six pictures, there is an overseas cap and a tie on the radiator behind the officers, but in the seventh those items are missing. This suggests a time lapse between the taking of those first six pictures and the seventh. It suggests that Ramey or his aide had removed them from the radiator before more members of the press arrived, and that Johnson was in and out before those others got there.

At 7:03 p.m. (EST, 6:03 CST), there was another "first lead" story from Washington, but this one hinted that there was nothing spectacular about the disc. It was now being identified as some sort of a meteorological device, or in other words, a weather balloon.[89]

Twelve minutes later, at 7:15 p.m. (EST, 6:15 CST), there was a bulletin that said General Ramey would make a statement on national radio. WBAP, a

---

[87] J. Bond Johnson, telephone interview by Kevin Randle, March 24, 1989.

[88] J. Bond Johnson, telephone interview by Kevin Randle February 27, 1989.

[89] "AP Wires Burn With 'Captured Disk' Story," *Daily Illini*, July 9, 1947, p. 5.

Dallas radio station, had arranged for the national hook up for Ramey[90]over the NBC radio network.

Then, at 7:29 p.m. EST (6:29 p.m. CST), came another new lead for the story. It said, "Procede [sic] Washington. Lead All Disk." This meant, simply, that the lead on the story that had been transmitted prior to this would be changed and the new lead substituted.

This was broken with another bulletin almost immediately. It said, "Fort Worth – Roswell's celebrated 'flying disk' was rudely stripped of its glamor by a Fort Worth army airfield weather officer who late today identified the object as a 'weather balloon.'"[91]

Johnson, in describing the situation at the *Star-Telegram* when he arrived from the Fort Worth Army Air Field, said:

> So, Cullen Greene, who was my city editor, said Bond, "Give us a wet print," which was not unusual. I normally operated on a very short time span at night or whatever… So I went in… they have brought a wire photo transmitter and had it set up there in the newsroom. There was some assorted people around there… these were technicians that had come over in the time that it had taken me to drive out to Carswell and interview General Ramey, get briefed and come back to the office. They had come from Dallas and set up this wire photo machine… I went in developed those two pictures and they were just identical almost. I came out with eight by ten wet prints and gave them to our photo people and they said thank you and by that time the telephone operator gave me a whole stack of messages that had come from all over the country.[92]

---

[90] Ibid.

[91] Ibid.

[92] J. Bond Johnson, telephone interview by Kevin Randle, March 24, 1989.

He would say that he wrote the first story about the recovery that appeared in an early edition of the newspaper. The last line of that story became important because it said, "After his first look, Ramey declared all it was was a weather balloon. The weather officer verified his view.[93]

Johnson had been given the solution, apparently before Newton arrived to provide the final, conclusive word. But that solution had been handed to the reporter for the *Dallas Morning News*, not by Ramey or Newton, but by Kirton, who was the intelligence officer. This would suggest that Newton was called in to provide some drama for the other reporters who did drive out to the airfield. But it would also suggest that a cover up was being created by Ramey and those at Fort Worth.

Thomas DuBose, who would eventually retire as a brigadier general, and who was in Ramey's office did talk about the press conference and the debris on the floor of Ramey's office. He said, "...actually it was a cover story, the balloon part of it for the remnants that were taken from this location and Al Clark[94] took it to Washington... That part of it was in fact a cover story that we were told to give to the public and the news and that was it."[95]

DuBose elaborated on this. He said, "That [balloon cover story] was the direction we were told. We were told this is the story that is to be given to the press and that is it... I don't know whether it was [Major General Clements]

---

[93] Johnson, claimed authorship of "'Disk-overy' Near Roswell Identified as Weather Balloon by FWAAF Officer," *Fort Worth Star-Telegram*, July 9, 1947. He would later deny that he had written the article.

[94] Colonel Al Clark was, in July 1947, the base commander at Fort Worth Army Air Field. He was subordinate to the Eighth Air Force commander, Ramey.

[95] Thomas J. DuBose, interviewed by Don Schmitt, August 10, 1990. Stan Friedman was also at the interview and asked questions as well. It is noted this way because Schmitt had arranged the interview, provided the recording equipment, and invited Friedman to attend.

McMullen[96] or [Colonel Alfred] Kalberer[97] or who, somebody cooked up the idea as a cover story we'll use this weather balloon."[98]

DuBose's confirmation of a cover up seems to confirm the suspicions that had arisen given the timing of various press stories. The timing of statements, such as that given by Kirton to the *Dallas Morning News*, suggest that the weather balloon identification had been made before Newton arrived at Ramey's office.

The photographs all show the remains of a rawin radar target in a very degraded state, and the blackened neoprene rubber of the balloon itself can be seen. Although Newton reported to various investigators and Air Force officers investigating the case in the 1990s that Marcel was attempting to convince him that some of the elements were strange, it seems that Marcel had not been fooled. According to Johnny Mann, a reporter for WWL-TV in New Orleans, he had interviewed Marcel in the early 1980s about what he had seen and what he had done in 1947. He showed Marcel the photographs that had been taken in Ramey's office and said, "Jesse, I've got to tell you that looks like a balloon."

Marcel, according to Mann said, "That's not the stuff that I brought from Roswell."[99]

That was not the only time that Marcel had been shown the pictures in *The Roswell Incident*. During the filming of *UFOs Are Real*, Marcel was shown the pictures and was asked, "This is not the material you found?"[100]

---

[96] In July 1947, McMullen was the Deputy Commander of the Strategic Air Command and was stationed in Washington, D.C.

[97] In July 1947, Kalberer was the Eighth Air Force Chief of Intelligence.

[98] Thomas J. DuBose, interviewed by Don Schmitt, August 10, 1990.

[99] Johnny Mann, telephone and personal interviews conducted by Kevin Randle, October, 1989; December, 1989.

[100] *UFO's Are Real*, shot script, May 1978, Tape 3, p. 1.

Marcel said, "Definitely not."

Even with the questions that should have been asked about how an experienced intelligence officer could mistake the remains of a flimsy weather balloon and an aluminum foil radar reflector for a flying saucer or why no one at Roswell could identify the wreckage for what it was the cover story was accepted. According to one story, "Brigadier General Donald M. Yates, chief of the AAF weather service, said only a very few of them are used daily, at points were some specific project requires highly accurate wind information from extreme altitudes."[101]

Newton, on the other hand, wasn't saying the rawins were quite so rare. The *Star-Telegram* reported, "Newton said there are some 80 weather stations in the United States using this type of gadget, and it could have come from any of them."[102]

The News Reports from Roswell

That wasn't the only point on which the stories disagreed. In earlier editions of the newspaper, it mentioned that "The disc landed on a ranch near Roswell sometime last week."[103] Later is was reported, "Brazell [sic], whose ranch is 30 miles from the nearest telephone and has no radio, knew nothing about the flying discs when he found the remains of the weather device scattered over a square mile of his property three weeks ago."[104]

---

[101] "AAF Finds 'Saucer', But Wishes it Hadn't," *The Boston Herald*, July 9, 1947, pp. 1 – 2.

[102] "New Mexico Rancher's 'Flying Disk' Proves to Be Weather Balloon-Kite," *Fort Worth Star-Telegram*, July 9, 1947, p. 1. This is from a later edition of the July 9, 1947, newspaper and adds details that were not reported in the earlier edition.

[103] This is from the United Press teletypes provided by Frank Joyce.

[104] "New Mexico Rancher's 'Flying Disk' Proves to Be Weather Balloon-Kite," *Fort Worth Star-Telegram*, July 9, 1947, p. 1.

That same day, in the *Albuquerque Journal*, Jason Kellahin reported, "Scattered with the materials over an area about 200 years across were pieces of grey rubber. All pieces were small."[105]

Interestingly, that is not the only fact that Kellahin reported that disagreed with the official story line. He wrote, "On July 4, after hearing about 'flying discs,' he took the find to Sheriff George Wilcox at Roswell who referred the discovery to intelligence officers at the Roswell field."[106]

This is an interesting statement because it is in conflict with almost all newspaper accounts that suggest Brazel had gone into Roswell on July 7. Marcel, in describing his activities, suggesting that they remained overnight on the Brazel ranch and went out the next morning, suggests a time line that put Brazel into Roswell on Sunday, July 6. After spending all day in the field, Marcel returned, arriving in Roswell, late on the evening or early in the morning of July 8, and then briefing Blanchard quite earlier.

By 10:00 p.m. (CST, 11:00 p.m. EST), the story was virtually over. Ramey had not appeared on NBC, but at 10:00 he was quoted on ABC's "Headline Edition," and the weather balloon story entered the public consciousness. The newspapers were reporting the error on the part of the officers in Roswell. The *Las Vegas Review Journal* reported:

> The excitement ran through this cycle:
>
> 1. Lieutenant Warren Haught [Walter Haut], public relations officer at the Roswell base, released a statement in the name of Colonel William Blanchard, base commander. It said that an object described as a "flying disk" was found on the nearby Foster ranch three weeks ago by W. W. Brazel and had been sent to "higher officials" for examination.

---

[105] Jason Kellahin, "NM Rancher Sorry He Said Anything About 'Disc Find,'" *Albuquerque Journal*, July 9, 1947, p. 2.

[106] Ibid.

2. Brigadier General Roger B. Ramey, commander of the 8th air force, said at Fort Worth that he believed the object was the "remnant of a weather balloon and radar reflector," and was "nothing to be excited about." He allowed photographers to take a picture of it. It was announced that the object would be sent to Wright Field, Dayton, Ohio, for examination by experts.
3. Later, Warrant Officer Irving Newton, Stessonville, Wisconsin, weather officer at Fort Worth, examined the object and said definitely that it was nothing but a badly smashed target used to determine the direction and velocity of high altitude winds.
4. Lieutenant Haught reportedly told reporters that he had been "shut up by two blistering phone calls from Washington."[107]
5. Efforts to contact Colonel Blanchard brought the information that "he is now on leave."[108]
6. Major Jesse A. Marcel, intelligence officer of the 509th bombardment group, reportedly told Brazel, the finder of the object, that it "has nothing to do with army or navy so far as I can tell."

---

[107] Haut, in a personal interview with Kevin Randle on April 20, 1989, said that he had received no calls from Washington, D.C. He said, "Well, I think that had I really gotten any calls from Washington, a first lieutenant getting calls from the big boys, I'd remember it." On the other hand, in an interview with Randle on December 5, 1990, George Walsh said, "He [Haut] said the just got a call from the War Department and [they] told him in two words, 'Shut up.'"

[108] As noted earlier, it was about 2:30 p. m. (MST) that Blanchard announced his leave. It seems to be a strange time to begin a leave, especially when it is about the same time that all the interest developed in the flying saucer story, based on the time lines published in 1947.

7. Brazel told reporters that he had found weather balloon equipment before but had seen nothing that resembled his latest find.

Those who saw the object said it had a flowered paper tape around it bearing the initials "D. P."[109]

While this story was winding down, with the principals unavailable for comment, that is Brazel being held by the military,[110] Marcel either in Fort Worth or enroute back to Roswell, and with Blanchard on leave, reporters were unable to gather any additional information. Sheriff Wilcox would say little. It was reported, "Wilcox said he did not see the object but was told by Brazell [sic] it was 'about three feet across.' The sheriff declined to elaborate. 'I'm working with those fellows at the base,' he said."[111]

There was a significant change in the world of flying saucers on July 9. The Associated Press reported, and the story appeared in newspapers around the country, that the Army and the Navy "...began a concentrated campaign to stop the rumors [of flying saucers whizzing through the sky]."[112]

The question generated by this is, "Why suddenly on July 9 did the military care about the flying saucers?" Prior to that point, they had admitted some

---

109 "Flying Disc Tales Decline As Army, Navy Crack Down," *Las Vegas Review-Journal*, July 9, 1947, pp. 1 – 2; Army, Navy Open Drive To Dry Up Saucer Talk," *Phoenix Gazette*, July 9, 1947, p. 4

110 Major Edwin Easley, Provost Marshal at the RAAF in July 1947, in a telephone interview with Randle, February 1990.

111 "Disc Mystery Is 'Solved' For Three Hours Until Roswell Find Collapses," *Albuquerque Journal*, July 9, 1947, pp. 1– 2.

112 "Services Try to Stop 'Disc' Talk," *Dallas Times Herald*, July 9, 1947; "Army, Navy Move on 'Flying Disc' Rumors, *El Paso Herald-Post*, July 9, 1947, p. 1; "Flying Disk Reports Fall Off Sharply," *Chicago Daily News*, July 9, 1947; "Flying Disc Tales Decline As Army, Navy Crack Down," *Las Vegas Review-Journal*, July 9, 1947, pp. 1 – 2; Army, Navy Open Drive To Dry Up Saucer Talk," *Phoenix Gazette*, July 9, 1947, p. 4.

confusion, had offered various explanations, and denied that they had anything in the inventory that would explain the sightings, or that there were no secret projects that might have given rise to the reports.[113]

To make such a change in policy, there must have some event to precipitate it. On July 7, the military didn't seem to care that people were seeing flying saucers, they didn't care that these were being reported, almost daily, on the front pages of newspapers across the country. Then, on July 9, the policy changed. Why did that happen?

Something was recovered near Roswell, and the military and the government no longer wanted people talking about flying saucers. They wanted the problem buried and began to work toward that end.

---

[113] "More 'Flying Saucers' Reported: Army Experts Puzzled," *Las Vegas Review Journal*, July 3, 1947, pp. 1 – 2; Army Probes 'Flying Saucer' Stories, *Des Moines Register*, July 4, 1947, p. 10; Flying Saucers Everywhere; New Tales Convert Skeptics," *Oregonian*, July 4, 1947, pp. 1, 24; "Army Shrugs Off 'Flying Saucers,'" *Carlsbad Current – Argus*, July 6, 1947, p 1.

Chapter 2:

In the Beginning

The story of the Roswell UFO crash became real for Don Schmitt and me when New Mexico rancher Bill Brazel told us, "My dad found this thing and told me a little bit about it. Not much ... because the Air Force asked him to take an oath that he wouldn't tell anybody in detail about it."[1]

He then told us, "I found a few bits and pieces later on. And the Air Force came out… they didn't confiscate it. They just put it to me in such a way that I should give it to them… And I had no use for them, and Dad had told me that he had taken an oath that he wouldn't tell anybody in detail. He told me I didn't need to know, and probably I'm better off."

His dad was William "Mack" Brazel, who in early July 1947 told the Chavez County Sheriff that he had found metallic debris on the ranch he worked near Corona, New Mexico.[2] Brazel took a sample of the debris into Roswell and then to the sheriff's office.[3]

---

[1] Bill Brazel, personal interview conducted by Kevin Randle and Don Schmitt, February 19, 1989 in Carrizozo, New Mexico. See also, Randle and Schmitt, *UFO Crash at Roswell*, p.127; Friedman and Berliner, *Crash at Corona*, p. 84 (which used the interview with neither credit nor attribution).

[2] "AAF Finds 'Saucer'," *Boston Herald*, p. 2; "NM Rancher Sorry," *Albuquerque Tribune*, p. 2; "Officers Say Disc," *Roswell Morning Dispatch*, p. 8. Carey, Tom and Donald Schmitt, *Witness to Roswell*, Pompton, Plains, NJ: New Page Books, 2009, p. 50.

[3] Bill Brazel, personal interview conducted by Kevin Randle and Don Schmitt, February 19, 1989 in Carrizozo, New Mexico.

According to Phyllis McGuire, the daughter of the sheriff, the military arrived almost immediately.[4] Although McGuire had been chased out of the office before the military arrived, she did remember that her father had dispatched two deputies to look for the area that Brazel had described.[5]

Deputy Sheriff B. A. Clark, who had the duty that day, listened to the story told by Brazel, but hesitated to call his boss. Although the sheriff and his family lived above the jail, Clark didn't want to interrupt the sheriff on a Sunday. Finally, unable to identify the material, he made the call.[6]

The Air Intelligence Officer of the 509th Bomb Group, Major Jesse A. Marcel, told researcher William Moore, "We heard about it on July 7 when we

---

[4] Phyllis McGuire, personal interview conducted by Kevin Randle and Don Schmitt, January 27, 1990 in Roswell, New Mexico.

[5] McGuire said the deputies returned having failed to find the debris field, but they did find an area where there was a large circular burn.

[6] Carey and Schmitt, *Witness to Roswell*, p. 50; "Flying Disk Transforms Sheriff's Office To International Newsroom," *Roswell Morning Dispatch*, July 9, 1947, p. 1

got a call from the county sheriff's office at Roswell.[7] I was eating lunch at the officers' club when the call came in that I should go out and talk to Brazel."[8]

It seemed to the sheriff that Marcel had arrived quickly, followed by the CIC man, which some have identified as an enlisted man but was probably Sheridan Cavitt, dressed in civilian clothes with no rank visible. Wilcox said afterward that it had seemed as if they were waiting for his telephone call because they had gotten there so fast.[9]

Marcel would later tell investigators that he had driven into town to talk with the rancher and then returned to the base to talk with Blanchard after he looked at the strange material. Brazel said that it looked as if something had exploded over one of the pastures. There was a lot of debris scattered around over a wide area. Brazel wondered who was going to clean up the mess.[10]

According to Marcel, he decided that he needed to tell the commanding officer of the 509th, Colonel (later general) William H. Blanchard about the debris. He said, "…we determined that a downed aircraft of some unusual sort

---

[7] Although this is a direct quote from Marcel, it is confusing. According to Stan Friedman, when he first talked to Marcel he wasn't sure of the date of the crash. It was only after Bill Moore had searched the newspaper files that they found the pictures of Marcel taken in Brigadier General Ramey's office on July 8 that the date was learned. By the time this interview was conducted by Moore, Marcel had either read the newspaper clippings or had learned the date in his discussions with Moore. Later in that same interview, Marcel uses the Sunday, July 6, date. Newspapers of the time, however almost universally use Monday, July 7 as the date. A few do report that Brazel went to the sheriff on Sunday. There is no definitive evidence for the date, though backtracking from the press announcement does support the July 6 date.

[8] Berlitz, Charles and William L. Moore, *Roswell Incident*, New York: Gosset & Dunlap, 1980, p. 63; Friedman and Berliner, *Crash as Corona*, p.9, Bob Pratt personal interview with Jesse Marcel, Dec 8, 1979.

[9] Phyllis McGuire, personal interview conducted by Kevin Randle and Donald Schmitt, Jan. 27, 1990.

[10] Tommy Tyree, personal interview conducted by Randle and Schmitt, Aug. 1989.

might be involved, so the colonel said I had better get out there, and to take whatever I needed to go. I and a CIC agent from west Texas by the name of Cavitt[11] followed this man out to his ranch…"[12]

Blanchard said, according to Marcel, "…take whatever you need with you but go. So I got one of my agents… so I took him. He drove a jeep carryall and I drove my staff car."[13]

According to Marcel, "We took off cross county [from the sheriff's office] behind this pickup truck this rancher had. He didn't follow any roads going out… so we got to his place at dusk. It was too late to do anything so we spent the night there in that little shack and the following morning we got up and took off."[14]

Later he would tell Linda Corley who reported it in *For the Sake of My Country*, "I went to his house. I followed him. We left Roswell in the afternoon and got there at dusk… so we couldn't do anything that evening. So we stayed at his house that night with Cavitt… [We] spent the night at his house. We were treated with a can of pork and beans and crackers."

In his interview with Moore, Marcel elaborated, saying, "We got there very late in the afternoon and had to spend the night with this fellow. All we had to eat was some cold pork and beans and some crackers."[15]

---

[11] According to Friedman, Marcel couldn't remember Cavitt's name, referring to him only as "Cav." See Friedman and Berliner, *Crash at Corona*, p. 10. Bob Pratt interviewed Marcel on Dec 8, 1979. In that interview Marcel said he thought the name of the CIC man was "Cabot."

[12] Berlitz and Moore, *Roswell Incident*, p. 63; Pratt, personal interview with Marcel, Dec 8, 1879.

[13] Jesse A. Marcel, personal interview by Bob Pratt, Dec 8, 1979.

[14] Ibid.

[15] Berlitz and Moore, *Roswell Incident*, p. 63.

Of the debris, Marcel said, "I'd never seen anything like that. I didn't know what we were picking up. I still don't know... And I brought as much of it back to the base as I could... It was strewn over a wide area, I guess maybe three-quarters of a mile long and a few hundred feet wide."[16]

Marcel told Moore, "We collected all the debris we could handle. When we had filled the carryall, I began to fill the trunk and backseat of the Buick."[17]

Marcel described the material that they had found, none of which sounded as if it was from anything conventional. He said, "I saw, well, we found some metal, small bits of metal but mostly we found some material that is hard to describe. I'd never seen anything like that..."[18]

Bob Pratt, the *National Enquirer* reporter who was interviewing Marcel asked, "It was something manufactured?"

Marcel said, "Oh, it definitely was, but one thing I do remember, I recall very distinctly, I wanted to see some of this stuff burn, but all I had, I had a cigarette lighter... I lit the cigarette lighter to some of this stuff and it didn't burn."[19]

In a similar description of the incident, Marcel told Moore, "I even took my cigarette lighter and tried to burn the material we found that resembled parchment and balsa, but it would not burn – wouldn't even smoke."[20]

---

[16] Jesse A. Marcel, personal interview by Bob Pratt, Dec 8, 1979; Karl Pflock, *Roswell: Inconvenient Facts and the Will to Believe.* Amherst, NY: Prometheus Books, 2001: p. 228.

[17] Berlitz and Moore, *Roswell Incident*, p. 67

[18] Jesse A. Marcel, personal interview by Bob Pratt, Dec 8, 1979.

[19] Ibid.

[20] Berlitz and Moore, *Roswell Incident*, p. 66

Once Marcel had filled his car, he drove back to Roswell and although it was late at night, he stopped at his home in Roswell early in the morning [Tuesday, July 8th] on the way back to the base. He took a box of the debris into the house to show both his wife and son.

Marcel spread the debris out on the kitchen floor. Jesse A. Marcel, Jr., who was eleven at the time, said there was something that looked like I-beams[21], oil, some sort of a wooden-like material and a few strands that looked like the fiber optics used in the modern world. There was nothing in the pile that any of them recognized, and according to Jesse, Jr., his father told him that he didn't know what it was.[22]

Jesse, Jr. noticed some writing on one of the beams and pointed it out to his father. He said it looked like purplish, geometric symbols, squares and circles and triangles, embossed on the inside of the beam.[23]

Jesse Sr. said that the writing "...was pink and purple. The main character might be pink and the tone behind it was purple. The others were purple and switched around... There's a lot of it."[24]

Marcel finally picked up the debris and with the help of his son, took the box back out to the car. He then waited until dawn before going to Blanchard's office. He waited for the colonel to come to work.

---

[21] Linda Corley had the opportunity interview Marcel in 1981 and he provided her with a drawing of the small beams as he remembered them. He didn't describe them as I-beams but as a "little member" that was rectangular in shape. This is a disagreement between father and son but the I section of the beam, according to Jesse Marcel, Jr., was very small.

[22] Jesse Marcel, Jr., personal interview with Kevin Randle, Aug 3, 1989; Randle and Schmitt, *UFO Crash at Roswell*, pp. 51 -52. It should also be noted that the timing of Marcel's return home is in dispute based on various interviews conducted with him See Chapter on The Jesse Marcel Conundrum.

[23] Jesse Marcel, Jr., personal interview with Kevin Randle, Aug 3, 1989.

[24] Linda Corley, *in MUFON Symposium 2000*, p.124.

It was some time after that, after Blanchard had seen the material, that First Lieutenant Walter Haut received a telephone call from Blanchard. He would say later that he didn't remember if Blanchard had dictated the press release to him or if he gave him the facts and Haut returned to his office to write the story.[25] He then took it into town to give to the four local media outlets. He said that he rotated the order of his deliveries to them so that one outlet didn't get all the releases first.[26]

Brazel and the Press

Frank Joyce[27], who was a reporter and an announcer for radio station KGFL in Roswell in 1947, said that Haut told him he was being given the press release first. Once Joyce read it, he contacted Haut again, asking him, "Walter, do you really mean to send this out like this?"[28]

---

[25] Haut would also say that he had gone to Blanchard's office to receive the information. In his 2003 Sealed Statement, Haut said that Blanchard dictated the press release to him over the phone. Carey and Schmitt, *Witness to Roswell*, 2007, p. 216.

[26] Walter Haut, personal interview with Kevin Randle and Don Schmitt, Apr 1, 1989; Randle and Schmitt, *UFO Crash*, p. 68

[27] Critics of Frank Joyce point out that he does tell some bizarre stories. According to him, on Wednesday, July 9, he, with George "Jud" Roberts and Walt Whitmore traveled to Brazel's ranch house. While they were inside, Joyce said that he remained outside, or in some kind of out building. He mentioned to Randle and Schmitt that he drank a grape soda on the way to the ranch and this seemed to confuse and confound him. He talked of not remembering much of the trip, other than that a "traveler" was with them who said nothing but seemed to communicate through telepathy. While Joyce waited in the out building, the "traveler" and others were inside. See Pflock, *Roswell Inconvenient Facts*, p. 123. Also see another version of this story in Carey and Schmitt, *Witness to Roswell*, 2007, p.169.

[28] Frank Joyce, personal interview by Kevin Randle and Don Schmitt, March 31, 1989

Joyce said that after the press release hit the news wires, the telephones went crazy. He even received a telephone call from a Colonel Johnson who said he was at the Pentagon. Johnson was screaming at Joyce, wanting to know who in the hell had told him to release that information. Joyce, finally having heard enough, told Johnson that he was a civilian and Johnson couldn't do anything to him.

Johnson responded, "I'll show you what I can do."[29]

Joyce had always been cagey about what he said. In the first interview conducted by Randle and Schmitt, Joyce told them that he knew a lot more than he wanted to say. He said, "I guarantee I could tell that story if I ever got in the mood and absolutely, positively connect everything and you couldn't catch me on one little deal. You couldn't find anything to say, 'Hey, you made that up.'"[30]

Over the years, Joyce would expand on what he had heard. After Tom Carey joined Don Schmitt, they visited Joyce on several occasions. He told them that he just didn't want to say any more, but once he retired from Albuquerque television station KOB, he revealed more of the story during a 1998 interview.

Joyce had said in earlier interviews that he would call the sheriff's office every day to see if anything had happened that might be of interest to his listening audience. He said that Wilcox mentioned that there was a rancher by the name of Brazel in his office at that very moment with an interesting tale and then handed the telephone to him. That would have been in the late afternoon of July 6.[31]

---

[29] Ibid

[30] Frank Joyce, telephone interview by Kevin Randle, March 30, 1989

[31] Frank Joyce, personal interviews by Kevin Randle and Don Schmitt, September 1991, videotaped March 1991, July 1992; see also Randle and Schmitt, *The Truth About*, pp. 33, 275, and Carey and Schmitt, *Witness to Roswell*, 2007, p.56.

Joyce said that Brazel was angry, wanting to know who was going to clean up the mess left in his pasture. Brazel wanted someone to come out and clean it all up.

Joyce wanted to know what stuff was. He asked Brazel, "What are you talking about?"

Brazel said, "Don't know. Don't know what it is. Maybe it's from one of them 'flying saucer' things."[32]

According to Joyce, at this point Brazel began to "lose it," and said, "Oh, God. Oh my God. What am I gonna do? It's horrible. Horrible. Just Horrible."[33]

Joyce asked, "What's that? What's horrible? What are you talking about?"

"The stench. Just awful."

Joyce was puzzled. "Stench? From what? What are you talking about?"

Brazel said, "They're dead."

"What? Who's dead?"

"Little people," Brazel said so quietly as to be nearly inaudible. "Unfortunate little creatures…"

Joyce now believed that Brazel was crazy but he kept asking questions. "What the…? Where? Where did you find them?"

"Someplace else."

---

[32] It has been suggested that such statements are untrue because the term, "flying saucer" wasn't being used then, but a survey of newspapers from June and July 1947 show that flying saucer was used interchangeably with 'flying disk' or 'flying disc.' *The Roswell Daily Record* for July 8, 1947 said, "RAAF Captures Flying Saucer On Ranch in Roswell Region."

[33] In earlier interviews, Joyce had always said that he didn't want to put words into a dead man's mouth and refused to provide any solid information. Slowly he began to open up and provide additional information. See *UFO Crash at Roswel*l, Randle and Schmitt, p. 135

Joyce said, "Well, you know, the military is always firing rockets and experimenting with monkeys and things. So maybe…

Brazel interrupted, shouting, "God dammit! They're not monkeys, and THEY'RE NOT HUMAN." He slammed down the phone.[34]

Joyce, in his earlier interviews with investigators, would only talk about Brazel when he was escorted to the radio station by the military on July 9, after he had given his new story to the *Roswell Daily Record*.[35] Brazel, it seemed, was there to straighten out the reports that had been made in the last few days. He was accompanied by at least two military officers. According to Joyce, Brazel was under a great deal of stress and said that it would go hard on him if he didn't cooperate.[36]

This time the story was quite a bit different than he had mentioned during the Sunday afternoon interview. Now Brazel was telling Joyce about a weather balloon he had found and saying that there just hadn't been all that much in the way of debris.[37] Joyce was confused by the changes, but Brazel stuck to the story until he got up to leave the office. With the military men out in the lobby waiting to escort Brazel to his next interview and unable to hear, Brazel

---

[34] Frank Joyce in a personal interview by Tom Carey and Don Schmitt, 1998; Carey and Schmitt, *Witness to Roswell*, p 56; Pflock, *Roswell: Inconvenient Facts*, p. 123. It should also be noted that the only living creatures that had been launched in the rockets at that time were insects and rodents. Primates would come in later experiments according to the records kept at the Space Museum in Alamogordo.

[35] "Harassed Rancher who Located 'Saucer' Sorry He Told About It," *Roswell Daily Record*, July 9, 1947, p. 1

[36] Randle and Schmitt, *UFO Crash at Roswell*, p. 136, Carey and Schmitt, *Witness to Roswell*, 2007, p. 61. Frank Joyce personal interview by Kevin Randle and Donald Schmitt, March 31, 1947

[37] "Harassed Rancher," *Roswell Daily Record*, p. 1

said, "You know how they talk about little green men? [38] Well, they weren't green."[39]

Joyce also said that he had attempted to gather up the various documents that related to the events, including the original press release. He said that someone had come into the office and cleaned out every bit of paper that had anything to do with the crash.[40] He did manage to save some teletype messages.[41] They added to the confusion by suggesting the material reported by Brazel had been found weeks earlier.[42]

According to these documents, "The disc landed on a ranch near Roswell sometime last week. Not having telephone facilities, the rancher, whose name

---

[38] The question, at that point became, when did 'little green men' enter the lexicon? In Edgar Rice Burroughs' John Carter on Mars series circa 1913, there was a race of green men but they were ten feet tall and had four arms. The first documented mention of Little Green Men, meaning aliens, is in 1908 in the *Daily Kennebec Journal*. They are referring to Martians. In either 1910 or 1915, there is a story from Puglia, Italy that a Little Green Man had been captured in his crashed spaceship. There was a Popeye cartoon, circa 1938, in which Popeye fights green aliens. After the Orson Welles *War of the Worlds* 1938 radio broadcast, Bill Barnard in the *Corpus Christi Times* wrote about "Thirteen little green men from Mercury..." Comic books from the 1930s and 1940s also regularly portrayed "Martians" as little, green, goblin-like creatures. In 1946, Harold M. Sherman wrote a science fiction novel called, *The Green Man: A Visitor from Space*. And finally, in July 1947, Hal Boyle wrote of his adventures with a green man from Mars which was a satire directed at the flying saucer reports. Of course it is unlikely that Mack Brazel would have been exposed to any of this.

[39] Frank Joyce, personal interview by Randle and Schmitt, Mar 31, 1989.

[40] Ibid., also personal interview with KGFL minority owner and fiddle player Bob Wolf who told Carey and Schmitt in a 1999 interview that he had been in an adjoining room at the station when the military came in to confiscate the papers.

[41] Joyce provided copies of the teletype messages to Randle, Schmitt and Carey, and later to Pflock. Pflock published some of them in his book, pp. 245 – 248.

[42] Teletype messages provided by Joyce; see Randle and Schmitt, *Truth About*, pp. 66 – 70; Pflock, *Roswell: Inconvenient Facts*, pp. 245 – 248.

has not yet been obtained, stored the disc until such time as he was able to contact the Roswell sheriff's office."[43]

But later in the sequence of teletype messages, it said, "Sheriff Wilcox (correct [meaning that this is the correct name of the sheriff]) of Roswell ways [sic] that the disc was found about three weeks ago by a rancher by the name of W. W. Brizell [sic] on the Foster ranch near Corona, about 75 miles north-west of Roswell near the center of New Mexico."[44]

In the course of these messages that came over the wire in a matter of about three hours, the time frame for the discovery of the debris changed. Remembering that these messages are dated on Tuesday, July 8, it can be suggested that "last week" could refer to Friday or Saturday. The reference that Brazel (Brizell) had found the debris three weeks earlier, changes the time frame but also suggests that he thought nothing of the strange debris for weeks and then decided to take it into the sheriff's office.

The real problem, however, is the mention of "Was it same ranch mentioned last week in flying disc hullabaloo?" which was a question that came from the Denver office.[45]

This reference, according to Terry Colvin, might be to a UFO landing near tiny Cliff, New Mexico on June 29, 1947. According to this, two "AAF Fliers," which might be interpreted as pilots, had attempted to learn more about this but all they could find was a "layer of stinking air."[46]

---

[43] Ibid.

[44] Ibid.

[45] Teletype messages provided by Joyce; Randle and Schmitt, *Truth About*, p. 65

[46] Terry Colvin, email dated September 9, 2011. The problem is that this reference can be traced no further. Colvin said that he found it while "trawling" the Internet. There is an article that appears in the *Albuquerque Journal* on June 30 that seems to be this particular incident. There are no other New Mexico sightings that fit the limited facts. It suggests, however, that this is probably not a reference to the Brazel

While this was going on in Roswell, the military, at least in Fort Worth, Texas was beginning to change the tone of the story. Marcel had been ordered to take some of the debris to Eighth Air Force headquarters at Fort Worth Army Air Field for the commanding officer there, Brigadier General Roger Ramey, to examine.[47] The story, having left Roswell, was now in Fort Worth.

According to Jesse Marcel, "…that next afternoon [48] [Tuesday, July 8th] we loaded everything into a B-29 on orders from Colonel Blanchard and flew it all to Fort Worth."[49]

Marcel continued, "Just after we got to Carswell [Fort Worth Army Air Field]in 1947] in Fort Worth, we were told to bring some of this stuff up to the general's office – which he wanted to take a look at it. We did this and spread it out on the floor on some brown paper."[50]

Reporter J. Bond Johnson was the first newsman to arrive at Ramey's office after learning the debris had been taken there. He was escorted in and allowed to photograph Marcel, posed near the debris as well as Ramey and

---

discovery because there is no indication that anything had been announced until July 8, 1947.

[47] "AAF Finds 'Saucer' But Wishes it Hadn't," *Boston Herald*, July 9, 1947; "Army Says New Mexico 'Disc' Wind Balloon," *Oregon Journal*, July 9, 1947; "'Disk' Near Bomb Test Site Is Just a Weather Balloon," *New York Times*, July 9, 1947, p. 1; "Mystery Object Found on Ranch Weather Device," *Syracuse Post Standard*, July 9, 1947, p 1.

[48] Given the timing, and the photographs that were taken of Marcel, along with a document giving a transmission time of 11:49 p.m., this is the afternoon of July 8, 1947.

[49] Berlitz and Moore, *Roswell Incident*, p. 67

[50] Ibid

then Ramey and his Chief of Staff, Colonel Thomas J. DuBose.[51] Once he got a look at it, Ramey said it was nothing more than a weather balloon.[52]

At this point the interest in the crash dissipated. The news went out over the wires that the disk had been identified as a weather balloon and a rawin radar target. Warrant Officer Irving Newton, called to Ramey's office to identify the debris, made the specific call. The information was mentioned on a radio broadcast on ABC's "Headline Edition" that said much the same thing.

Back in Roswell, on that Wednesday [July 9th], Brazel was back in town, giving new and different interviews. A number of his friends saw him in town, escorted by military officers.[53] The Roswell Army Air Field Provost Marshal, Major Edwin Easley said that Brazel had been held at the Guest House on the base,[54] and a neighbor, Marian Strickland, who with her husband, remembered that Brazel sat in their kitchen complaining about being held in jail.

Bill Brazel, who in July 1947 was living in Albuquerque, said that he had seen his dad's picture in the newspaper. He said, "I got the paper and I looked at it and here's my dad's picture.[55] I said to Shirley [Bill Brazel's wife], 'What

---

[51] J. Bond Johnson, personal interview conducted by Kevin Randle, Mar 24, 1989.

[52] "'Disk-overy' Near Roswell Identified As Weather Balloon by FWAAF Officer," *Fort Worth Star-Telegram*, July 9, 1947; Telephone interview with Irving Newton by Kevin Randle, Mar 24, 1990, Pflock, *Roswell: Inconvenient Facts*, pp. 162 – 164

[53] Bill Jenkins, personal interview by Don Schmitt; September 1991, Randle and Schmitt, *Truth About UFO*, p. 36; Carey and Schmitt, *Witness to Roswell*, 2007, p. 64-65; Berlitz and Moore, *Roswell Incident*, pp. 83– 85.

[54] Edwin Easley, personal interview by Kevin Randle, Feb 2, 1990, Randle and Schmitt, *Truth About UFO*, p 35, Bill Brazel, personal interview by Randle and Schmitt, Feb 19, 1989.

[55] In the first stories, in both Albuquerque newspapers, the *Albuquerque Tribune* and the *Albuquerque Journal* were on the front page. They discussed Brazel's discovery but no photograph. His picture did appear in other newspapers around the country. There is a front page photograph of him in the *Roswell Daily Record* for July 10, 1947.

the hell did he do now?' So I proceeded out to the ranch. I think it was two or three days before my dad showed up. I said, 'What did you get into?' and he said, 'Oh, I found this thing out there.'"[56]

Brazel said that he asked his father, once in a while, what had happened, and his father would say little about it. He made it clear that he had promised that he wouldn't talk about it, and to Mack Brazel, it meant he wouldn't talk to anyone about it. He did, however, give Bill a clue now and then.

Bill Brazel said, "I'd ask him a question and he told me a little about it, naturally. But if I asked very many questions, he'd just say, 'Well, I told them I wouldn't tell and you don't need to know.' That's just the way he was."[57]

That doesn't say much about the nature of the find. But Brazel had more to say about that. He said that he had found some pieces, not more than a dozen and maybe only eight. He said, "There were only three items involved. Something on the order of balsa wood and something on the order of heavy gauge monofilament fishing line and a little piece of… it wasn't really aluminum foil and it wasn't really lead foil but it was on that order. A piece about the size of my fingers with ragged edges."[58]

The items were not, however, those things. The wood-like debris was just light weight, the way balsa wood is, but much stronger, according to Brazel. He said that the monofilament fishing line was strange. You could shine a light in one end and it came out the other, or, in other words, it was fiber optics.

Describing the foil, he said, "I couldn't tear it. Hell, tin foil or lead foil is easy but I couldn't tear it. I didn't take pliers or anything. I just used my fingers. I didn't try to cut it with my knife… It was pliable. Real pliable. I would bend it over and crease it and if you straightened back up there wouldn't be a

---

[56] Bill Brazel, personal interview by Randle and Schmitt, February 19, 1989; Randle and Schmitt *UFO Crash*, p.128

[57] Ibid

[58] Ibid

crinkle in it. Nothing. It would flatten out it was just as smooth as ever. Not a crinkle or anything."[59]

Brazel told Bill Moore, "There were also several bits of a metal-like substance, something on the order of tinfoil except this stuff wouldn't tear and was actually darker than tinfoil – more like lead foil, except very thin and extremely lightweight. The odd thing about this foil was that you could wrinkle it and lay it back down and it immediately resumed its original shape. It was quite pliable, yet you couldn't crease it or bend it like ordinary metal. I don't know what it was, but I do know that Dad once said the Army had told him that they had definitely established it was not anything made by us."[60]

Brazel said that he showed his father the bits of debris he'd found. His father said, "Oh, yeah, that's some of the thing I found."[61] This, of course, ties the material that Bill Brazel found to the discovery by his father. It was all from one source.

As happens so often Brazel no longer had the debris. He said that he had been visited by the military, he thought there were four of them, and the officer made it clear that they weren't going to leave without the debris.

He said, "I still am not positive, but I'm almost positive that the officer in charge, his name was Armstrong. A real nice guy… They came out to the ranch and they were talking to me and they said, 'We understand your father found a weather balloon,' and I said, 'Yeah.' And they said, 'We understand you found some bits and pieces… we would like to take it with us.'"[62]

---

[59] Ibid

[60] Berlitz and Moore, *Roswell Incident*, p. 79

[61] Bill Brazel personal interview by Randle and Schmitt, Feb 19, 1989.

[62] Ibid

Brazel said that he had no use for it, and it was clear to him that they were not going to leave without. He surrendered it to them. As he did, he was asked, "How well did you examine it?"[63]

Brazel said, "Well enough to know that I don't know what it is."

They asked him not to talk about it, told him that they were going to look over the field, and then left the ranch. Brazel, though he looked, never found any additional debris.[64]

This was but the beginning of the journey. It would take more than twenty-five years, involve hundreds of interviews, and yield little in the way of documentation and more than a few interviews where it became clear that the "witness" was less than candid. But the bare bones were there and there were questions that needed answers. One of them was how the most highly trained unit in the world could be so badly fooled by nothing more than a weather balloon and a radar reflecting target?

---

[63] Ibid

[64] ibid

Chapter 3:

Finding the Metallic Debris

William "Mack" Brazel has been described as an "old fashioned western cowboy"[114] who didn't say much. According to his son, Bill Jr., "My dad knew a lot of things that he wouldn't talk about. He might know who stole the calf, but he... didn't do a lot of talking. His philosophy in life was if you can't say anything good about a person, don't say anything."[115]He was "a very self-sufficient country man who was very dependable, self-confident. He was not a stupid man. He was not a highly-educated person."[116]

The conventional story is that Brazel was alone at the ranch he managed for the Fosters when he heard an explosion inside a thunderstorm that didn't sound like normal thunder to him.[117] But it seems that Bill Brazel had told Bill Moore that his sister Bessie and brother Vernon had been at the ranch during that thunderstorm. He told Moore, "Dad was in the ranch house with two of the younger kids late one evening when a terrible lightning storm came up."[118]

---

[114] Bill Brazel, Jr. personal interview with Kevin Randle and Donald Schmitt, February 19, 1989; see also, Randle, Kevin and Don Schmitt, *UFO Crash at Roswell,* New York: Avon Books, 1991: p.129.

[115] Ibid. p. 129

[116] Marian Strickland, personal interview with Kevin Randle, Donald Schmitt and Don Berliner, September 27, 1991, Roswell New Mexico.

[117] Marian Strickland, personal interview with Kevin Randle and Donald Schmitt, September 27, 1991; Randle and Schmitt, *UFO Crash at Roswell*, p. 129

[118] Berlitz, Charles and William Moore, *Roswell Incident*, New York: Grosset & Dunlap, 1980: p. 77. Brazel added, "...so the next day he rounded up the two kids and took off for Roswell by way of Tularosa, where he stopped off and left the kids with Mother." p. 78.

Marian Strickland described that storm as well. She said, "...all the ranchers go to out on the porch to see where it's raining and to see who got some. He [Lyman Strickland, her husband] had to stand out there to see where it's raining... there was so much thunder and lightning that I begged him to come into the house and finally there was this terrible thunderclap and he came in and said, 'Boy that was something.' You know how it will have a different sound. It has a hollow sound."[119]

The next day Brazel was in the pastures, checking to see where the rain had fallen. On horseback, with young neighbor Timothy Proctor, called Dee, he was planning to move the livestock to the fields where the rain fell because the grass would be softer and greener.[120] South of the ranch headquarters, in a field near a windmill and watering station, they came to a pasture filled with metallic debris. It ran from a small circle of hills, down an arroyo, up another, gentle slope, and disappeared on the reverse side.[121]

The debris looked to be metallic. Some of it was shiny, but most of it was a dull gray color. The debris was so densely packed that the sheep refused to cross it and had to be driven around it to get them to water.[122] Bessie Brazel

---

[119]Marian Strickland, videotaped interview with Randle, Schmitt and Berliner, September 27, 1991.

[120] Loretta Proctor, personal interview with Kevin Randle and Donald Schmitt, April 20, 1989. The question that comes to mind is why was Dee Proctor with him rather than either of his two children. Although we were told that Dee was with him, no one suggested that Vernon was with them and Bill had said that Mack had taken the children to Tularosa to be with their mother.

[121] Bill Brazel took both Randle and Schmitt out to where he had found some of the bits of metallic debris and showed them the lay of the land. He also described a gouge that ran down the center of the field that was tapered at both ends and expanded to a place near the center that was about ten feet wide. He told Bill Moore that the wreckage was scattered over "a patch of land about a quarter mile long or so and several hundred feet wide." Berlitz and Moore, *Roswell Incident*, p. 77.

[122] Bill Brazel, interview by Randle and Schmitt, March 31, 1989; Tommy Tyree, interview by Randle and Schmitt, August 12, 1989 in Corona, New Mexico.

would later suggest that the shiny metal blowing in the wind would frighten the sheep.[123]

With young Proctor, Brazel rode down into the field. There was debris but no sign of any occupants or people killed in a crash. Later, during his interrogation by the Army, he was told that "it wasn't anything made by us."[124]

Mack Brazel examined the material closely and although he was never interviewed by investigators, he did provide observations to family and friends and he did show them bits of the debris. According to Tommy Tyree, Brazel told him that the debris was so light that it stirred in the wind. It was so strong that pieces of it would not flex.[125] Brazel attempted to cut it with his knife and couldn't do it. Bill Brazel would later say that he attempted to whittle on a small piece to see if there was any stratification in it but was unable to raise a shaving.[126]

Mack Brazel picked up some small pieces of the debris and then took young Proctor home. There he showed some of it to Floyd and Loretta. According to an interview conducted in June 1979, Floyd Proctor said:

> He wanted me to come over with him and look at it, and described it as "the strangest stuff he had ever seen." I was tired and busy and just didn't want to bother going all the way over there right then… He described the stuff as being very odd.

---

[123] Bessie Brazel Schreiber affidavit, see Pflock, Karl. Roswell: *Inconvenient Facts and the Will to Believe*, Amherst, NY: Prometheus Books, 2001: p. 277.

[124] Carey, Tom and Don Schmitt, *Witness to Roswell*, Pompton Plains, NJ New Page Books: pp. 210 – 211.

[125] Various interviews including that with Tommy Tyree on August 12, 1989; Loretta Proctor on April 20, 1989, Bill Brazel, Jr. February 19, 1989; Sallye Tadolini, signed affidavit September 27, 1993.

[126] Bill Brazel, personal interview by Kevin Randle and Donald Schmitt, February 19, 1989

> He said whatever the junk was, it had designs on it that reminded him of Chinese and Japanese designs. It wasn't paper because he couldn't cut it with a knife, and the metal was different from anything he had ever seen. He said the designs looked like the kind of stuff you would find on firecracker wrappers… some sort of figures all done up in pastels, but not writing like we would do it.[127]

Loretta Proctor was there during the conversation. In later interviews, she would say that they didn't go down there because gas cost money and tires cost money and it would take time and "…everything was real short."[128] She did describe what she had seen when Brazel came by. She said:

> … he did bring a little sliver of a wood-looking stuff up but you couldn't burn it or you couldn't cut it or anything. I guess it was just a little sliver of it, about the size of a pencil and about three or four inches long… I would say it was kind of brownish tan but you know that's been quite a long time. It looked like plastic, of course there wasn't any plastic then but that was kind of what it looked like… We didn't [try to cut it]. He did and he was telling us about the other material that was so lightweight and that was crinkled up and then would unfold out. He said there was more stuff there, like a tape that had some sort of figures on it… He never did say what he had. But anyway, he went into Roswell and reported it and they flew him back out there to show them the location. They took him

---

[127] Berlitz and Moore, *Roswell Incident*, p. 83

[128] Loretta Proctor, telephone interview by Kevin Randle, April 20, 1989.

> back to Roswell and kept him down there until they got it all cleaned up.[129]

During the July 4th weekend, Brazel talked to friends and family about the debris he had found. Norris Proctor said that Brazel should try for the rewards being offered for proof that the flying saucers were something real.[130] Others told him to contact the sheriff or the government because it was probably one of the government's experiments, maybe from White Sands Proving Ground (later White Sands Missile Range). Brazel again approached the Proctors, but they still didn't want to take the time to drive down to see the strange debris. They did suggest that the military might want to know about it.[131]

The evidence suggests that Brazel went into town on Sunday, July 6,[132] and drove to the sheriff's office. Wilcox, of course, suggested that they call out to the air base. While they waited for someone to drive out to the sheriff's office, Frank Joyce, of radio station KGFL called, asking Wilcox if anything

---

[129] Ibid.; See also, Randle and Schmitt, *UFO Crash at Roswell*, p. 148; Loretta Proctor, telephone interview by Kevin Randle, April 20, 1989.

[130] It seems unlikely that Brazel had heard about these rewards which totaled three thousand dollars. He had no radio and no newspaper. Most of the stories mentioning the rewards were published after the weekend and after Brazel had gone into Roswell. One of them had expired at sundown on Sunday. There seems to be no evidence that a reward was the motivation for Brazel going to Roswell.

[131] Norris Proctor, personal interview by Kevin Randle and Don Schmitt, August 12, 1989.

[132] There is a point of contention here. Most newspaper accounts from July 1947 suggest that Brazel came into town on Monday, July 7. Although Berlitz and Moore report in *The Roswell Incident* that Marcel said it was July 7, the earliest interviews with Marcel, those conducted by Friedman and Len Stringfield, Marcel was unsure of the date and said only that it was 1947 while he was assigned to the base in Roswell. In his interview with Bob Pratt on December 8, 1979, Marcel said, "I don't remember the exact date. It was in July 1947." Finally, a time line constructed working backward from the press release suggests that Marcel and Cavitt followed Brazel out later Sunday afternoon, arriving at the ranch about dusk. They spent most of Monday in the field, putting them back in town late in the afternoon of July 7.

was going on. Wilcox was unimpressed with Brazel's tale but then handed the phone to Brazel saying, "Here's something that you might find interesting."[133]

Joyce, however, offered a slightly different version of this in an interview conducted on March 31, 1989. He said, "...he [Brazel] called me on the phone and the reason he called was because I was a radio personality in the day... on a little hundred-watt radio station... I am sure I figured it out he heard me reading this story about the guy in Washington who had just seen thirteen or fourteen of them."[134]

Joyce listened to what Brazel had to say, but exactly how he treated it in his radio report has been lost. Joyce only said that he had interviewed Brazel while Brazel was at the sheriff's office and before the military had arrived.[135] He was not very impressed with what Brazel had to say. Joyce said, "I have never told anyone exactly what W. W. [Mack Brazel, as Joyce insisted on calling him] told me. I have and the reason was that after all he's dead, and it would be like putting words in a dead man's mouth."[136]

When the military returned to the sheriff's office, Brazel agreed to take them out to the debris field. Although Marcel told Moore that they had arrived at the ranch late in the evening and remained overnight, Sheridan Cavitt, at the time a captain and the senior Counterintelligence Corps officer at the base, later disputed this in an interview conducted in 1994 by Colonel Richard

---

[133] Frank Joyce, personal interview by Kevin Randle and Don Schmitt, March 31, 1989, Albuquerque, New Mexico.

[134] Frank Joyce, telephone interview by Kevin Randle, March 30, 1989. It should be noted here that Mack Brazel had neither telephone nor radio at the ranch and the signal of KGFL would not have reached to the ranch anyway. Joyce's suggestion that he had called the sheriff's office looking for anything interesting is a more likely scenario.

[135] Karl Pflock, in *Inconvenient Facts*, p. 121, noted, "Joyce, the Radio KGFL announcer and United Press stringer who might have had a scoop had he taken Mack Brazel seriously the first time he spoke with him..." This suggests that Joyce might not have even put anything on the air, which makes sense when other facts are considered, especially the timing of the events after Brazel found the debris.

[136] Frank Joyce, telephone interview with Kevin Randle, March 30, 1989.

Weaver.[137] In the morning Brazel saddled two horses, Cavitt, climbed onto one horse and Brazel took the other. Marcel drove.[138] Once they were out at the debris field, Brazel left them alone. Cavitt said there wasn't much to see and they picked up some of the debris quickly.

Walt Whitmore, Sr., then the majority owner of radio station KGFL, had heard about the Brazel story and wanted to learn more. George "Jud" Roberts was the minority owner. He said, "I don't know who tipped us off but the rancher came into town… That's how Mr. Whitmore… got in touch with him and we decided that we had us a hell of a scoop. So, as a matter of fact, he kind of made it possible for the rancher to stay overnight at his house instead of going back to the ranch."[139]

Walt Whitmore, Jr., confirmed that Brazel had stayed at the house. Whitmore said that he had been in law school in Colorado and had come home late to find Brazel in his bed. Whitmore had to sleep on the couch.[140] Brazel, up early to make coffee, woke Whitmore and they talked about why he was in town and what he knew. He told Pflock:

> …Brazel sketched a map for me, showing which roads to take and how to find the site. I drove there… a distance of 65 or 70 miles. No one was there when I arrived. I do not remember

---

[137] Richard L. Weaver and James McAndrew, *The Roswell Report: Fact vs. Fiction in the New Mexico Desert*, 1995, section 18.

[138] Cavitt, of course, disputed this, but did say that the description, of the one officer as a "good, old West Texas boy," who could ride sounded like him, but he didn't remember Marcel going out to the site with him. Cavitt, personal interview with Kevin Randle and Donald Schmitt, June 25, 1994. As a side note, Cavitt denied that he had been interviewed by anyone, though it is clear from the timing that Weaver had been there about a month earlier.

[139] Jud Roberts, videotaped personal interview by Kevin Randle and Don Schmitt, January 31, 1990, Roswell, New Mexico.

[140] Karl Pflock, *Inconvenient Facts*, pp. 153 – 154; Walt Whitmore, Jr. personal interview with Kevin Randle, September 1993.

seeing any sign anyone had been on the site, and I saw no one else while I was there…

The debris covered a fan- or roughly triangle-shaped area which was about 10 or 12 feet wide at what I thought was the top end. From there it extended about 100 to 150 feet, widening out to about 150 feet at the base. This area was covered with many, many bits of material.[141] Many pieces had been blown out of the main area, and I could see them stuck to bushes as far as a city block away.

Most of what I found was white, linen-like cloth with reflective tinfoil attached to one side. Some of the pieces were glued to balsa-wood sticks, and some of them had glue on the cloth side, with bits of balsa stuck to it. Most of the pieces were no larger than four or five inches on a side, although I found one or two about the size of a sheet of typing paper… One of the larger pieces of foiled cloth, measuring about 8 by 12 inches, had writing on the cloth side. Someone had used a pencil to do some figuring, arithmetic. There were no words, only numbers. I did not see any writing or markings on any of the other debris.

I collected some of the foiled-cloth material, including the pieces with the writing on it, and a few sticks, filling a large, 9 by 12 envelope with it… I still have the material… It is… stored in a safe and secure place.[142]

---

[141] This would confirm the information by Tommy Tyree that the debris was thickly strewn over the field in a dense pattern. Tyree personal interview by Kevin Randle and Donald Schmitt, August 12, 1990.

[142] Karl Pflock, *Inconvenient Facts*, p. 154. Whitmore told Randle in September 1993 that the debris had been in his safe deposit box until he took a trip to Europe and had

This tale told by Whitmore in recorded conversations with Karl Pflock in 1992 and 1993 does not square with the stories that Whitmore had told earlier to Bill Moore.[143] In this version, according to Moore:

> Whitmore, Jr., said that while he did not see the actual crash site until after the Army Air Force [sic] had "cleaned it up," he did see some of the wreckage brought into town by the rancher. His description was that it consisted mostly of a very thin but extremely tough metallic foil-like substance and some small beams that appeared to be either wood or wood-like. Some of the material had some sort of writing which looked like numbers that had been either added or multiplied. He recalls that his father went out to the site in a Buick but was turned back by armed MPs who had set up a road block…
>
> Several days later Whitmore, Jr., ventured out to the site and found a stretch of about 175 – 200 yards of pastureland uprooted in a sort of fan-like pattern with most of the damage at the narrowest part of the fan… He added that the largest piece of this material that he saw was about four of five inches square, and that it was very much like lead foil in appearance but could not be torn or cut at all. It was extremely light weight.

---

to store, other, more valuable items in it. The debris was moved to Whitmore's "junk room," and from there it disappeared. No one ever saw it, and though he promised to provide it to Max Littell at the International UFO Museum and Research Center upon his death, no one, in various searches of his junk room and home, was able to find the debris.

[143] Pflock reported, "I asked Whitmore about this during our October 31 [1992] interview, and he emphatically stated that Berlitz and Moore had misquoted him. He told me he pointed this out to Moore in detail after the book [*Roswell Incident*] was published in 1980, but no corrections were made in subsequent printings or for the 1988 paperback edition." See page 166, footnote 31, in *Inconvenient Facts*.

Brazel, after breakfast in the Whitmore house, was taken to the radio station for an interview. Roberts said, "…we had those old RCA wire recorder which weighed 85 pounds which we used to haul around."[144] Roberts said that the interview with Brazel lasted about an hour, but that he hadn't sat in on it, so he didn't know what was said.

That interview never aired. "We had the material but we were tipped off by Washington that this was a closed in matter and that anyone that didn't play ball, why we could start hunting for our licenses. It was as simple as that."[145]

The Army was apparently searching for Brazel while he was staying with Whitmore. Once the radio interview was conducted, Whitmore took Brazel out to the base. The question is: why did the interview not air prior to Brazel returning to military control and before anyone could learn what he might have said?

The answer seems to be the nature of the wire recorder. According to Roberts, it took as long to rewind as the interview had taken to record. In other words, if the interview was an hour long, it would take that long for the wire to rewind. They simply might have not have had the time.

Brazel, then, on Tuesday, July 8 was in Roswell and spent the night with Whitmore. From that point on, he was held by the military, and according to various sources, prevented from leaving the base without escort.

There is, however, other testimony that disputes this. Bessie Brazel Schreiber, who had apparently been on the ranch during the thunderstorm said she, her brother and mother had accompanied Mack into Roswell. She said that it was a day or two later that the military arrived at the ranch.[146] She said

---

[144] Jud Roberts, videotaped personal interview by Kevin Randle and Don Schmitt, January 31, 1990, Roswell, New Mexico.

[145] Ibid

[146] Berlitz and Moore, *Roswell Incident*, pp. 86 – 87; Pflock, *Inconvenient Facts*, pp. 23 – 24, 162 – 163, 277 – 278; personal interview by John Kirby and Don Mitchell, March 8, 1995; Randle, Kevin D. "Bessie Brazel's Story," *International UFO Reporter*, 20,3 (May/June 1995): pp. 3 – 5, 24.

that her father hadn't gone back into Roswell nor had he been kept there for a number of days.[147]

There were a number of people who saw Brazel in the company of the military. Floyd Proctor was in Roswell with L. D. Sparks when "...we were in town one time, and he was all surrounded by military men, at least a half dozen, and walked past us like he didn't even know us."[148]

Had that been the lone testimony about it, the idea of Brazel in military custody might be ignored. Marian Strickland remembered Brazel sitting at her kitchen table, talking about his time in Roswell. She said, "He was held for, I believe, it was three days and three nights... I think he called Bill... They literally threw him in jail. It was two or three days later Lyman [Strickland] and Floyd Proctor had some business with the Bureau of Land Management (BLM) which had an office... They saw Mack being escorted by two Army men..."[149]

Bill Brazel, of course, said that his father had been gone from the ranch a number of days. According to Bill, he learned of his father's plight in the newspaper on July 10. He said that he didn't arrive at the ranch until two or three days later, and that his father arrived two or three days after that.[150] Given the timing, it seems that Brazel was held for six to eight days.

---

[147] This belief by Schreiber is contradicted by the newspaper articles that appeared around the country and the story published on July 9, 1947 in the "Harassed Rancher" article in the *Roswell Daily Record.* Clearly Mack Brazel was in Roswell on that day for interviews by the reporters dispatched from Albuquerque and El Paso. Bill Brazel also said that he had gone to the ranch after seeing a story about his father in the newspapers because he knew no one would be there. His father didn't return for several days after Bill arrived there.

[148] Berlitz and Moore, *Roswell Incident*, p. 84

[149] Marian Strickland, personal videotaped interview with Kevin Randle, Donald Schmitt and Don Berliner, September 27, 1991 in Roswell, New Mexico.

[150] Bill Brazel, personal interview with Kevin Randle and Donald Schmitt, February 19, 1989; Randle and Schmitt, *UFO Crash at Roswell*, p. 43; see also Carey and Schmitt, *Witness to Roswell*, p. 54

Even with a number of Brazel's neighbors mentioning that he had been in military custody, there is one other source. In July 1947 Major Edwin Easley was the provost marshal at Roswell. He was in charge of the military police and the base security.[151] Easley said that Brazel had not been held in the stockade, but had been put up at the guest house on base.[152] While not the same as being in jail, he was not allowed to leave without an escort.[153]

Brazel made the rounds of the local media. At the *Roswell Daily Record* office, he gave an interview to Jason Kellahin and Robin Adair of the Associated Press in Albuquerque[154]and to a local reporter who provided information about what Brazel had found. In that article, Brazel said that they had found the debris on June 14 and called it a "large area of bright wreckage made up of rubber strips, tinfoil, a rather tough paper and sticks."[155]

Brazel provided many other details about the wreckage. The newspaper noted that:

---

[151] Unit History, 509th Bomb Group for June, July and August, 1947; 1947 Roswell Army Air Field Yearbook prepared by Walter Haut; Edwin Easley personal audiotaped interview by Kevin Randle, January 11, 1990.

[152] Edwin Easley, personal interview by Randle February 4, 1990.

[153] Karl Pflock, *in Inconvenient Facts*, p. 170, complained that the interview with Easley that mentioned that fact had not been recorded, though he was aware of the circumstances surrounding the interview and that Randle had taken notes. Had this been the only reference to Brazel being held in Roswell by the military, it could probably be dismissed. There are several other references, as outlined, including Brazel's son Bill, Floyd Proctor, Lyman and Marian Strickland and Loretta Proctor. Clearly Brazel was kept in Roswell for several days and was seen in the company of military men on, at least, one occasion.

[154] "Send First Roswell Wire Photos from Record Office," *Roswell Daily Record*, July 9, 1947, p. 1

[155] "Harassed Rancher who Located 'Saucer' Sorry He Told About It," *Roswell Daily Record*, July 9, 1947, p. 1.

> Brazel said he did not see it fall from the sky and did not see it before it was torn up, so he did not know the size or shape it might have been, but he thought it might have been about as large as a table top. The balloon which held it up, if that was how it worked, must have been 12 feet long, he felt, measured by the size of the room in which he sat. The rubber was smoky gray in color and scattered over an area about 200 yards in diameter… That when gathered up the tinfoil, paper, tape, and sticks made a bundle about three feet long and 7 or 8 inches thick, while the rubber made a bundle about 18 or 20 inches long and about 8 inches thick. In all, he estimated, the entire lot would have weighed maybe five pounds… There were no words to be found anywhere on the instrument although there were letters on some of the parts. Considerable scotch tape and some tape with flowers printed upon it had been used in the construction.[156]

Skeptics, in their analysis of the Roswell case, often quote this part of the interview, suggesting that the materials described seem to be a strange way to construct a spaceship. They believe that this description points to a weather balloon, as was suggested by Brigadier General Roger Ramey at the Headquarters, 8th Air Force in Fort Worth.[157] They ignore the second to the last paragraph in which it was reported, "Brazel said that he had previously found two weather observation balloons on the ranch but that what he found this time did not in any way resemble either of these."[158]

---

[156] Ibid.

[157] "Gen. Ramey Empties Roswell 'Saucer," *Roswell Daily Record*, July 9, 1947, p. 1

[158] "Harassed Rancher who Located 'Saucer' Sorry He Told About It," *Roswell Daily Record*, July 9, 1947, p. 1.

Brazel was also taken to the officers of KGFL where he was again interviewed by Frank Joyce. This interview was different, with Brazel giving Joyce the story of the balloon-like debris. Joyce pointed this out to Brazel but Brazel said that it would go hard on him if he didn't stick to the new story. And Joyce said, "I told him, what you're saying is not what you were saying the other night. And then he turned around and admitted that he had been told to come in or else."[159]

Joyce would say that there were two officers with Brazel when he came by the KGFL office.[160] Joyce said, about the military officers waiting in the front room, "I knew they were there but I didn't know who they were."[161]

As Brazel was leaving, according to Joyce, he turned and then asked about the little green men. After telling Joyce they weren't green, he left but not without opening a door.[162] That one statement implied something that hadn't been discussed to that point. Nowhere in the first interviews, various press releases, or newspaper articles, had anyone talked about alien bodies. This was Brazel telling Joyce that he had seen something more than just the metallic debris, and telling Joyce that the story of the balloon was a cover for the real event.

Jesse Marcel, Sr., had not talked of bodies,[163] nor had Bill Brazel. This was the one area that required more protection. Metallic debris, no matter how ex-

---

[159] Randle and Schmitt, *UFO Crash at Roswell*, p. 165, based on personal interviews with Joyce on March 31, 1989 and January, 20 1990.

[160] Frank Joyce, personal interview with Kevin Randle and Donald Schmitt, March 31, 1989.

[161] Ibid.

[162] Randle and Schmitt, *UFO Crash at Roswell*, Karl Pflock, *Inconvenient Facts*, Carey and Schmitt, *Witness to Roswell.*

[163] Jesse Marcel, Jr. said repeatedly that his father never mentioned bodies to him, however a relative, Norris Marcel had suggested that he heard about bodies from the

otic, was metallic debris. It could be explained as experimental, usual, and impossible to find in the civilian world, but bodies of alien creatures could not be explained as easily.

At what point, then, did Brazel discover the bodies of the alien flight crew. Was it while he was still riding the range with Dee Proctor, or was it sometime between the point that Marcel and Cavitt left and Walt Whitmore arrived to take him back to Roswell?

Dee Proctor never really talked to investigators about the event though I spoke with him twice, both times briefly and both times by accident.[164] He did confirm that he had been with Brazel and that military authorities had talked to him in the days that followed the discovery. In fact, it was clear that he had not only been there, but he had taken some friends out there with him. He would not say what the military told him, nor would he say much about what he had seen, other than to say that it was a field with metallic debris and the remnants of a craft. It was clear that these experiences with the military left a lasting impression on him which guided what he said for the remainder of his life.

Proctor died in January, 2006 at age 65. He had always been a somewhat reclusive man, quick to anger and reticent to talk about these events. In 1996, he took his mother to a bluff about ten or so miles from their ranch house and about two and a half miles from the debris field. He told her that was the field in which more than just debris had been found.[165] Any trace of the craft or its impact was long gone in 1996.

---

elder Marcel. Given the relationship between father and son, had Marcel, Sr. seen bodies and told anyone about them, it would have been Jesse Jr. The circumstances suggest that Marcel, Sr. had not seen bodies and had been cut out of the loop at that point.

[164] Each of those telephone calls was the result of attempts to speak with his mother. In neither case was it expected that he would be at his mother's home. In the second of those, he was caught off guard as well, but did answer some benign questions, at least confirming that he had been with Brazel and he had seen both debris and remnants of the craft. Since the first crash site was only metallic debris, this means that Proctor had been at another site where bodies were found.

[165] Carey and Schmitt, *Witness to Roswell*, pp. 138 – 139.

He also said that the craft was of extraterrestrial origin, though those words came from the older man fifty years later and not the seven-year-old boy in 1947.[166] Such is the contamination in the world of today.

It would have been on the afternoon of July 8, 1947, after Marcel had returned to tell Colonel Blanchard about what he had found, and after there had begun a recovery effort, that bodies were located. Men of the 509th Bomb Group who had not been involved in other activities or whose specific duties could wait were ordered to assist in the recovery operation. One those would have been Melvin Brown. Although he died before any UFO researcher spoke with him, his family was aware of the story. Timothy Good interviewed Beverley (Brown) Bean who said her father who had been stationed at Roswell in 1947.[167] Brown's photograph appears in the Roswell *Yearbook*, verifying that he was there in July 1947. Further verification came on one of the documents offered by Bean to prove her father served in Roswell. This was an order with several names on it including that of Major Jesse Marcel.[168]

In a videotaped interview conducted in England on January 4, 1991, Bean said, "Dad used to tell us this story and he didn't tell us often."[169] He said he "had to go out into the desert. All available men were grabbed and they all went out into the desert in trucks where a crashed saucer had come down."

---

[166] Dee Proctor, unrecorded conversation with Randle, 1994.

[167] Good, Timothy. *Alien Contact: Top Secret UFO Files Revealed.* New York: William Morrow and Company, 1991: pp. 99 – 101, 276 – 277. See also, Randle, Kevin D. *Roswell Revisited.* Lakeville, MN: Galde Press, Inc. 2007: pp. 25 – 28.

[168] Randle, *Roswell Revisited.* p. 25.

[169] Beverley Bean, videotaped interview in her home in England, January 4, 1991, copies supplied to the Fund for UFO Research and the J. Allen Hynek Center for UFO Studies.

Brown, and another soldier whose name he never gave to his daughter and whose name he might not have known, were pulled aside for guard duty.[170] They were told not to look under the tarp in the truck, but Bean said, laughing, that the minute someone tells you that, the first thing you do is take a look. She said that her dad told her, "He and this other guy lifted up the tarpaulin or something..."

She said, "I can remember my dad saying he couldn't understand why they wanted refrigerated trucks. And him [sic] and another guy had to sit on the back of a truck to take this stuff to a hangar. They were packed in ice. And he lifted up the tarpaulin and looked in, and saw two, possibly three bodies."[171]

She said that she and her sister now argue about the number of alien creatures under the tarp. Bean says it was two, but her sister insists that it was three. No matter now. The point is that Brown described the creatures for them.

According to her, "He said they were smaller than us, not more than four-foot-tall... much larger heads than we have. Slanted eyes and [the skin was] yellowish."[172] She said, "They looked Asian… but had no hair. He said they could have passed for Chinese."[173]

Bean wondered if he had been scared but he said that he wasn't. He thought they had nice faces and they looked as if they would have been friendly. According to Bean, he repeated that as often as he told the story, which, over the years was fewer than a dozen times.

Bean, of course, sometimes pestered him for more information. After the release of *Close Encounters of the Third Kind* in 1977, she asked him about

---

[170] It is not unusual for soldiers not assigned as MPs or to security to be tasked for guard duty when the numbers needed exceed those available. Part of all soldiers' basic training is training in performing guard duties.

[171] Good, Timothy. *Alien Contact:* p. 99.

[172] Good. *Alien Contact*. p. 99. Also see Randle, *Roswell Revisited*. 26; Videotape interview with Bean, January 4, 1991.

[173] Ibid.

the movie and how authentic it might be. He said that it was the biggest load of crap he'd ever seen and not like the real thing at all. When she tried to learn more, he told her, "That's all I can tell you. I can't tell you anymore."

The late Karl Pflock, in his book, *Roswell, Inconvenient Facts and the Will to Believe,* complained that Bean's story was second hand and that neither her sister nor her mother would comment on it.[174] Pflock had to know that both the mother and the other daughter had confirmed the tale because he had access to the video tapes of those 1991 interviews. He is right about this being a tale told by the daughters and wife of the man who lived it. By the time his name surfaced in the investigation, he had died from complications of various lung diseases, but it is not true that his wife or other daughter refused to talk about this or that they didn't corroborate what Bean was saying.

Ada Brown added little to the complex tale told by Beverly Bean when she was interviewed on videotape in January 1991. She merely confirmed that she too had heard about the crash over the years and that it was something from another world.[175] She seemed a little uncomfortable sharing a secret left by her husband.

Bean's sister, Harriet Kercher, on January 4, 1991, was also interviewed on videotape. She had heard her father tell his tales a couple of times when Beverly was there, but there was one incident when Beverly was absent and her father gave her just a little more information to her sister.

Kercher, in her early teens said that she was with friends when she saw something flash by. Her friends saw it too, and then, in the distance, that something reappeared and seemed to be coming at them. Kercher said they were frightened by that shiny object but they weren't far from her house so they ran there, slamming the door behind them.

---

[174] This is the same criticism that Kal Korff raised in his book, but the facts do not bear this out. Both Bean's sister and her mother, in videotaped interviews on January 4, 1991 reported to have heard these stories from Melvin Brown.

[175] Ada Brown, videotaped interview, January 4, 1991. Copies supplied to the Fund for UFO Research and the J. Allen Hynek Center for UFO Studies.

Her father met them and asked them why they seemed to be in such a panic. Kercher said that her father, after hearing the tale of the shining object, told her, "It's nothing to be frightened about."[176]

The friends didn't understand, exactly, what he meant and he told them about the crashed flying saucer, saying that there were a few bodies on it. He provided few new details. He just made it clear that there was something about the creatures that suggested to him that they were not to be feared.

It was also on the afternoon of July 8 that an officer from the base visited the Roswell Fire Department. Frankie Rowe had said that her father, a fire lieutenant with the Roswell department in 1947 had been to the crash site.[177] She said that her father had come home after his shift at the fire station (which lasted about twenty-four hours) and had something important to say. He then told them, according to Rowe, that they had gone about thirty miles outside of Roswell and then a few miles back to the west on the dirt roads.

He said there had been some kind of a crash and that he had called it a spaceship or a flying saucer or something. Then she said one of the most important things. According to her, "I remember him saying that some of them helped pick up some pieces of the wreckage. He said he saw two bodies in bags and one that was walking around."[178]

She said, "...he said he was sure that there were bodies because the third one would go over to them... he talked about [how] this third one would go

---

[176] Harriet Kercher, videotaped interview, January 4, 1991.

[177] Frankie Rowe, personal interview by Kevin Randle and Donald Schmitt, November 1990, September 1991. Interviews by Randle, January 2, 1993 and January 27, 1993. Also, interview with J. L. Smith, telephone interview 2009. See also: ufor.blogspot.com/2009/03/**roswell-fireman**-confesses-it-was-flying.htm.

[178] Frankie Rowe, multiple interviews with Randle and Schmitt. See also, Pflock. *Inconvenient Facts*. p. 46 – 47; Randle and Schmitt, *Truth About UFO Crash*. p. 16 -17; Randle, *Roswell Revisited*. p. 21 – 24.

back and forth between different parts of the wreckage and was walking around dazed. He didn't say if anyone tried to talk to this person."[179]

The creatures were, according to what Rowe remembered her father had said, about the size of a ten-year-old, meaning that they were smaller than a human adult. He said "the face looked like that little bug, Child of the Earth [more commonly called the Jerusalem Cricket]."[180] It was sort of copper colored or maybe a sort of dark brown.

Unfortunately, as has happened so often in this case, no researcher had a chance to talk with Rowe's father to get his first-hand observations. He died long before the investigation began. But I did have the opportunity to talk with her sister, Helen Cahill. She was married in 1947 and living in California at the time of the crash, but had heard some discussion about the events during a visit to New Mexico in 1960. Although her information wasn't as complete as that of Rowe, it confirmed, for what it's worth, that Rowe did not invent the tale of the crash. Of course, it does little to validate it, except to suggest that Rowe's father was talking about a UFO crash long before the reports of the Roswell events came to light and at a time when few people thought of UFOs as being from other worlds.[181]

There has been, however, some corroboration for Rowe's story. Tony Bragalia and I have been in contact with a former member of the Roswell Fire Department. This man was interviewed by Karl Pflock and Pflock cited him as saying the Roswell Fire Department didn't make a run outside the city to the crash site.[182]

---

[179] Frankie Rowe in various interviews with UFO researchers. See also: Pflock. *Inconvenient Facts*. p. 46 – 47; Randle and Schmitt, *Truth About UFO Crash*. p. 16 -17; Randle. *Roswell Revisited*. p. 21 – 24.

[180] Frankie Rowe, personal interview with Kevin Randle, January 2, 1993.

[181] Helen Cahill, telephone interview with Randle, September 1993: See also: Randle and Schmitt, *Truth About UFO Crash*. p. 17.

[182] Pflock. *Inconvenient Facts*. p. 63 – 64, see also Footnote 41, p. 66 for more details on Pflock's interviews.

For Pflock, this disproved Frankie's story.[183] And the man, J. L. Smith, said the same thing in 1993. The fire department didn't make a run to the crash site. But then the retired fire fighter said something else. He said that a colonel had come out from the base and told them not to go out there. That they, the military, would handle anything on that site.[184]

When asked if he knew Dwyer and he said that he did. When asked about Dwyer making a run outside the city he said that "colonel" from the base had advised against it. The Fire Department didn't make the run but that Dwyer, in his personal car did drive out to the crash site.[185] Dwyer and not the fire department, which explains why there is no record of it and why other fire fighters didn't remember it.[186]

---

[183] Pflock also claimed that a former city councilman had said that the Fire Department didn't make runs outside the city limits. The councilman was Max Littel and he wasn't on the city council in 1947. He didn't serve until the 1950s. Records in the Fire Department do show runs outside the city in June 1947 but no record exists for a run on the dates in question.

[184] J. L. Smith, telephone interview with Randle, 2009. See also: ufor.blogspot.com/2009/03/**roswell-fireman**-confesses-it-was-flying.htm.

[185] In *Witness to Roswell*, Carey and Schmitt report on Lee Reeves who apparently served in various capacities for the city including that as a fire fighter. According to what Reeves' son told them, when the call came in that there had been a crash north of town, Dwyer and Reeves were dispatched in what was described as the fire station's tanker, a pick-up truck with a tank in the back. They arrived before the military so that they had an opportunity to see that what had crashed was not an aircraft but some sort of egg-shaped vehicle. The description of the events in the book is confusing as is the source of the Reeves' tale. One quote that is clearly from Frankie Rowe is attributed to a source who requested anonymity but was the son of Dr. Foster's housekeeper according to them. The evidence and documentation available does not support this story and while Reeves was a member of the fire department at some point, in 1947 he is listed as working in private industry. Had he and Dwyer made a run in a fire department vehicle that should have been properly noted in the fire log. This tale is, at best, second hand but without a named source, is virtually useless.

[186] J. L. Smith, telephone interview with Randle, 2009. See also: ufor.blogspot.com/2009/03/**roswell-fireman**-confesses-it-was-flying.htm.

The retired fire fighter was quite clear about these points. They had been visited by an officer from the base, they had been told not to go out there, but Dwyer, in his personal car, did. From his observations out on the site, he told his family what he had seen and he told some of his fellow firefighters what he had seen.

By the end of the day on July 8, the situation had changed. The story had shifted from Roswell and the report of the flying saucer crash was hidden by other news. Brigadier General Roger Ramey had claimed there was nothing to the report.[187] The Army and the Navy began a "campaign to stop the rumors" of flying saucers.[188] While that was going on, the clean up outside of Roswell continued.

---

[187] "General Ramey Empties Roswell Saucer." *Roswell Daily Record*. July 9, 1947. p. 1

[188] "Army, Navy Open Drive To Dry Up Saucer Talk." *Phoenix Gazette*, July 9, 1947; "Services Try To Stop 'Disc' Talk." *Dallas Times Herald*, July 9, 1947. Dozens of other newspapers printed similar stories that were sent over the news wires by the United Press.

Chapter 4:

Another Crash Location and Examining the Metallic Debris

Over the years there have been those who have suggested that Don Schmitt and I had not found the debris field. However, we were taken to that spot by Bill Brazel, who parked his truck and pointed at the ground saying that here was where he had found some of the strange metallic debris. He would describe three types of debris. One he suggested was as light as balsa wood but extremely tough, one was like fiber optics, and one was a heavy foil-like material that could be crumbled into a ball which would then unfold itself.[189]

Bill Brazel showed the debris to his friends at the time. Sallye Tadolini, the daughter of Lyman and Marian Strickland said that she had an opportunity to see and touch the debris.[190] Her signed affidavit said:

> (4) In July 1947, I was nine years old and lived with my parents, Lyman and Marian Strickland... The neighboring ranch was the Foster place, which was managed by William W. ("Mac") [sic] Brazel...

---

[189] Complete descriptions of the material appear earlier int his work. See also, Randle, Kevin D. and Donald R. Schmitt. *UFO Crash at Roswell*. New York: Avon Books, 1991: pp. 52 – 53, 129 – 133; Bill Brazel, personal interview by Kevin Randle and Don Schmitt, February 19, 1989

[190] Randle, Kevin D. and Donald R. Schmitt. *The Truth about the UFO Crash at Roswell*. New York: M. Evans and Company, 1994. pp. 138, 140, 173. Signed affidavit collected by the Fund for UFO Research, published in Pflock, Karl. *Roswell: Inconvenient Facts and the Will to Believe*. Amherst, NY: Prometheus Books, 2001. p. 285.

(5) I remember my parents talking about Mac Brazel finding a lot of unusual debris in one of his pastures and that there was a great deal of excitement about it among the neighbors. I recall the adults at first thought it was some kind of newfangled weather balloon, then deciding there was no way it could be anything like that. I also recall that, later, the neighbors talked about how badly Mac Brazel had been treated and that when he came back to the ranch, he never again wanted to talk about what he had found.

(6) A week or so after the excitement, Mac's son, Bill, who was quite a bit older and married, stopped by our house. He had someone with him, and while I am not absolutely certain, I think it was his brother Vernon, who was my age. We – my father, brothers, myself and possibly my mother[191] – sat at the kitchen table with them. Bill showed us a piece of the thing his father had found, and he asked us not to say anything about it.

(7) What Bill showed us was a piece of what I still think of as fabric. It was something like aluminum foil, something like satin, something like well-tanned leather in its toughness, yet it was not precisely like any of one of those materials. While I do not recall this with certainty, I think the fabric measured about four by eight or ten inches. Its edges, which were smooth, were not exactly parallel, and its shape was roughly trapezoidal. It was about the thickness of very fine kidskin glove leather and a dull metallic grayish silver, one side

[191] In various interviews with Marian Strickland, she made it clear that she was on the periphery of these events and rarely participated in them. In those interviews she never indicated that she had seen or held any of the debris that had been found.

> slightly darker than the other. I do not remember it having any design or embossing on it.[192]
>
> (8) Bill passed it around, and we all felt of it. I did a lot of sewing, so the feel made a great impression on me. It felt like no fabric I have touched before or since. It was very silky or satiny, with the same texture on both sides. Yet when I crumpled it in my hands, the feel was like that you notice when you crumple a leather glove in your hand. When I released it, it sprang back into its original shape, quickly flattening out with no wrinkles. I did this several times, as did the others. I remember some of the others stretching it between their hands and "popping" it, but I do not think anyone tried to cut it or tear it.
>
> (9) While all I saw was the piece of fabric, I remember hearing discussions about what must have been part of the frame, which was said to be somehow very different. I also remember Mac Brazel referring to – and I think these were his exact words – "all that junk all over out there." These recollections make me think there must have been more than just a lot of fabric there.[193]

Though not as impressive, other civilians did have the chance to see debris. Tommy Tyree wasn't exactly a ranch hand for Brazel but did work with him

---

[192] In the Air Force report written by Richard L. Weaver and James McAndrew, *The Roswell Report: Fact vs Fiction in the New Mexico Desert* (page 23 of the Executive Summary), they quote from this paragraph of Tadolini's affidavit. They ignore the paragraph that follows which provides more information and more description of the material she handled.

[193] Pflock, *Inconvenient Facts*. p. 285.

on occasion. He did not see the debris field before it was cleaned but he had heard Brazel complain about it. The debris was scattered over a large area and the sheep refused to cross it, meaning that Brazel had to drive them around it to get them to the water.[194]

Tyree also said that while he had not handled the debris himself, he was told that it was so light that it was stirred by the wind.[195] More importantly, Tyree said that while on the range one day, Brazel pointed at a sinkhole that had water standing in the bottom of it. Floating on top was a small piece of the metallic debris. To both of them at that time, the sinkhole was too deep and it was too much trouble to attempt to recover the material.[196]

Dr. John Kromschroeder wasn't on the scene in 1947, but did say that he was shown a fragment some years later. Pappy Henderson, who was one of the pilots who allegedly flew debris onto Wright Field, apparently had a piece of the metallic debris. Kromschroeder, who had an interest in metallurgy, said that he had never seen anything like it.[197]

The metal was gray and resembled aluminum but was harder and stiffer. He said that he was unable to bend it, but had to be careful handling it because the edges were sharp. He said that it seemed to have a crystalline structure, and

---

[194] Tommy Tyree, personal interview with Kevin Randle and Don Schmitt, August 12, 1989 in Corona, New Mexico.

[195] Bessie Brazel Schreiber said that the reason the sheep refused to cross the field was they were spooked by the blowing, flashing debris. See Pflock, Inconvenient Facts for Schreiber's affidavit, pp. 277 – 278.

[196] Ibid. Although that idea seems preposterous in today's world, at that time, in 1947, neither of them understood the overall importance that such a scrap might present. The story had faded from the public arena by the time they saw the debris.

[197] Dr. John Kromschroeder, personal interview by Kevin Randle and Stan Friedman, January 29, 1990. See also Randle and Schmitt. *Truth about UFO Crash.* p. 139; Friedman, Stanton and Don Berliner. *Crash at Corona.* New York: Paragon House, 1993: pp. 126 – 127.

based that on the fracturing of it and based on an examination under magnification. He said that it hadn't been torn.[198]

According to Kromschroeder, Henderson had told him the metal was part of the lighter material lining the interior of the craft. He said that when properly energized, it produced perfect illumination. It cast a soft light with no shadows.[199]

Henderson's debris apparently came from Major Ellis Boldra. He had subjected it to a number of tests. He said that it was thin, incredibly strong, and dissipated heat in some manner. Boldra had used an acetylene torch on the metal but couldn't get it to melt. It barely got warm and didn't glow. Once the heat source was removed, it cooled rapidly and could be handled in seconds.[200]

Other members of the military also had the chance to see the metallic debris. Major Jesse Marcel, Sr. said that it was clear to him "something... must have exploded above the ground and fell."[201]

Marcel said that he was able to determine "which direction it came from and which direction if was heading. It was in that pattern... You could tell that it was thicker where we first started looking and it was thinning out as we went southwest."[202]

---

[198] Ibid. Also, Dr. John Kromschroeder, telephone interview with Kevin Randle, July 15, 1990.

[199] Ibid.

[200] Greg Boldra, telephone interview with Kevin Randle, January 1992. See also, Randle and Schmitt. *Truth about UFO Crash*. p. 139.

[201] Jesse Marcel, Sr. personal interview conducted by Bob Pratt, December 8, 1979. Full transcript available in Pflock, Karl. *Inconvenient Truth*. pp. 225 – 233. Pflock interpreted the interview from Pratt's notes and there are places in which the placement of a comma alters the meaning of the sentence. Full photographs of that interview available at http://kevinrandle.blogspot.com/2013/09/jesse-marcel-sr-bob-pratt-and-interview.html.

[202] Pratt interview with Marcel; Pflock, Karl. *Inconvenient Truth*.

He said, "We found some metal, small bits of metal. We picked it up." He also said that it was something that had been manufactured and that they had found some small, beam-like structures. These didn't bend and they couldn't break them but they didn't look like metal. "They were, as I recall, perhaps three-eighths of an inch by one quarter of an inch, and [came] in just about all sizes, none of them very long... You couldn't even tell you had them in your hands."[203]

Marcel and Cavitt loaded as much as they could into the vehicle driven by Cavitt and Marcel sent him back to the base. He then loaded more into his car and drove home. Once they had returned from the Brazel ranch, Lewis Rickett, the NCOIC[204] of the Counterintelligence Office (CIC) in Roswell was taken out to one of the crash sites. According to Rickett, "There were four or five military vehicles at the crash site. The MPs checked our IDs. All of them had .45s and some of them had Thompsons or old grease guns... The MPs, four or five of them in the first group were close to the gouge. There were twenty-five or thirty others scattered around the perimeter. The Provost Marshal didn't want anyone just wandering up on it."[205]

Rickett explained that his boss, Captain Sheridan Cavitt, had wanted him to see the metallic debris. Rickett described the trip out to the field which didn't seem to be as far away as others had placed it. Rickett said, "I remember it took us about forty-five minutes. I said, 'How far are we going out here?' and he [Cavitt] said, 'It's just over here – I just want you to see it instead of

---

[203] Ibid.

[204] The Noncommissioned Officer in Charge (NCOIC) is the highest ranking NCO in the office.

[205] Lewis Rickett, personal interview with Mark Rodegheir January 1990; personal interview with Don Schmitt, October 29, 1989 and October 17, 1992; See also Randle and Schmitt. *UFO Crash at Roswell*, pp. 40, 61 – 63, 65, 159 – 163; Carey, Tom and Donald R. Schmitt. *Witness to Roswell*. Pompton Plains, NJ: New Page Books. 2009: pp. 198 – 201.

just talking about it.' [Marcel and the Provost Marshal Edwin Easley]... we were all gathered around."[206]

Rickett's suggestion that the crash site was less than an hour from Roswell indicated another location. That has been corroborated, after a fashion by others who were there. Chester Barton said that he had been ordered by his boss, Major Edwin Easley, to inspect the field. According to Barton, the crash site was about forty-five minutes from the base.[207]

Although Barton thought that what had crashed was a B-29[208] he saw no wreckage to suggest that. According to Barton, "...no part of the 'airplane' was recognizable as such." Barton saw a large burn area the size of a football field and burned debris scattered around.[209]

Dan Dwyer, the Roswell fire fighter lieutenant, told his daughter, Frankie Rowe, and fellow fire fighter J. L. Smith that he had driven out about thirty miles, or less than an hour north of Roswell to the crash site.[210]

One other person mentioned the location of this particular site. In 1947 George Erl was a staff sergeant who was assigned as the NCOIC for Entry Control at the main gate. He suggested that the area around Ramon and Mesa,

---

[206] Lewis Rickett, personal interview with Don Schmitt, October 29, 1989. It should be noted that based on Rickett's description, he was not at the debris field found by Brazel but on a site closer to Roswell. Others who visited the field later suggested that the trip was much shorter than it would have been to get to the debris field.

[207] Chester Barton, telephone interview by Joe Stefula, August 1995; information available at http://www.roswellproof.com/Barton.html (accessed December 3, 2007).

[208] An Air Force search of the records found no aviation accident that fit the time frame. See Richard L. Weaver and James McAndrew, *The Roswell Report: Fact vs Fiction in the New Mexico Desert* (page 19 – 20 of the Executive Summary).

[209] Chester Barton, telephone interview by Joe Stefula, August 1995; information available at http://www.roswellproof.com/Barton.html (accessed December 3, 2007).

[210] Frankie Rowe, personal interview with Kevin Randle and Don Schmitt January 2, 1993 and January 27, 1993; J. L. Smith, audiotaped telephone interview with Kevin Randle, 2009.

some thirty or forty miles north of Roswell would be near the location for the crash site.[211]

There is also the testimony of William Woody who was 12 in 1947. He said that he had seen something in the air and watched as it crossed the sky, falling somewhere north of Roswell. He said that he and his father, two or three days later went looking for the object that had fallen. He said, "We headed north through Roswell on U.S. 285. About 19 miles north of town, where the highway crosses Macho Draw, we saw at least one uniformed soldier stationed beside the road. As we drove along we saw more sentries and Army vehicles… they were armed, some with rifles, others with sidearms."[212]

Once out on the field Rickett said that Cavitt had thought that it was important for someone else in the CIC office to see the debris. Rickett said:

> …I said it looks like metal and then I asked if it was hot [radioactive]. My boss [Cavitt] told me no, go ahead and pick it up… There was a slightly curved piece of metal, real light. It was about six inches by twelve or fourteen inches. Very light. I couched down and tried to snap it. My boss laughs and said, 'Smart guy. He's trying to do what we couldn't do.' I asked, 'What the hell is this stuff made out of?' It didn't feel like plastic and I never saw a piece of metal this thin that you couldn't break.'
>
> As we walked around, my boss said, "You and I were never here. You don't see any military people out here.' And I said, 'Yeah, that's right. We never left the office.'"[213]

---

[211] George Erl, personal letter to Kevin Randle, June 28, 1994.

[212] William Woody, personal interview with Kevin Randle, September 30, 1990. See also Pflock, Karl. *Inconvenient Facts,*" for affidavit, p. 291.

[213] Ibid. Randle and Schmitt. *UFO Crash at Roswell*, pp. 40, 61 – 63, 65, 159 – 163. Lewis Rickett, taped interview by Don Schmitt, February 13, 1990.

Cavitt, when interviewed by others, denied that this took place. He was interviewed by Colonel Richard L. Weaver and said:

> …I couldn't swear to the dates, but in that time, which must have been July [1947] we heard that someone had found some debris not too far from Roswell and it looked suspicious; it was unidentified. So, I went out and I do not recall whether Marcel went with Rickett and me; I had Rickett with me. We went out to this site. There were no, as I understand, check points or anything like that (going through guards and that sort of garbage) we went out there and we found it. It was a small amount of, as I recall, bamboo sticks, reflective sort of material that would, well at first glance, you would probably think that it was aluminum foil, something of that type. And we gather up some of it. I don't know whether we even tried to get all of it. It wasn't scattered; well, what I call, you know, extensively. Like, it didn't go along the ground and splatter off some here and some there. We gathered up some of it and took it back to the base and I remember I had turned it over to Marcel. As I say, I do not remember whether Marcel was there or not on the site. He could have been. We took it back to the intelligence room… in the CIC office.[214]

Cavitt, however, had said in an interview that he had never participated in the recovery of a balloon.[215] He said that he was too busy to waste time on

---

[214] In the Air Force report written by. Weaver and McAndrew, *The Roswell Report: Fact vs Fiction in the New Mexico Desert.* Section 18 "Interview. Col Richard L. Weaver with Lt. Col. Sheridan D. Cavitt, USAF (Ret). May 24, 1994.

[215] Sheridan Cavitt, personal interview with Kevin Randle and Don Schmitt in Sierra Vista, Arizona, January 29, 1990.

picking up balloons. He also said that those who put him on the site were wrong.[216]

When Cavitt was interviewed by Weaver, he would say, "I thought a weather balloon."[217]

He then said, "This is all over my head. When I saw it was to [sic] flimsy to be anything to carry people or anything of that sort. It never crossed my mind that it could be anything but a radio sonde."[218]

Cavitt also said that he had only made one trip out there. Given the circumstances, it is quite possible that he only made a single trip to the Brazel ranch and that would have been with Marcel. The other sites were closer to Roswell, as described by Rickett when he thought it hadn't been much more than thirty miles to the site.

Rickett did supply additional information. He said that the metal wasn't heavy, just that it was hard. He said, "It was light, but just like foil. It wasn't real brilliant."[219] He said that he later ran into a friend who seemed to have more information about it but all he was told was, "Our metal experts don't know what it is. They can't even cut it."[220]

The debris was taken from the field, to the base in Roswell for air transport to Wright Field. One of those responsible preparing and transporting the material was Sergeant Robert Smith. He said, "We started out in the morning, say

---

[216] Cavitt, in a personal interview with Kevin Randle and Don Schmitt held in Sequim, Washington, on June 25, 1994, just months after Weaver's interview, Cavitt would deny that he had been out there, seen anything and couldn't understand why people thought he had been out to the field with Marcel.

[217] Weaver and McAndrew, *The Roswell Report: Fact vs Fiction in the New Mexico Desert.* Section 18 "Interview. Col Richard L. Weaver with Lt. Col. Sheridan D. Cavitt, USAF (Ret). May 24, 1994.

[218] Ibid.

[219] Lewis Rickett, audiotaped personal interview with Don Schmitt, October 29, 1989.

[220] Ibid.

eight or nine o'clock. Seems like it lasted up to nearly four o'clock with a break for lunch. The reason I say it took so long is the weight and level guy worked to determine how much each crate weighed. He had to counterbalance the aircraft."[221]

The crates were made of plywood with one-inch furring strips on the around the edges. They varied in size from four to six feet wide, eight to ten feet long and three to four feet high. The problem wasn't the weight but the size. Smith said that they were light. Some were difficult to handle because of the size. He also said the crates were marked but he couldn't remember what the markings said, other than to suggest they were top secret and there were armed guards watching them.

When they finished with the loading, the three aircraft were sealed. He said that he and some of the other NCOs were discussing the nature of the cargo. Smith said:

> We were talking about what was in the crates and so forth and he (another of the NCOs) said, 'Oh do you remember the story about the UFOs.'[222] Or rather the flying saucers. That was what we called them back then. We thought he was joking, but he let us feel a piece and he stuck it back into his pocket. Afterwards we got to talking about it and he said he'd been out there helping them clean this up. He didn't think taking a little piece would matter.

---

[221] Robert Smith, videotaped personal interview with Kevin Randle and Donald Schmitt, January 19, 1990 in Waco, Texas. See also, Randle and Schmitt. *UFO Crash at Roswell*. pp. 83 – 86.

[222] Clearly Smith is applying the modern term to the phenomenon. In 1947, they were referred to as flying disks or flying saucers. Although there was periodic mention of unidentified flying objects, the term, "UFO" was created in the early 1950s.

> It was just a little piece of metal or foil or whatever it was. Just small enough to be slipped into a pocket. I think he just picked it up for a souvenir.
>
> It was foil-like, but it was a little stiffer than foil that we have now. In fact, being a sheet metal man, it kind of intrigued me, being you could crumple it and it would flatten back out again without any wrinkles showing up in it. Of course we didn't get to look at it too close because it was supposed to be top secret. He just popped it out there real quick and let us feel it and so forth while everybody was doing something else.
>
> I know there was a pretty good gathering of them [soldiers out in the field] and they went across the field about a couple of feet apart, and they just kept going across picking up the pieces. They were loading them in wheelbarrows and things like that.[223]

When the Fund for UFO Research was gathering affidavits about the crash, Smith provided additional information about the debris.[224] He said, "The piece of debris I saw was two – to – three inches square. It was jagged. When you crumbled it up, it then laid back out; and when it did, it kind of crackled making a sound like cellophane, and it crackled when it was let out. There were no creases."[225]

---

[223] Robert Smith, videotaped personal interview with Kevin Randle and Don Schmitt, January 19, 1990. See also Smith's affidavit in Pflock. *Inconvenient Facts*. pp. 283 – 284.

[224] These affidavits were all signed by the witnesses and officially notarized. These are sworn statements and while not as important as testimony given in a court, it does rise above the level of witness testimony.

[225] Robert Smith, affidavit signed on October 10, 1991. Affidavits are filed at FUFOR and CUFOS. For Smith's affidavit see Pflock, *Inconvenient Facts*. pp. 283 – 284.

Smith also wrote, "The largest piece was roughly 20 feet long; four-to-five feet high; four-to-five feet wide. The rest were two-to-three feet long; two feet square or smaller."[226]

Although he didn't seem to be among those tasked to go out to clean the fields, his affidavit does suggest that the debris came from outside. Smith said, "There was a lot of farm dirt on the hangar floor. We loaded it on flatbeds and dollies; each crate had to be checked…"[227]

Smith's information was corroborated and added to by Staff Sergeant Earl Fulford. He said that he had been in the mess hall eating breakfast when his superior entered and told him to join a group assembling outside. They were all ordered onto a bus and then driven out into the country. The rides had taken hours. Fulford said:

> The ride took about two hours. The site was northwest of Roswell. We went north up Highway 285, then west on the Corona road… I remember seeing a little house, which was not far from where we were going… A major was in charge and armed MPs ringed the site…Sergeant Rosenberger then handed each of us a burlap bag and told us to "police up" the site…[228]

In interviews conducted by Tom Carey and Don Schmitt Fulford suggested that the field had already been cleared by another crew because he could see tire tracks from heavy trucks all over it. Fulford told them:

---

226 Ibid.

227 Ibid.

228 Carey and Schmitt. *Witness to Roswell.* pp. 113 – 116.

> I picked up small, silvery pieces of metallic debris, the largest of which was triangular in shape, about 3 to 4 inches wide by about 12 to 15 inches long. It looked like thin, light, aluminum foil that flexed slightly when I picked it up, but once in the palm of your hand, you could wad it up into a small ball. Then, when you let it go, it would immediately assume its original shape in a second or two – just like that. That was the only type of debris I saw that day. I thought to myself, "Hey this is neat. I'm going to keep a piece for myself." But they searched us so thoroughly when we got back to make damned sure that none of us had anything… We didn't see any other type of debris or pieces of debris with writing on them and we didn't see any bodies. We also did not see any balloons or balloon material… When we got back to the base, everything that we picked up was taken to Hangar 3. We were lined up and told one by one by our first sergeant in no uncertain terms that we didn't see anything, and we didn't say anything; and if we did from that point forward, we might be court martialed…[229]

There were discussions of the debris by some high-ranking officers who had some knowledge of the debris. In 1947, Arthur Exon was a lieutenant colonel at Wright Field and he retired as a brigadier general after a tour as the base commander at Wright-Patterson Air Force Base in the 1960s.[230] Exon would

---

[229] Ibid. It should also be noted that while skeptics don't believe that the soldiers were threatened, the sort of thing that Fulford reported to Carey and Schmitt would be a fairly standard warning about sharing classified information with those not cleared to receive it. Almost every classified briefing begins with a warning about the possible penalties for sharing classified material with those not authorized to have it. No matter the source of the Roswell debris, if those in authority thought the recovery was classified, they would warn the soldiers about telling it to friends and family.

[230] Biography of General Exon can be found at: http://www.af.mil/AboutUs/Biographies/Display/tabid/225/Article/104819/brigadier-general-arthur-ernest-exon.aspx; see also Randle and Schmitt. *Truth about UFO Crash*. pp. 62 – 63.

say that in 1947, he had flown over the crash site. He said, "[It was] probably part of the same accident, but [there were] two distinct sites. One, assuming that the thing, as I understand it, as I remember flying the area later, that the damage to the vehicle seemed to be coming from the southeast to the north-west, but it could have been going in the opposite direction, but it doesn't seem likely. So the farther northwest pieces found on the ranch, those pieces were mostly metal."[231]

In describing the scene, as he flew over it sometime later, Exon said, "I remember auto tracks leading to the pivotal sites and obvious gouges in the terrain."[232]

Exon, who was a student at the Air Force Institute of Technology in July 1947, he was also assigned to the headquarters staff at Wright Field in Ohio. Exon said:

> We heard the material was coming to Wright Field... Everything from chemical analysis, stress tests, compassion tests, flexing. It was brought into our material evaluation labs. I don't know how it arrived, but the boys who tested it said it was very unusual... [Some of it] could be easily ripped or changed... There were other parts of it that were very thin but awfully strong and couldn't be dented with hammers. It was flexible to a degree... some of it was flimsy and was tougher than hell and the other was almost like foil but strong. It had them pretty puzzled... They knew they had something new in their hands. The metal and material was unknown to anyone I talked to. Whatever they found, I never heard what the results

---

231 Arthur Exon, audio taped personal interview by Don Schmitt, June 19, 1990 in Riverside, California.

232 Arthur Exon, letter to Kevin Randle, dated November 24, 1991.

> were. A couple of guys thought it might be Russian, but the overall consensus was that pieces from space.[233]

There is other evidence that the material, the metallic debris, was sent to Wright Field. In the days that followed the cleanup. Captain O. W. "Pappy" Henderson flew a C-54 loaded with the metallic debris to Wright Field. He didn't mention anything about the flight to anyone including his wife until after 1980 when the information about the Roswell case appeared in a tabloid newspaper. He told his wife, "I guess I don't have to keep it a secret anymore."[234]

Henderson told his wife that he had wanted to tell her about the trip for years. He knew that the flying saucers were real because he had flown a plane load of material to Wright Field. He didn't say much about it other than it was strange.[235]

She wasn't the only one to hear about the strange flight. Four years before Henderson died in 1986; he was at a reunion with his flight crew from the Second World War and told them about flying the wreckage of a spacecraft from Roswell to Wright Field. He also mentioned the bodies that had been recovered as well.[236]

---

[233] It should be noted here that Exon was explaining what he had heard from those he knew who had been involved in the analysis of the material. Arthur Exon, audiotaped telephone interview by Kevin Randle, May 19, 1990. See also Randle and Schmitt. *Truth about UFO Crash*. pp. 62; Carey and Schmitt. *Witness to Roswell*. pp. 219 – 219.

[234] Sappho Henderson, audiotaped telephone interview with Kevin Randle, August 3, 1989. See also Randle and Schmitt. *UFO Crash*. pp. 82 – 83. As an aside, Henderson was wrong about being free to talk about the UFO crash. Although reported in the tabloid, the information was still considered classified. Henderson's obligation was to deny knowledge of it when asked.

[235] Ibid.

[236] Vere McCarthy, letter, July 12, 1989.

In a strange coincidence, Robert Porter, brother of Loretta Proctor, was also a member of the 509th Bomb Group in 1947.[237] He said that he was on an aircraft that carried some of the debris to Fort Worth. There were only four small packages on the flight and they were handed up to him through a hatch. The largest of the pieces was two and a half to three feet across and three to four inches thick. It was triangular shaped and sounds like it was part of a rawin reflector when folded together.[238] The other three packages were about the size of shoe boxes but seem as if they were empty. All packages were wrapped in brown butcher's paper.[239]

Porter did suggest that things were strange when they arrived in Fort Worth. In his affidavit he said, "After we landed at Fort Worth, Col. Jennings told us to take care of the maintenance and that after a guard was posted, we could eat some lunch."[240]

Another of those who knew something about the flights out and the retrieval of the metal debris was Milton Sprouse.[241] He said that a number of his friends were taken out to the debris field and participated in the clean up there. Sprouse suggested that about 500 soldiers were on that field, moving shoulder to shoulder and picking up everything they could find.[242]

---

[237] Master Sergeant Porter's picture does appear on pages 1, 124, 126 of the Yearbook created by Walter Haut in 1947. He was a member of the 830th Bomb Squadron.

[238] Porter didn't describe it as such, but photographs of the rawins, before they are opened up for flight look like the article Porter described.

[239] Robert Porter, telephone interview with Kevin Randle and Don Schmitt, April 8, 1990. See also Randle and Schmitt, *UFO Crash*. pp. 81 – 82; Pflock, Karl. *Inconvenient Facts*. p. 270.

[240] Pflock, Karl. *Inconvenient Facts*. p. 270

[241] Staff Sergeant Milton Sprouse appears on pages 129 and 139 in the Yearbook, establishing his presence in Roswell at the right time.

[242] Milton Sprouse, audio taped telephone interview with Kevin Randle, October 2, 2007.

He said that these men hadn't seen the bodies because they had already been removed, and they didn't see a craft. He did say, "There were big pieces... up to twelve inches or bigger all over that ranch. But they didn't recognize what any of it was."

He said that they laughed about the weather balloon explanation. He said, "I've seen weather balloons. They've been launched all around my airplane there at Roswell many times."

One of his friends, a barracks buddy, had been a medic and had, one night, been called to the hospital. Sprouse said, "I lived in the barracks. I was single at the time... One of the barracks buddies was a sergeant and he worked in the Medics. He lived in the barracks. And he got a call to report to the hospital and he went up and when he come back he said, 'You wouldn't believe what I been through and what I've seen.' He didn't have too much to say about anything because they told him not to talk about everything and we didn't get much from him but later on he did say he was one of the few enlisted there and there were two doctors and two nurses in there. And of course he left right after that incident and never said good-by to us or nothing."

There are a few problems with Sprouse's story. First is that he can't remember any of the names of those who went out onto the debris field for the retrieval. He said that the names found in the Yearbook for his crew were those who had joined him later, after these events. He said that his aircraft commander was Colonel William H. Harrison, and Harrison was certainly an aircraft commander at Roswell in the right time frame. But, when Sprouse called Harrison, he was told that he, Harrison, remembered very little. In fact, according to Sprouse, Harrison said, "'I don't remember, Milt, who you are.' And he and I were friends because I was his crew chief for a heck of a long time. I flew all over the world with him... but I didn't get to talk to his wife or nothing because he hung up on me after a while and I've never called him again because it was a waste of time because he didn't know who I was and we had nothing to talk about in common."

Second, while I have heard tales of newspapers printing flying saucer stories and that someone made an effort to recall all those newspapers, this is not something that had been associated with the Roswell case until now. It wasn't effective because we have all seen the newspaper. I suppose they might have wanted to get it off the base to inhibit the soldiers talking about it and didn't care what was left in the civilian community, but that makes no real sense.

Finally, we have been unable to discover who the medic was. He's just another unnamed source who might have been a staff sergeant at the time, and who might have been promoted to what was known as a technical sergeant (now known as either a platoon sergeant or a sergeant first class) but that doesn't help us much. There are eleven men in those two grades in the Yearbook. Yes, it will take some time to check them out with no guarantee that anyone of them will be the right guy. Haut said that fifteen to twenty percent of the soldiers at Roswell were not included in the Yearbook for a variety of reasons.

This is a second-hand story that might provide some clues about the Roswell case.

Chapter 5:

Talk of the Bodies

If what had fallen at Roswell was an alien spacecraft, there was one man who would have been privy to the whole episode and that was Colonel William Blanchard.[243] He was the commanding officer and the communications with higher headquarters would have gone through his office. He would make all the decisions about what had to be done, how to do it and who all would have to be involved. Like so many of the witnesses, he was never interviewed about this, though some others have suggested that he mentioned the crash and recovery to them.

Robert Hastings was interviewing Chester Lytle concerning UFO sightings related to nuclear facilities. During an interview, Lytle diverted the conversation to the Roswell case. Hastings wrote:

> Unexpectedly, Lytle told me that he had once heard – from a high-credible source – that the object recovered near Roswell in 1947 was indeed a crashed extraterrestrial spacecraft. That source was none other than William H. Blanchard…
>
> According to Lytle, both he and Blanchard, was what by now a general, had been visiting Eielson AFB, Alaska, in mid-February 1953… Lytle's wife was in Chicago, about to give birth to a son and Lytle was desperate to get home. Blanchard, who was 'a very close friend'… offered to personally fly him in a

---

[243] Later Blanchard would be promoted to general and would be the vice chief of staff when he suffered a heart attack at the Pentagon. He was one of the top officers in the Air Force and had a very distinguished career after having commanded the 509th Bomb Group in Roswell.

> bomber to an Air Force base in Illinois... From there, Lytle could take a short commercial flight to Chicago...
>
> During the long flight... the subject of UFOs can up... Suddenly the general mentioned the Roswell Incident. Lytle, who held top secret clearances relating to his work with the AEC, was informed by Blanchard that a crashed alien spacecraft had indeed been recovered in July 1947. The general said that four dead humanoid beings had been aboard.[244]

Art McQuiddy, who was the editor of the *Roswell Morning Dispatch,* corroborated the story after a fashion. In an interview conducted over the telephone, he said, "The commanding officer out there was a great guy named Butch Blanchard. He and I were good friends. All I know is that he said they put the wreckage or whatever it was they found in a B-25 and flew it to Fort Worth and it went right from there immediately to Wright-Patterson.... He was the one who told Walter Haut to put out the press release."[245]

Bill Brainerd, a former mayor of Roswell talked to Blanchard at a reunion dinner in 1965, added to this. He said, "...I was at a table with the general and several other locals when they were interrogating him about the 1947 incident and he declined to answer any of the questions except he did comment that it was the damnest thing he'd ever seen."[246]

---

[244] Hastings, Robert H. *UFOs and Nukes*. Bloomington, IN: Author House, 2008. pp. 510 – 512.

[245] Art McQuiddy, audio taped telephone interview with Kevin Randle, January 19, 1990. See also, Randle, Kevin D. and Donald R. Schmitt. *UFO Crash at Roswell*. New York: Avon Books, 1991. p. 71.

[246] Quote from a BBC Roswell UFO crash documentary found at: www.youtube.com/watch?v=0WuqJolib5k. This is a third source for Blanchard commenting on the events which meets certain journalist standards for reporting.

Marion M. Magruder

This isn't the only example of a second party revealing what a former high-ranking military officer had said. Some of the most comprehensive information about the bodies came from the family of former Marine colonel Marion M. Magruder, who during World War II had developed the Marine night fighter capability.[247] In the summer of 1947, Magruder was assigned as a student to the Air War College and it was during that that year that he saw the alien creature that had been at Roswell.[248] According to what he told his children:

> The Air War College students were led into a room where they were told about the crash outside of Roswell, New Mexico earlier the same month, and then they were shown debris from the crash. They were also brought into first-hand contact with one of the craft's inhabitants, the only one alive after the crash. Magruder described the being in specific detail to his sons, but kept repeating that the being seemed "squiggly" to him, a remark that he repeated as he lay dying 40 years later.
>
> "He said the creature was not like the aliens you see depicted on television," Mike Magruder remembered. "Except for the large head. Actually instead of a gray color, the being had a flesh tone. And it resembled a human more than anything else.

---

[247] Noory, George. "Coast to Coast AM: Once a Hero." *UFO Magazine* 21,4 (June 2006) pp. 26 -27, 75; Birnes, William J. with Mark Magruder, Merritt Magruder and Natalie Magruder. "Squiggly." *UFO Magazine* 21,4 (June 2006) pp. 32 – 39; Carey, Tom and Donald R. Schmitt. *Witness to Roswell.* Pompton Plains, NJ: New Page Books, pp. 161, 237 – 239; Information about the Air War College gathered from Maxwell Air Force Base and other official sources in August 2008 by Kevin Randle.

[248] Birnes, William J. with Mark Magruder, Merritt Magruder and Natalie Magruder. "Squiggly." *UFO Magazine* 21,4 (June 2006) pp. 35.

> It had a large head, it had large eyes – larger than usual – and it had only a slit for a mouth and hardly any nose. In fact, my father said it didn't really have a nose the way human beings have noses. The creature was human-like. Yet, despite its humanoid appearance, it was clearly, my father said, not from this planet. It was not like any human being he had ever seen."
>
> ...It was five feet tall or smaller, a little creature, but it did not look like a small adult. It was clearly something different. ...it was wearing a coverall that reminded him of a flight jumpsuit.[249]

Magruder expanded on the description telling his granddaughter, Natalie, that the creature had long arms, large eyes and an oversized, hairless head. It had a slit-like mouth, no nose or ears, just small orifices. He told her that there was no question that it came from another planet.[250]

There is a problem with the timing of all this, however. Although it was originally suggested that Magruder and his class had traveled to Wright Field in August 1947, the class didn't begin until September. Magruder had a delay in route which is a way of giving him a leave without charging it as leave. According to the school records, Magruder did travel to Wright Field as part of his class work, but that wasn't until April 1948.[251]

---

[249] Ibid. p. 35

[250] Carey and Schmitt. *Witness to Roswell.* p. 239

[251] Class rosters, class curriculum, and other information provided by the Air War College archives and Maxwell Air Force Base in August 2008.

More Talk of Alien Bodies

Edwin Easley, who retired as a colonel was the provost marshal at Roswell in July 1947.[252] He had little to say to UFO researchers, though he did confirm that he believed the craft to be alien. When asked, specifically if we, Don Schmitt and I, were following the right path, Easley said, "Let me put it this way, it's not the wrong path."[253]

Easley was reluctant to talk about the events in 1947, saying, repeatedly, "I can't talk about it. I was sworn to secrecy."[254]

He did, however, share some things with family members. Dr. Harold Granik, who practiced in several Fort Worth hospitals, reported that just before Easley died, his granddaughter asked him about the Roswell case. According to the information, Easley said, only, "Oh, the creatures."[255]

Retired Brigadier General Arthur Exon also mentioned the bodies. He said, "There was another location… apparently the main body of the spacecraft was… where they did say there were bodies… they were all found, apparently outside the craft itself but were in fairly good condition. In other words, they weren't broken up a lot."[256]

---

[252] Easley's position at the RAAF is verified through the Yearbook published in 1947, the Unit Histories published in June, July and August, 1947 and in interviews with other members of the 509th Bomb Group who knew him and saw him at the base in 1947.

[253] Edwin Easley, unrecorded telephone interview with Kevin Randle, February 2, 1990.

[254] Edwin Easley, telephone interview with Kevin Randle, October 1989; January 11, 1990. See also Randle, Kevin D. and Donald R. Schmitt. *The Truth about the UFO Crash at Roswell*. New York: M. Evans and Company, 1994; 14. Letter from Shelly Easley Perkins to Kevin Randle, January 13, 1992; Note written by Easley on package of UFO related material received after the *Unsolved Mysteries* broadcast.

[255] Dr. Harold Granik, telephone interview by Mark Rodeghier, April 14, 1992.

[256] Arthur Exon, telephone interview by Kevin Randle, May 19, 1990. It should be made clear that Exon did not see the bodies himself. He was reporting what he had

He added, when asked about the specifics, "Well, that's my information. But one of them was that it went to the mortuary outfit… I think at the time it was in Denver [Lowery Air Force Base] where these people were being identified. But the strongest information was that they were brought into Wright-Pat."[257]

Captain (later Colonel) Darwin E. Rasmussen while at a family barbeque said that he had no doubt that flying saucers were real because he had helped recover the bodies from a crash. There were four bodies and as soon as they were loaded into the truck, they were driven to Roswell.[258] He also mentioned that within a matter of hours they were told that they hadn't seen anything and the suggestion was that they not talk about it.

High-ranking officers aren't the only ones who told family members about the crash. Carlene Green is the daughter of Sergeant Homer G. Rowlette, a member of the 603rd Air Engineering Squadron who was stationed in Roswell in 1947. According to her, she learned of her father's part in the crash retrieval just days before he died. He was on a Gurney and about to be wheeled into an operating room when he asked her to come closer so that he could speak to her.[259]

He told her that he had been at the base in Roswell when the "spaceship" crashed and that he had seen it. He said that the craft was rounded; "somewhat

---

been told by his colleagues at Wright Field in 1947 who he said were in a position to know.

[257] Arthur Exon, telephone interview by Kevin Randle, May 19, 1990. Follow up interview in person by Don Schmitt, June 18, 1990. Both interviews recorded on audio tape.

[258] Elaine Vegh, audiotaped telephone interview with Kevin Randle March 1, 1990. Vegh was Rasmussen's cousin and overheard the conversation between her father and Rasmussen. See also Carey and Schmitt, *Witness to Roswell*, p. 83. Although Rasmussen's picture does not appear in the Yearbook, his name and rank along with a telephone number does appear in the RAAF Base directory for 1947.

[259] Carlene Green, unrecorded personal interview conducted by Kevin Randle July 2, 2011 conducted in Roswell, New Mexico.

circular" and that he had seen three "little people" with large heads and suggested that one of them had survived the crash.[260]

He apologized for not telling her sooner but that he had told her brother the story sometime earlier. Finally, he cautioned her to keep it all to herself, or else.

This all happened in 1988, according to what she said. While the story of the Roswell crash wasn't as well known then as it is today, there was a book published about it, there were television shows that touched on it, there were documentaries about it and some magazine articles that told of it. In other words, this information didn't appear in the vacuum that existed prior to 1980. And to make it worse, like the stories that preceded it, all there were second hand.

There were also some civilians who had information about alien bodies and the events around Roswell. Barbara Dugger, granddaughter of Sheriff George Wilcox, said that she had talked to her grandmother about the crash. Dugger said, "Someone came and told my grandfather about this incident that happened outside Roswell. My grandfather went out there and when he got there, there was a big burned area… He saw debris."[261]

Dugger said that her grandfather said that there were four bodies.[262] She said, "They were, like gray and granddaddy said their heads were large and the little suits they had on were like silk or something… They were gray…. The little people were lying on the ground…. She said I think one of them was alive."[263]

---

[260] Carey and Schmitt. *Witness to Roswell Revised.* 233 – 234.

[261] Barbara Dugger videotaped personal interview with Kevin Randle and Donald Schmitt, March 4, 1991. See also, Randle, Kevin D. and Donald R. Schmitt. *The Truth about the UFO Crash at Roswell.* New York: M. Evans and Company, 1994; 34 – 35.

[262] It should be made clear that Dugger heard nothing directly from her grandfather. The comments were made by her grandmother, which, unfortunately, makes the information third hand at best.

[263] Barbara Dugger videotaped personal interview with Kevin Randle and Donald Schmitt, March 4, 1991.

The crash site visited by Wilcox, according to Dugger was about thirty miles outside of Roswell. There was a big burned area and there was debris scattered around.[264]

Dugger said that her grandmother had written something about her experiences in the late 1940s. It was part of an article about the time George Wilcox was the sheriff of Chaves County and she was working in the jail as the matron. She wrote, (reproduced here just as she typed it but without the strikeovers):

> One day a rancher North of town brought in, what he called a "FLYING SAUCER", there had been many reports all over the United States by people who claimed they had seen a FLYING SAUCER. the rumors were in many variations, The saucer was from a different planet, and the people flying on it, were looking us overf. The Germans had invented this strange contraption, a fromible weapon. Other tales, that one had landed and strange looking people all seven feet tall or more walked from it, but quickly departed on sighting any on looker. All the papers played the stories up, and many people searched the skies at night to catch sight of one. Since no one had seen a flying saucer, Mr Wilcox called headquarters at Walker Air Force Base, and reported the find. Beofre he hung up the telephone almost, an officer walked in. He quickly loaded the object into a truck and that was the last glimps any one had of it.
>
> Simultaneously the telephone began to ring, long distance calls from News papers in New York, England, France Government officials, Military officials, and the calls kept up for 24 hours straight. They would speak to no one but the Sheriff.

---

[264] Ibid.

> However the Officer who picked up the suspicious looking saucer, admonished Mr Wilcox to tell as little as possible about it and refer all calls to Walker Air Force Base. A secret well kept, for to this day, we never found out if this was really a FLYING SAUCER.[265]

The trouble here is that this was an addition to the original article which suggests it was added at a later date, though that is certainly speculation. According to the available records, Inez Wilcox died a few years after the publication of *The Roswell Incident* which means she could have added the information about the flying saucer after the publicity of the crash began in 1980. While there is nothing to suggest that Wilcox's addition was confabulation based on that information, there just is no way to verify the information in it or that it was written prior to the publication of the book.[266]

There is also information that the discovery was filtering up through the highest levels of the New Mexican state and federal elected officials. Jud Roberts, minority owner of KGFL in 1947 said that Walt Whitmore, Sr., the majority owner had found Mack Brazel and recorded an hour long interview with him. Before they could air it, they received a call from one of their United States senators who told them not to broadcast it. Roberts was told, "if we released it our license would be gone the second day."[267]

---

[265] Inez Wilcox. "Four Years in the County Jail." Unpublished manuscript. The Author. No date attached to it. Also see Barbara Dugger, videotaped personal interview by Kevin Randle and Donald Schmitt March 4, 1991. Randle and Schmitt. *UFO Crash at Roswell.*

[266] According to Dugger, Inez Wilcox told her that she had something to tell her that she had never mentioned to anyone else and warned her not to discuss it with anyone. This would suggest that Inez had kept the secret for a long time before revealing it to Dugger. Although there is no real date on this conversation, it probably took place after the publication of the Berlitz and Moore book.

[267] George 'Jud' Roberts, telephone interview with Kevin Randle January 19, 1990; personal videotaped interview with Don Schmitt and Kevin Randle, January 31, 1990 in Roswell, New Mexico.

This, of course, does not mean that the senator who called the station knew what was going on; only that he was doing a favor for the Army. But there is other evidence. Ruben Anaya, who, in 1947, worked as a civilian contractor at the base, suggested that then Lieutenant Governor Joseph Montoya had called requiring a ride off the base. Anaya was told by the senator, "Get your car Ruben and pick me up. Get me the hell out of here."[268]

The Whole Story Told by the Anaya Brothers

Anaya gathered his brother Pete and a family friend Moses Burrola[269] and headed out to the base. Anaya worked at the base and had a sticker on his car showing that it had been registered so that he was waved through the gate. Montoya had told them not to go near the base headquarters because there were too many people there. He would be waiting near the water tower on the east side, near the hangars.

Montoya was waiting where he said he would be. According to Anaya, Montoya was pale and shaking. Montoya was very scared. Montoya got in the car and said, "Get me the hell out of here. I want to go."[270]

During the ride back into Roswell, Montoya didn't say much. He was staring out the window.[271]

---

[268] Ruben Anaya, audiotaped personal interview by Kevin Randle and Don Schmitt, November 17, 1991. Also, interviews with Pete and Mary Anaya by Kevin Randle and Don Schmitt on February 2, 1992.

[269] Moses Burrola died before researchers had a chance to interview him. His widow did confirm the account, saying that her husband had mentioned the Montoya story to her. She was interviewed by Kevin Randle and Don Schmitt, February 2, 1992.

[270] Ibid.

[271] Pete Anaya, audiotaped personal interview by Kevin Randle and Don Schmitt, February 2, 1992.

Inside the house, with a glass of Scotch in his hand, Montoya said, "You all are not going to believe what I've seen. If you ever tell anyone, I'll call you a damned liar. We don't know what it is... They say it moves like a platter. It's a plane without wings... It's not a helicopter... I don't know where it's from... It could be from the moon... We don't know what it is."

The Scotch wasn't enough and according to Anaya, "The man took half a quart just, boom, boom, boom. He was very scared."[272]

Montoya told them there had been "four little men... [They were] short... skinny-like with big eyes... [The] mouth was real small, like a cut across a piece of wood."[273]

Pete Anaya remembered the less of the description or maybe he heard less of it. He recalled only that Montoya said that the beings were small with big heads. That was the thing he remembered the most, the discussion of the over-sized heads.[274]

According to what Ruben Anaya remembered, Montoya said they were so skinny that they simply did not look human. Although he didn't get a very good look at them, he did see that they had no hair, that the skin was white, and each wore a silver, tight fitting, one-piece suit that Anaya described as similar to those worn by navy deep sea divers. They had four long, thin fingers, were bald and the eyes were larger than normal.[275]

According to Montoya, the bodies had been in the hangar stretched out on long tables like those found in the mess hall. He thought that one of them might

---

[272] Ibid.

[273] These descriptions are second hand. Neither of the Anayas saw the creatures but were repeating what Montoya had told them. Interviews were conducted by Kevin Randle and Don Schmitt on November 17, 1991 and February 2, 1992.

[274] Pete Anaya, personal interview with Kevin Randle and Don Schmitt, on November 17, 1991 and February 2, 1992. See also, Carey and Schmitt, *Witness to Roswell*, pp. 88 – 90; Randle and Schmitt. *Truth about UFO Crash*, pp. 17 – 21.

[275] Ibid.

have been alive because he thought it was moaning. It was moving a hand, barely and had its knees drawn up.

After a short time, the bodies were transferred to the base hospital. He said that soldiers brought in some of the debris at that point. It was just pieces of metal. Nothing that resembled a large ship.[276]

Ruben Anaya could see that Montoya was becoming more agitated by the questions and stopped asking them. With that Montoya began to relax and said that he needed to take a nap. Eventually he fell asleep, but then would jerk awake suddenly which would indicate that he was still under some kind of stress.

Before he left them, Montoya told them not to discuss what they had heard. Threats had been made and for decades they rarely discussed it, even among themselves.[277] Karl Pflock, in his book, suggested that he has spoken to friends of Montoya and to Montoya's widow but none of the remember having heard Montoya speak of these experiences. Montoya's widow said that she did not remember her husband being in Roswell at that time nor did she remember a panicked telephone call from him at that time.[278]

All this is interesting because it hints at alien bodies and many of those who told the tales were important people who would have no reason to invent such colorful stories. Recent history has shown that even those in high profile positions and who seem to be credible sources will sometimes deviate from the truth. Proof of this made national news in February 2015 when Brian Williams, the anchor of the *NBC Nightly News* was caught embellishing a tale about having been on a helicopter that was shot down in Iraq in 2003. He was not on that helicopter and his apology days later only made it worse when it

---

[276] Ruben Anaya, personal interview by Kevin Randle and Don Schmitt, November 17, 1991.

[277] Randle and Schmitt. *Truth about UFO*. p. 20; Carey and Schmitt. *Witness to Roswell*. pp. 88 – 92

[278] Pflock, *Inconvenient Facts*, p. 63

was suggested that his helicopter had been hit by ground fire but had not been forced to land. This too was not true, though he might have been in a helicopter near one that had taken ground fire.

More Witnesses to the Bodies

It could be argued that the men telling stories of seeing the alien bodies from the Roswell crash were just making an interesting story better. It could be said that family members misunderstood what was being said at those times. Second-hand sources are simply not as important as those who were on the scene and made the observations. It is the first-hand witnesses who are the important ones.

One of those was Thomas Gonzales, a sergeant in the transportation section stationed in Roswell in July 1947.[279] Gonzales had been caught when officers swept through the base looking for soldiers available for various tasks including standing guard and providing security around a hangar and on one of several sites that were off base.

According to Gonzales, the craft he saw looked more like an airfoil than it did a saucer. He said that the bodies were like "little men," but not like the "Greys" of the abduction literature.[280]

When interviewed, Gonzales was 78 years old and living with his son. He was not an articulate man and his descriptions ran to one or two words for most questions. He did say that shortly after these events, he was transferred with little or no notification. Research has shown that a large number of those who

---

[279] Documentation available including the Yearbook produced for the 509th Bomb Group in 1947 proves that Gonzales was stationed in Roswell at the proper time. The Yearbook went to the printers in August 1947 and was returned to the base in November.

[280] Ecker, Don. "The Tale of a New Roswell Witness." *UFO* 12,9 (1994): 12- 13. See also, Randle, *Roswell Revisited*, 32 – 33. Thomas Gonzales, audiotaped telephone interview with Randle, November 1994.

were directly involved were transferred from the base in the months after the UFO crash.[281]

Another former member of the 509th to have claimed to see the bodies was PFC Eleazar Benavidez whose story was first told in *Witness to Roswell*. He was identified as Elias "Eli" Benjamin in the book.[282] According to the book, Benavides and his wife visited the International UFO Museum in Roswell in 2002. Although Benavidez didn't want to reveal what he knew, his wife talked with museum director Julie Shuster. They arranged for a private interview.[283]

According to what he told Carey and Schmitt, and later told to Richard Dolan and later still for the Roswell Slides presentation in Mexico in a videotaped interview,[284] Benavidez was told to grab his weapon and run over to one of the hangars for guard duty.[285]

He said that he drew his weapon, reported to the hangar and then searched for the Officer in Charge. There was an officer there, whose name Benavidez could not remember, who was supposed to be in charge of transferring the

---

281 See the 509th Bomb Group Unit History for June, July and August. Major Edwin Easley, in his section of the report complained that nearly three quarters of the soldiers assigned to him had been transferred over the previous months. The overall number of soldiers assigned to the base didn't change much because other soldiers were transferred in as those at the base were transferred out.

282 The real name was deduced by a number of UFO researchers. Don Schmitt confirmed the name for Kevin Randle at the 6th Annual Crash Conference in Las Vegas on November 7, 2008. Benavides has been interviewed for a number of documentaries which also used his name.

283 Carey and Schmitt. *Witness to Roswell Revised.* 151 - 156.

284 Benavidez interview can be found here:
https://www.youtube.com/watch?v=p25nbISLvps,

285 There has been discussion about how some of these soldiers were dragged from their normal duties to participate in guarding the crash sites or hangars. Nearly every soldier has completed some form of basic training and part of that training is instruction on performing duties of an armed guard. These claims are not in conflict of Army doctrine in 1947.

bodies of the alien creatures to the base hospital. Benavidez believed that the officer had been to the crash site and then, in the hangar had seen the bodies. This caused him undue stress. With the officer unable to finish his task, another, higher ranking officer told Benavidez to take charge of the detail and ordered that the bodies, all on Gurneys, taken to the hospital.

The bodies, according to Benavidez, were grayish. They had swollen, hairless heads that were oversized for the bodies. He said that they had slanted eyes, holes for the nose and a slit for a mouth. He thought that one of them might have been alive.[286]

Although Benavidez had been concerned about his pension, he did participate in a number of additional interviews including one for a Canadian documentary in which his name was revealed as well. Finally, in 2015, Benavidez was interviewed for the short clip for the trailer for a documentary, Kodachrome.[287]

Although many soldiers said they had seen something strange, had seen bodies of alien creatures and feared for their pensions, there is no evidence that any of them lost those benefits. No one has come forward to say that after he spoke about what he had seen, his pension and benefits had ceased.[288] Not only would that open the military to a lawsuit, but it would give a platform to that

---

[286] Carey and Schmitt. *Witness to Roswell Revised.* 151 - 156.

[287] The trailer for Kodachrome can be found here: https://www.youtube.com/watch?v=jL0MvHpieaE. In it, Benavidez is pressed into saying that the body seen in the Roswell Slides looked like what he had seen. A full discussion on the Roswell Slides can be seen in Chapter 9; The Roswell Slides.

[288] An exception might be seen as Robert B. Willingham who claimed he was denied a pension for revealing what he had seen at the site of a recovery operation in 1948 or 1950 or 1954. He offered all these dates. The best evidence available however is that Willingham had not served long enough to receive a pension. His tale of a case in the Del Rio region of Texas cannot be corroborated and the best evidence is that he invented it for the notoriety.

soldier in a legal environment. The government would have to defend its position and to do so, would have to reveal there had been some sort of alien event in Roswell.

In 1994, a civilian witness, Anna Willmon was discovered who claimed that in 1947, she had seen the craft and the alien bodies.[289]. She talked of a crash site that was closer to Roswell and closer to the highway leading to the west out of town.[290] She said that she was returning to Roswell with her first husband, W. I. Witcamp when they spotted something shiny off the highway. They were about twenty or thirty miles outside of Roswell when they saw it and stopped to find out what it was.

According to Willmon, they moved through the bush until they came to an object shaped like an overturned washtub.[291] She didn't think it was very large, only twelve to fifteen feet in diameter. She did see the bodies of the flight crew, or the "little guys" as she called them.

There were two bodies. One was lying face down in the dirt and the other in the shade of some cedar trees, as if he had crawled over there before he died. The skin looked like burnt rubber. It was grayish brown, she said but she also said that it was hard to define.

According to Willmon, "...that other one was laying up kind of towards this brush and the other was out back... like he had been flung out of the thing. And this other little guy looked like he'd got out and went off and laid down....

---

[289] Anna Willmon, interview conducted and videotaped in her home in Roswell, New Mexico by Kevin Randle, May 5, 1994. See also Randle, *Roswell Revisited*. 28 – 30.

[290] See "Map of New Mexico Key Points," in Carey and Schmitt, *Witness to Roswell*. 269.

[291] This is nearly the same description of the craft provided by Johnny McBoyle when talking to Lydia Sleppy who was manning the telephones at the radio station. See Berlitz and Moore, *Rowell Incident*. 14- 16.; Randle and Schmitt, *Truth About UFO*. 141. McBoyle, unrecorded telephone interview by Randle, December 17, 1990.

And he wasn't very big. He was about as big as a little five-year-old kid. A little one."[292]

She said that one of them was slim, skinny, with short arms and little hands and feet. The other one, the one that she thought might have survived the crash only to die a little later was chubby. His arms were short too and his feet looked like human feet. The only real difference between them was that one was a little heavier.

The surface of the craft was very shiny, almost mirror-like. It was in two pieces. One of those pieces sat up on four short, stubby legs and the other, smaller piece looked as if it had been knocked from the top of the craft during the crash. That was sitting a short distance away. She insisted on calling it a flying saucer and said that her husband had called it that repeatedly when they talked about it.

She said that they went into Roswell and called the sheriff to let him know what they had seen. They then returned to the site to await the arrival of the police. She said that she later talked to a colonel from the base and he requested that they not talk to anyone about what they had seen. She made it clear that it hadn't been an order but merely a request.[293] This seems to suggest that she was treated more as someone who had seen something classified and had been asked not to talk about it. This is in contrast with what some of the other civilians had said.

Martin Jorgenson walked into the International UFO Museum and Research Center in Roswell and mentioned, off-hand, casually, that he had been around Roswell in 1947. Don Burleson, a Roswell resident, happened to be in the museum at the time and, according to Burleson, "[I] tackled the guy (almost literally)."[294]

---

[292] Anna Willmon, interview conducted and videotaped in her home in Roswell, New Mexico by Kevin Randle, May 1994. See also Randle, *Roswell Revisited.* 28 – 30.

[293] Ibid.

[294] Information supplied by Don Burleson in email to Kevin Randle, April 16, 2002.

Sometimes called "Tex" by researchers,[295] Jorgenson said that he was a civilian working for the military and that he, along with his colleagues, stumbled onto the site while searching for a drone jet aircraft that had been tested on July 3 but that had crashed somewhere near Roswell. He, along with several NCOs had been told to find it.

As they neared the site, Jorgenson said that he saw pieces of bright silver metal. The metal appeared to be very light and later he would learn quite hard. He said that he saw some sort of writing, that is "hieroglyphics," on it. The craft, he said, was stubby with curved-back wings. It was a small object, about twenty feet long and only about 12 to 14 feet wide. It had the look of some of the rocket ships in the old Buck Rogers serials.[296]

There were three creatures. One was dead, one, though alive, was slumped over and the third was standing in the canopy looking at the men. They were all small, with a grayish, greenish skin, and dark slanted eyes. They had large heads and small noses.[297]

Unlike Willmon, but like others, Jorgenson said they were wearing uniforms, or what might have been uniforms, but he wasn't sure. He didn't notice their hands or their feet.

Jorgenson said that he left the site shortly after the recovery began. He returned to the base and there he was warned not to talk about what he had seen. He said that he had to sign some kind of an oath or a non-disclosure agreement. He thought, by the time he said anything to Burleson and others, that the agreement had expired.

There is one real concern about this story. Jorgenson said that he was assigned to the base where this jet drone project was located and it was a test

---

[295] For another take on this information see Klass, Philip, *Skeptics UFO Newsletter*. May 2000: p. 4. Carey, Tom, "New Roswell Witness Says He Saw Alien Craft, Bodies," copy emailed to Kevin Randle from Karl Pflock April 9, 2002.

[296] Carey, Tom, "New Roswell Witness Says He Saw Alien Craft, Bodies."

[297] Information supplied by Don Burleson in email to Kevin Randle, April 16, 2002.

flight that went astray. There are no records to corroborate this claim. And his discussion of using radar to track it is also a worry, given the nature of the terrain and the locations of the various radars in 1947. Had this test gone astray from White Sands Proving Grounds or from Alamogordo Army Air Field, the radars there wouldn't have been able to track it much below 10,000 feet as it traveled to the north, which meant it could have crashed almost anywhere. The search probably would have been conducted by aircraft and not guys in a jeep.[298]

Physical Evidence of the Bodies

The problem is that all these stories surfaced after the publication of *The Roswell Inciden*t. Although it could be argued that the book was a national publication, there is no evidence that it was big news in Roswell or anywhere else outside of the UFO community. It could be argued that many of those who told of seeing the bodies had not heard about the book when they told their tales. It could also be argued that many of those stories were the result of people influenced by that book.

But that is all there is to this aspect of the case. We just have the stories told about the bodies, few of them told by first-hand witnesses. Those who have claimed to be first-hand-witnesses have nothing to prove they were there or that they saw anything extraordinary. There have no diaries, no journals, no letters from that time which would add a note of credibility to what they have claimed.

This is not to mention that many of those who originally claimed to have seen bodies have had their credibility challenged. Some of them have been caught in forging documents, changing the story to fit the new evidence as it is uncovered, and making up their involvement. There are no solid first-hand

---

[298] There has been no positive corroboration for Jorgensen's tale. It matches, to some extent those offered by others, but he is a self-identified witness and told the story at the International UFO Museum and Research Center in Roswell, so this tale must be examined in that light.

tales complete with some form of documentation. All we have are the tales often with no corroborating testimony to go with it, or that all those who might have seen the same thing no longer available for interviews.

Almost the only exception to this is the article written by Inez Wilcox about her time as matron of the Chaves County Jail. The problem is that there is no date on the document, it is clear that this was an addition to the original story and was probably added after the publication and publicity about *The Roswell Incident*. This doesn't mean what she wrote is untrue, only that it can't be dated prior to the publicity for the case. Could that be done, it would have been a bit of corroboration. As it stands, it is merely an interesting footnote to what we know.

Many of those telling second-hand tales such as that told by Vern Maltais say that he learned of the alien spacecraft crash and the bodies of the flight crew in the early 1950s. The exact date and the exact information about the location wasn't readily available. Even with those problems, there is no reason to suspect that Maltais is lying about what Barney Barnett said, which is not the same thing as saying that the story is the truth. The story was important because it mentioned not only a crash but alien bodies had been recovered. Those relating these stories probably did hear them from the source as they claimed, but that doesn't mean that the stories are grounded in reality.

In fact, with the Barney Barnett tale, there is a diary. Ruth Barnett kept it for the year 1947, but there is nothing in it to suggest that Barnett had seen anything strange. There is no hint that he stumbled onto the scene of a UFO crash.[299]

Other sources, such as the late Brigadier General Arthur Exon related what he had heard while assigned to Wright Field in 1947. He made it clear that he had seen little himself but he had not seen bodies. He'd heard that they had been sent to Wright Field, but given his position, he had no need to see them or to be involved in any research conducted on them.

---

[299] For complete details on the Barney Barnett sighting and an analysis of the Plains of San Agustin crash, see Appendix B.

And this is the trouble we now have. While there are some tantalizing hints about the retrieval of alien bodies, there is nothing solid on which to base a conclusion. No documentation, no photographs, no autopsy reports, only the tales told by some who claim involvement and more stories offered by relatives and friends of those who might have been involved.

The bottom line on this is that while it might be intriguing, there is no evidence that it is true other than the testimonies offered. No matter what we think might be true we can't prove any of it. These tales are exciting but in the end, that is all they are… tales.

Chapter 6:

The Cover-Up

Everyone agrees that something fell at Roswell in 1947. They just don't agree on what it was. Everyone agrees there was a cover up and even the Air Force admitted that in their 1994 investigation. These two facts make the Roswell case extraordinary in the realm of the UFO. But the story didn't begin with the Roswell case, nor with the Arnold sighting of June 24, 1947. The modern era of the UFO began during the Second World War.

The Foo Fighters were a phenomenon observed by the members of the military in all branches of the service. They were described as balls of fire, blobs of light and a form of St. Elmo's fire. Some of these sightings were of solid objects and on several occasions they were fired on. The witnesses said they saw the bullets absorbed by the objects or saw them bounce off, suggesting that the objects were dense metal.[300] The Allied intelligence services never debated the reality of the reported objects, only what their purpose might be. They feared an enemy weapon that could change the course of the air war over Europe or in the Pacific, and they needed to find a way to combat it. When the war ended, the intelligence imperative ended because hostilities had ended. Besides, it was clear that the Foo Fighters, whatever they were, did not belong to the Axis Powers.

There is one item from all this that is overlooked but is of significant importance. One of the American intelligence officers involved was Colonel Howard McCoy.[301] He was working at the highest levels to determine the

---

[300] For a comprehensive analysis of the Foo Fighters, see Keith Chester's *Strange Company* from Anomalist Books, published in 2007. See also, Randle, Kevin. *Government UFO Files*. Canton MI: Visible Ink Press, 2014: pp. 1 – 18.

[301] Keith Chester. *Strange Company*. pp. 80, 215.

source and purpose of the Foo Fighters. When the war ended, he returned to the United States but remained in the Army.

Although the height of the wave was in the summer of 1946, the Scandinavian Ghost Rocket sightings began earlier in the year. According to documents found in various official files, a letter written by Colonel William E. Clingerman for Colonel Howard McCoy asked Lieutenant Colonel George Garrett for all the files on the Ghost Rockets.[302]

In December 1946, according to information developed by Wendy Connors and Michael Hall, McCoy received instructions from Air Materiel Command Commander, Lieutenant General Nathan F. Twining, to establish an unofficial investigation into these strange aerial phenomena that were now being reported over the United States. They set up their investigation in a single locked room at Wright Field. Access was limited to only a few people but when the Arnold sighting was made on June 24, 1947, and more importantly reported in newspapers around the country, the unofficial investigation evolved into one with official status.[303]

At this point there was no real cover up. Information, especially that developed during the war, was, quite naturally classified. The same might be said for the communications between the embassy staffs in various Scandinavian countries and the intelligence sections back at Wright Field. Given that one of the theories developed was that the Soviet Union was responsible for the Ghost Rockets in an attempt to intimidate the Swedish government, a certain amount of secrecy can be expected.[304] That doesn't lead to cover up, but standard intelligence policies.

---

[302] Randle, Kevin. *Government UFO Files: The Conspiracy of Cover-Up*. Detroit: Visible Ink Press, 2014; pp. 39 – 40. Original data uncovered by Connors and Hall. See also Hall, Michael D. and Wendy A. Connors. *Alfred Loedding & the Great Flying Saucer Wave of 1947*, Albuquerque: Rose Press, 1998.

[303] Randle, *Government UFO Files*. pp. 49 – 51. Connors, telephone communication by Kevin Randle.

[304] With the collapse of the Soviet Union it became clear to various UFO researchers and reporters that the Soviets had nothing to do with the Ghost Rockets.

McCoy surfaces again in September 1947 when Twining issued his letter that suggested that the "...phenomenon reported is something real and not visionary or fictitious."[305] This document doesn't suggest a cover up, but does created a classified project to investigate the reports of the flying saucers. While all the information wouldn't be shared, a great deal of it would be and given the nature of the world in 1947, this made some sense. They also announced the creation of Project Saucer which was the public face of the UFO project.[306]

This all dealt with the flying saucers as they were generally perceived in 1947. There is another aspect of this however. In the days that followed the reporting of Kenneth Arnold's sighting, the newspapers were filed with stories about the flying saucers and the flying disks.[307] There were explanations for the sightings, serious reports and humorous articles. There were limited tales of alien beings but curiously, there were a number of tales of UFO crashes, all of which received a great deal of publicity. Preliminary research and follow up stories showed that nearly every one of those reports were hoaxes.[308]

---

[305] Randle, Kevin D. *Project Moon Dust*. New York: Avon Books, 1998. pp. 8 – 10. The entire letter has been reprinted in a number of books. See Good, Timothy. *Above Top Secret*. New York: William Morrow and Company, 1988. pp. 476 – 478; Dolan, Richard M. *UFOs and the National Security State*. Charlottesville, VA: Hampton Roads Publishing Company, 2002. 43 – 44; Swords, Michael and Robert Powell. *UFOs and Government*. San Antonio, TX: Anomalist Books, 2012. pp. 476 – 478. Twining's Letter can be seen here: http://www.nicap.org/twining_letter.htm (Accessed February 20, 2015.

[306] Clark, Jerome. *The UFO Encyclopedia: Second Edition*. Detroit: Omnigraphics Inc. 1998, pp. 746 – 747; see also Randle, Kevin. *Government UFO Files*. pp. 92 – 95; http://www.project1947.com/fig/projsauc.htm (Accessed February 20, 2015).

[307] For a comprehensive list of the sightings in 1947, see Bloecher, Ted. *Report on the UFO Wave of 1947*. Washington, D.C.: privately printed, 1967.

[308] For a comprehensive list of these UFO crashes, see Randle, Kevin. *Crash: When UFOs Fall from the Sky*. Franklin Lakes: NJ: New Page Books, 2010. pp. 67 – 106. Newspaper stories can be seen in Bloecher, Ted. *UFO Wave of 1947*.

As noted, on July 8, 1947, it was announced throughout the world, that a flying saucer had been discovered near Roswell. Within hours that story had been changed to nothing more than a weather balloon and interest in that story faded. But on July 9, newspapers carried a new story from the United Press said, "Reports of Flying Saucers whizzing through the sky fell off sharply today as the Army and Navy began a concentrated campaign to stop the rumors."[309]

This wasn't part of an attempt to classify the event until a proper investigation could be completed. This was the government shutting down the reporting on something that hadn't bothered them for two weeks. Surveys of the news stories ran from just sightings to landing to the crashes. Scientists were speculating about the cause of the sightings. Military organizations were denying that the flying saucers belonged to them or that they were some new technology. Newspapers continued to print the stories because they were selling a large number of newspapers. At this point, there was no prohibition about all of that.

The Cover Up Begins

Then on July 9, the attitude changed. Suddenly, the military didn't want the stories to circulate. They wanted the interest to end.[310] The question seems to be why they would suddenly care. True, the various Scandinavian governments had resorted to a suppression of the information in the summer of 1946

---

[309] "Army, Navy Move on 'Flying Disc" Rumors." *El Paso Herald Post*, July 9, 1947: pp. 1, 11; "Services Try to Stop 'Disc' Talk." *Dallas Times Herald*, July 9, 1947: p. 1. These are just two of the examples of the story that was carried the day after the Roswell crash case.

[310] Information in Bloecher's analysis of the Wave of 1947 demonstrates the sudden and rapid decrease in the number of stories reported and printed. It is traceable back to the July 9, 1947, report.

and that had slowed down the reporting.[311] It also seems that the cover up here began in General Roger Ramey's office with the announcement that the object recovered near Roswell was a regular weather balloon.

The Eighth Air Force Chief of Staff at the time, then Colonel Thomas Dubose said, "...actually, it was a cover story, the balloon part of it for the remnants that were taken from this location and Al Clark took it to Washington... That part of it was in fact a cover story that we were told to give the public and the news."[312]

In 1994, the Air Force confirmed the balloon cover story but assigned it a new culprit. James McAndrew wrote, "This article appeared to have been an attempt to deflect attention from the Top Secret MOGUL[313] project by publicly displaying a portion of the equipment and offering misleading information. If there was a 'cover story' involved in this incident, it is this article, not the actions of statements of Ramey."[314]

In that same report, in the Executive Summary written by Colonel Richard Weaver, it said, "Thirdly, when such claims are made, they are often attributed to people using pseudonyms or who otherwise do not want to be publicly identified, presumable so that some sort of retribution cannot be taken against them

---

[311] Randle, Kevin. *Government UFO Files*. pp. 30 – 31; see also Svahn, Clas and Anders Liljegren. "Close Encounters with Unknown Missiles." *International UFO Reporter* 19,4 (July/August 1994): 11 – 15; Berliner, Don. "The Ghost Rockets of Sweden," *Official UFO* 1,11 (October 1976): 30 – 31, 60 – 64.

[312] Thomas J. Dubose, personal videotaped interview by Don Schmitt and Stan Friedman, August 10, 1990.

[313] The secrecy of the Mogul Project is argued in Appendix C. It is clear that the secrecy revolved around the purpose of Mogul and not the activities in New Mexico in 1947. Even the name was known to the civilians working on the New York University Constant Level Balloon Project which suggests that the level of classification claimed for the activities in New Mexico simply did not exist.

[314] McAndrew, James, in *The Roswell Report: fact vs Fiction in the New Mexico Desert*, Section 32," Synopsis of Balloon Research Findings." Washington, D.C.: Government Printing Office, 1995.

notwithstanding that nobody has been shown to have died, disappeared, or otherwise suffered at the hands of the government during the last 47 years."[315]

While it is true that no one's pension seemed to have been denied or revoked, and there is no evidence of anyone being jailed or fined, it is also true that there is a large body of testimony to suggest that threats, some mild and some very specific, were made. Edwin Easley, the Roswell Provost Marshal, in his first interview said, repeatedly, "I can't talk about that. I was sworn to secrecy."[316] That implied a warning that discussing classified material with those not authorized to hear could result in a stiff fine and a prison sentence.[317]

Earl Fulford reported much of the same sort of thing. He said, "We were lined up and told one by one by our first sergeant in no uncertain terms that we didn't see anything, and we didn't say anything; and if we did from that point forward, we might be court martialed…"[318]

In an even milder case, Thomas Dubose said that he had been told by Major General Clement McMullen not to discuss it with anyone. He wasn't supposed to even mention it to Ramey. There were no threats implied here, only an order not to talk about it with anyone.[319]

With others the threats became more specific and more deadly. These, it seems, were directed at the civilians who had somehow become involved. Barbara Dugger, granddaughter of Sheriff George Wilcox, said that she was told

---

[315] Weaver, Richard L. and James McAndrew. *The Roswell Report: Fact vs Fiction in the New Mexico Desert*, "Executive Summary." Washington, D.C.: Government Printing Office, 1995. pp. 30 – 31.

[316] Edwin Easley, telephone audiotaped interview by Kevin Randle, January 11, 1990.

[317] See 18 U.S.C. 793 et. seq. The unauthorized disclosure will result in fine or imprisonment or both.

[318] Carey, Tom and Donald Schmitt. *Witness to Roswell*. Pompton Plains, NJ: New Page Books, 2009: pp. 113 – 116.

[319] Thomas J. Dubose, personal videotaped interview by Don Schmitt and Stan Friedman, August 10, 1990.

that the military police had gone to the jail house. She said that her grandmother said, "... [They] told George and I [that is we] ever told about the incident, talked about it in any way, not only would we be killed, but they would get the rest of the family."[320]

Frankie Rowe also recalled threats made to her and the family. She said that she was in the house with her mother and brothers and sisters when the military arrived. "They sent Susy and Donnie and Pat outside and they asked my mother who was the child at the fire station."[321]

These men were armed and one of them stood outside with his rifle at port arms. Another stood beside Rowe and her mother as another questioned them. It was clear that the military was in charge and they didn't want anyone talking about the events.

In her affidavit, she said, "A few days later, several military personnel visited the house, telling my younger brothers and sisters to wait outside. My mother and I were told to sit at the dining room table where I was questioned about the piece of metal I had seen. I was told that if I ever talked about it, I could be taken out into the desert never to return, or that my mother and father would be 'Orchard Park', a former POW camp."[322]

Helen Cahill remembered her father's (Dan Dwyer) intense rage about the menacing talk. Cahill also said that soldiers had threatened the fire fighters[323]

---

[320] Barbara Dugger, videotaped interview by Kevin Randle and Don Schmitt, March 4, 1991. Affidavit published in Pflock, Karl. *Roswell: Inconvenient Facts and the Will to Believe*. Amherst, NY: Prometheus Books, 2001: p. 258

[321] Frankie Rowe, personal audiotaped interviews with Kevin Randle and Don Schmitt, January 2, 1993.

[322] Affidavit published in Pflock, Karl. *Inconvenient Facts*. p. 276.

[323] Although Cahill said that the fire fighters were warned, it seems that the warning was little more than an officer from the base telling them that they didn't have to worry about it. That the fire fighters at the base had the situation under control.

but more importantly, she confirmed that something of enough significance had happened that soldiers visited them to warn them.[324]

All this suggests a cover up. Reminding soldiers of their obligations when exposed to classified information doesn't necessarily prove a cover up. Threats made to civilians certainly suggest an attempt to keep the information bottled up but it could be argued that there was a good reason for that. But other information leads to the cover up.

Cover Up at Roswell

In July 1947, Major Patrick Saunders was the base adjutant[325] which meant that he dealt with all the personnel actions and the paperwork that went with running a military base.[326] Given his position at the time, he would have been involved with the activities surrounding the recovery.

When first contacted and asked about the UFO crash he said, "I can't specify anything."[327]

But that wasn't his last word on the subject. He did begin to talk to family and friends about what he had seen and done. The flyleaf to *The Truth about the UFO Crash at Roswell* said:

> Damage Control
>
> Files were altered. So were personal records, along with assignments and various codings and code words. Changing serial

---

[324] Helen Cahill, telephone interview by Kevin Randle, September 1993.

[325] Patrick Saunders' position at Roswell confirmed by the Yearbook prepared by Walter Haut and the 509th Bomb Group Unit History for June, July and August 1949. He confirmed the information during a June 14, 1989 interview with Kevin Randle.

[326] In today's world, the adjutant is known as the S1 or the Personnel Officer.

[327] Patrick Saunders, telephone interview with Kevin Randle, June 14, 1989.

> numbers ensued that those searching later would not be able to locate those who were involved in the recovery. Individuals were brought into Roswell from Alamogordo, Albuquerque, and Los Alamos. The MPs were a special unit constructed of military police elements from Kirtland, Alamogordo, and Roswell. If the men didn't know one another, or were separated after the event, they would be unable to compare notes, and that would make the secret easier to keep.
>
> After the impact site was cleaned, the soldiers debriefed, and the bodies and craft removed, silence fell. It would not be broken for almost forty-five years.

This wouldn't be all that important because it could be argued that a copywriter had created the paragraph to promote a book. Patrick Saunders read what had been written and sent a photocopy of the page to a friend and wrote at the top, "Here's the truth and I still haven't told anybody anything!" He then signed it.[328]

There was a similar experience with *UFO Crash at Roswell*. The flyleaf of that book said:

> Top Secret
>
> Rickett, the senior counterintelligence man, and the Provost Marshal walked the perimeter of the debris field examining the wreckage scattered there. Most of the pieces were small, no more than a few inches long and wide, but some measured a couple of feet on a side.

---

[328] Information supplied by Evie Smith, August 29, 1996 to Kevin Randle.

> The came to one piece that was about two feet by two feet. According to Rickett, it was slightly curved. He locked it against his knee and tried to bend or break it. The metal was very thin and very lightweight. Rickett couldn't bend it at all.
>
> As they prepared to leave the crash site, the senior CIC agent turned to Rickett. "You and I were never out here," he said. "You and I never saw this. You don't see any military people or military vehicles out there either."
>
> "Yeah," Rickett agree. "We never even left the office."[329]

This page came with the inscription, "Susan – You were there!" it was signed by Saunders.

In a letter from his daughter, Saunders was quoted as saying, "At one point he bragged to me about how he had covered the 'paper trail' associated with the clean up!"[330]

Smith, in one of her letters, also wrote, "…You have to realize the situation, the background first. Los Alamos was there. Roswell had the 509th with the crews and bombs. Sandia Caves held the bombs secretly. The Trinity Site – all were there in the area. New Mexico was the center were our greatest capability lay… Suddenly we were confronted with a capability greater than ours. 'They' had our skies. We were powerless. There was no idea of 'their' intentions. Their technology was more advanced than ours… The government was there reluctant to release anything about them… Our first responsibility

---

[329] For the complete interview see Randle, Kevin D. and Donald Schmitt, *UFO Crash at Roswell*, New York: Avon Books, 1991: pages 61 – 63.

[330] Letter from Susan Simmons, daughter of Patrick Saunders to Kevin Randle, April 20, 1997.

was to secure the citizens. Losing control would make them look bad and possibly panic the people."[331]

That wasn't his only comment on the topic. In a letter dated January 22, 1997, Saunders' daughter wrote, "...he [Saunders] felt the threats to people who talked were very real."

To that she added, "I know he struggled with his conscious over what he could say."[332]

While most of this information is second-hand, it does reflect the statements of Saunders to various people. It was reinforced in his statement, "I can't specify anything." He was saying that he had been sworn to secrecy but without giving away much in his statement.

Other elements of the cover up were even more subtle. According to the information available, on the afternoon of Tuesday, July 8, 1947, after having alerted the world to the crash, Colonel Blanchard went on leave.[333] Joe Briley, who became the Operation Officer in the middle of July 1947, said that Blanchard had gone out to the debris field.

One of the problems with this is that Blanchard apparently decided to begin his leave on a Tuesday afternoon. Blanchard, as the commanding officer was in a critical position on the base. He was not as easily replaced as many of the other officers so his leaves had to be planned unless it was an emergency leave. Given the timing, it would seem that some sort of emergency dictated that he leave the base in the middle of the day. There are no indications that such was the case which means that Blanchard's leave would have been planned and if that was true, then he would have been able to schedule it to

---

[331] Evelyn Smith, letter to Kevin Randle, September 12, 1996.

[332] Letter from Susan Simmons, daughter of Patrick Saunders to Kevin Randle, January 22, 1997.

[333] Randle, Kevin D. and Donald Schmitt. *Truth about UFO Crash*. New York: Avon Books, 1994: p. 167; Pflock, Karl. *Inconvenient Truth*. pp. 100 – 101; "Commander on Leave." *Albuquerque Journal*, July 10, 1947.

take advantage of the three day weekend. He would not have scheduled it in the middle of the day meaning that he would lose a day of leave by doing so.

Another factor in this is that on July 9, 1947, Blanchard had been scheduled to meet with New Mexico governor Tom Mabry to have a proclamation signed for "Air Force day." It seems odd that Blanchard would begin his leave on a Tuesday afternoon, especially when he was scheduled to meet with the governor the next day. Briley said, "The story was changed and hushed up immediately... Frankly it was just hushed up so quickly… and so completely that nothing was ever said about it."[334]

Chester Barton, who was stationed in Roswell in 1947, added to this. Barton was assigned as a crypto officer in 1947 in the 509th Communications Unit.[335] According to Barton, he saw no communications traffic about the incident and even though he originally believed the wreckage was the result of a B-29 crash, he was surprised that it had not generated any encrypted traffic. Barton said that in July 1947, given his job, he would have seen this traffic if anything had been forwarded to a high headquarters.[336]

Barton also made it clear that he didn't accept the balloon explanation. Barton knew the balloon theory was complete nonsense, which is a way of suggesting a cover up. He also said that he did not make a written report, but briefed Easley, the provost marshal on what he had seen. This, of course, means that there is no documentation of his experiences on the crash site.[337]

---

[334] Joseph Briley, telephone interviews with Kevin Randle, October 20, 1989 and April 9, 1990. See also Carey and Schmitt. *Witness to Roswell.* pp. 88 – 89; Pflock, Karl. *Inconvenient Facts*. p. 249.

[335] Barton's picture appears in the Yearbook as a member of the 1395th MP Company. Information about the Communications Unit comes from Joe Stefula interview with Barton.

[336] Chester Barton, telephone interview with Joe Stefula, retired Army Warrant Officer and MUFON member, August 1995.

[337] Ibid.

As with the idea that something fell at Roswell, everyone agrees that there had been a cover up. Even the Air Force admitted that in their massive report issued in 1995. In the executive Summary it said, "He [Charles Moore] stated, 'It appears that there was some type of umbrella cover story to protect our work with MOGUL.'"[338]

The idea was that while the equipment was unclassified and the research conducted in New Mexico was unclassified, the ultimate purpose and the name were highly classified. According to the Air Force, "Subsequently, a 1946 HQ AMC memorandum surfaced, describing the constant level altitude balloon project and specified that the scientific data be classified Top Secret Priority 1A. Its name was Project MOGUL."[339]

This all demonstrates that there was a cover story. The reason for it is in question, yet the activities by the New York University balloon project in New Mexico, were printed in the newspapers of the time. Moore even said that the step ladder in one of the pictures of a balloon launch was one that he had bought with petty cash. This suggests that there was nothing happening with the Mogul equipment and arrays that demanded a Herculean effort to cover it up. There was no reason to go to any lengths, and in fact, they didn't make much of an effort to recover all of the balloons launched.

It was clear, however, from the testimony of so many of those involved, that there was an effort to cover up the debris found by Brazel and at the other locations. There was a cover up and the question that this fact generates is what was so important that they went to such extremes to cover up the recovery.

---

[338] Weaver, Richard L. and James McAndrew. *The Roswell Report: fact vs Fiction in the New Mexico Desert*, "Executive Summary." Washington, D.C.: Government Printing Office, 1995. p. 27

[339] Ibid. p. 25. For the Mogul story see Appendix C.

Chapter 7:

The Air Force Investigation

It could be argued that the Air Force's public investigation of the Roswell UFO Crash began in 1994[340] when New Mexico Congressman Steven Schiff asked the General Accounting Office to look into the information about the case.[341] The Air Force undertook a search of their own. According to the Executive Summary of their study:

> Although the GAO effort was to look at a number of government agencies, the apparent focus was on the United States Air Force (USAF) SAF/AAZ, as the Central Point of Contact for the GAO in this matter, initiated a systematic search of current Air Force offices as well as numerous archives and records centers that might help explain this matter. Research revealed that the "Roswell Incident" was not even considered a UFO event until the 1978 – 1980 time frame. Prior to that, the incident was dismissed because the AAF originally identified the debris recovered as being that of a weather balloon. Subsequently, various authors wrote a number of books claiming that not only was the debris from an alien spacecraft recovered, but also the bodies of the craft's alien occupants. These claims continue to evolve today and the Air Force is not

---

[340] Date for the beginning of the GAO investigation can be found in Weaver, Richard L. and James McAndrew. *The Roswell Report: Fact vs. Fiction in the New Mexico Desert*. Washington, D.C.: Government Printing Office, 1995.

[341] "Results of a Search for Records Concerning the 1947 Crash Near Roswell, New Mexico." General Accounting Office, July 1995, p. 1.

> routinely accused of engaging in a "cover-up" of this supposed event.[342]

The purpose of the investigation by the GAO wasn't to look into the evidence for the crash of an alien craft, but to look for records of the event.[343] The Air Force, which had been tasked with the investigations of UFOs until 1969,[344] followed a similar instruction. In other words, this was not an investigation into the UFO crash, but an attempt to find any relevant documents about it.[345] A great deal of the study concerned Project Mogul, the allegedly top secret attempts to create a constant level balloon.[346] Of those interviewed in the study, the vast majority of them were the civilians who worked on Mogul and the officers who were involved with support of it at Wright Field. Although the taped interviews of various officers and enlisted soldiers from the 509th Bomb Group were offered to the Lieutenant McAndrew as he worked on the investigation, he expressed no interest in them.[347] The only interview they conducted with any of those who had been in Roswell in 1947 was retired

---

[342] Weaver and McAndrew, *Fact vs. Fiction*, Executive Summary, p. 9.

[343]. "Results of a Search for Records Concerning the 1947 Crash Near Roswell, New Mexico." General Accounting Office, July 1995, p. 1.

[344] Project Blue Book, the official USAF investigations of UFOs began in 1948 with the creation of Project Sign, was ended in 1969 at the conclusion of the University of Colorado UFO study sponsored by the Air Force.

[345] According to Colonel Richard Weaver in an email to Randle, "The USAF was not 'motivated' on its own to "investigate the Roswell". It was to be a review of the "DOD policies and procedures for acquiring, classifying, retaining and disposing of official government documents dealing with weather balloon [sic] aircraft and similar crash incidents."

[346] For the history of Project Mogul, see Appendix C.

[347] Personal conversations with McAndrew by Kevin Randle during the course of their investigation.

Lieutenant Colonel Sheridan Cavitt who had been assigned as the CIC OIC (Counterintelligence Corps Officer in Charge) in June 1947.[348]

The Project Blue Book (Project Sign at the time) files contain a single mention of the Roswell UFO crash and that is in the third paragraph of a four paragraph newspaper clipping about another UFO sighting. The mention is that Walter Haut had been chewed out for issuing the press release.[349] The story, as seen, was front page news around the country on July 8 and 9 but nothing about it made it into the Blue Book files. Other reports of UFO crashes, those with plausible explanations and those which were obvious hoaxes had been investigated and reports had been made to the Army (later the Air Force). One of those, from Shreveport, Louisiana, provided an illustration of this.

The Shreveport UFO Crash

According to a report in the Project Blue Book files dated 23 July 1947, the intelligence office at Barksdale Field near Shreveport, had received information that a flying disc had been located, or crashed, in Shreveport.[350] A witness told the Army officers that he, a man identified only as Harston in the report, "had heard the disc whirling through the air and had looked up in time to see it when it was approximately two hundred feet in the air and coming over a sign board adjacent to the used car lot where he was standing... [witness]

---

[348] Information about Cavitt came from personal interviews with him by Kevin Randle and Don Schmitt on January 29, 1990; March 27, 1993; June 25, 1994; Cavitt's statement to Colonel Richard Weaver dated May 24, 1994; Carey, Tom. "Will the Real Sheridan Cavitt Please Stand Up?" *International UFO Reporter*, Fall 1998: pp. 14 – 21; Maccabee, Bruce. "Did Sheridan Cavitt Visit the Same Crash Site? Or "Cavitt Emptor!" http://brumac.8k.com/Roswell/CavittEmptor.html.

[349] Project Blue Book files, July 9, 1947, case number 53. Clipping labeled "Saucers, If Any, Fading Away to Blue Yonder."

[350] Project Blue Book files, July 7, 1947, case number 41.

stated that smoke and fire were coming from the disc and that it was traveling at a high rate of speed and that it fell into the street..."[351]

The following day, the Army investigators talked to another witness who told them in an interview "that he had made the disc in order to play a joke on his boss and that the starter had been taken from a florescent light and two condensers from electric fans."[352]

With that the Army closed their investigation. But the FBI was still interested and the importance of that interest would be learned later. The FBI document covers the same information. The report added one detail. According to the document, "Harston stated that the disc made a sound when traveling through the air similar to a policeman's whistle and that the smooth side was toward the earth while in flight."[353]

Here is where this case becomes important. In July 1947 the military asked the FBI for assistance in investigating the flying saucers. J. Edgar Hoover responded to the request which came from the Army Air Forces' Brigadier General George F. Schulgen, at the time Chief of Requirements Intelligence Branch, by writing, "I would do it but before agreeing to it we must insist upon full access to discs recovered. For instance in the La. case the Army grabbed it and would not let us have it for cursory examination."[354] This statement, which seems to suggest the recovery of a flying saucer, did, for many years, become the focus of a dispute as to exactly what Hoover meant when he wrote

---

[351] Project Blue Book files, July 7, 1947, letter dated 23 July 1947 from the officer of the ACofS A-2 (Assistant Chief of Staff, Intelligence), Barksdale Field, LA and sent to the Commanding General, Army Air Forces, Washington, D.C.

[352] Ibid.

[353] Project Blue Book files, July 7, 1947, copy of document found in the files dated July 8, 1947 that includes the handwritten statement, "10 July 47 - Major Carlau (US Br of ID) says FBI advises this was a hoax."

[354] Handwritten notation on FBI Document dated July 10, 1947 by D.M. Ladd for E. G. Fitch with J. Edgar Hoover's note written on the second page.

it in 1947 and was used to suggest that the FBI had been left out of the Roswell investigation.

The "La." is written in such a way that it could just as easily be read as "Sw." or "gov." or "Sov." or even "2a." Hoover's handwriting was sloppy enough that any of these interpretations could be accurate and all have been offered as a solution for the handwriting. However, in another document completed at the end of July 1947, the statement had been typed and it was interpreted by the typist in 1947 as "La." Since this document was an FBI creation, and that it was typed at the FBI headquarters by a secretary familiar with Hoover's handwriting, it would seem that the interpretation is accurate.[355]

This demonstrates that some of the UFO crash stories had thick files at Project Blue Book. This case did not make the national news the way the Roswell crash did but there is a file. For Roswell there is nothing and if the Air Force believed in 1947 that the logical response was a weather balloon and radar reflector, then why no similar report.

The Spitzbergen UFO Crash

There were other reports of UFO crashes in the Project Blue Book files. In 1952, a series of messages passed through the Air Force communications system that reported a crashed disk had been recovered on the Swedish controlled island of Spitzbergen.[356] The information about the case is sketchy but it was

---

[355] Letter from E. G. Fitch to D. L. Ladd (both with the FBI) dated July 24, 1947.

[356] For the most comprehensive analysis of the Spitzbergen crash, see Braenne, Ole Jonny. "Legend of the Spitzbergen Saucer." *International UFO Reporter*, 17,6 (November/December 1992): 14 – 20. See also, Randle, Kevin D. *Crash: When UFOs Fall from the Sky*. Franklin Lakes, NJ: New Page Books. 2010: 146 – 152: Wood, Ryan. *Majic Eyes Only*, Broomfield, CO: Wood Enterprises, 2005. 102 – 104; Edwards, Frank. *Flying Saucers – Serious Business*. New York: Bantam Books, 1966: 44 – 48.

covered in the Project Blue Book files that included several pages of teletype messages that gave details but offered nothing much in the way of a solution.[357]

The story as told is that an article had appeared in the German newspaper, *Saarbrucker Zeitung* on June 28, 1952, that reported a "silvery disc with a dome of plexiglass [sic] and 46 jets on the rim" had been found on Spitzbergen Island and it was believed to be of Soviet manufacture. Norwegian military officers landed near the disc and began a detailed study of it. According to this report, the saucer was 48.88 meters in diameter, was made of metal of an unknown alloy, had instruments in Russian, seemed to have a combat radius of 30,000 kilometers, and could fly at an altitude of 160 kilometers.[358]

That article set off a chain of events that resulted in several classified messages being sent from Europe to Wright-Patterson Air Force Base. One of those dated August 19, 1952, recounts the information that had been published in the *Berlin Volksblatt* that repeated the information that has been circulating. The classified message said, "Request check on validity of this information or on any occurrence which may have formed the basis for this rumor."[359]

The next day, in another message to the Chief, ATIC, it was noted that the story was getting a big play in the German media and that the Germans had been running the story "continuously since shortly after 9 July." They were again asked to check on the validity of the story.

On August 22, 1952, in a classified message that was directed to the "USAF WASH DC ATTTN: [sic] AFOIC 2A2 [the Intelligence Office]," it said, "Ref your 56013 DTG 192143Z RNAF states this information is definitely false." It was signed "Reminton."[360]

---

[357] Project Blue Book files, July 9, 1952, case number 1411.

[358] Braenne. "Spitzbergen Saucer."

[359] Project Blue Book files, July 9, 1952, case number 1411.

[360] Ibid.

The Project Blue Book files contained a great deal of information about this hoax, including the messages that had been sent through the Air Force communications system and the directives sent first, accidentally to the Army Attaché and then on to the Air Force Attaché at the embassy in Oslo. Or, in other words, they processed and filed the information on this crash, labelling it a hoax in the documentation but suggesting it had been caused by an airplane in the final listings of the Blue Book sightings.[361]

This certainly suggests that there should have been a great deal of information about the Roswell case in the Project Blue Book files. That it had been "identified" as a weather balloon would be no reason to exclude it, given the extensive coverage of other reports of flying saucer crashes.

More On the History of the Roswell Crash

The Roswell case had been relegated to the hoax category. One of the few mentions of it appeared in the "Hoaxes and Mistakes" section of Ted Bloecher's *Report on the UFO Wave of 1947*. That entry said:

> While newspapers still carried a few apparently genuine UFO reports – often buried among the mish-mash of superficial nonsense – the kind of stories that made the headlines after July 8th were the sort a [sic] reader found impossible to take seriously. If a report wasn't an out-and-out hoax, it was an embarrassingly obvious mistake. One of those mistakes, given the widest possible publicity, had its original near Roswell, New Mexico, when a farmer names William W. ("Mac") Brazel[362] discovered the wreckage of a disc on his ranch near Corona, early in July. After hearing news broadcasts of flying saucer reports, Brazel, who had stored pieces of the disc in a

---

[361] Project Blue Book Index, Roll No. 1.

[362] Name altered or added in 2005 edition of the report.

barn, notified the Sheriff's Office in Roswell, who, in turn, notified Jesse A. Marcel, of the Roswell Army Air Field intelligence office. The remnants of the disc were taken to Roswell Field for examination. Though a series of clumsy blunders in public relations, and a desire by the press to manufacture a crashed disc if none would obligingly crash of itself, the story got blown up out of all proportions that read "Crashed disc Found in New Mexico."

According to AP on July 8th, public information officer Walter Haught [sic] made an announcement of the discovery: "The many rumors regarding the flying disc became a reality yesterday when the intelligence office of the 509th Bomb Group of the Eighth Air Force, Roswell Army Air Field, was fortunate enough to gain possession of a disc through the cooperation of one of the local ranchers and the sheriff's office of Chavez [sic] County." The effect of this reckless statement was equal to an atomic detonation; results were immediate. While newspaper [sic] deluged the air base for additional information, a search party was sent out to scour the landing site for additional fragments; the collected remains of whatever it was that had crashed on Brazel's ranch were taken to Eighth Air Force headquarters in Fort Worth, Texas. There, Brigadier General Roger M. Ramey tried to clarify matters by first explaining that no one had actually seen the object in the air; that the remains were of flimsy construction; that it was partially composed of tinfoil; and, finally, that it was the wreckage of "a high altitude weather device." Warrant Officer Irving Newton, a weather forecaster at the Fort Worth Weather Station, had identified the crashed "disc" as the remains of weather equipment used widely by weather stations around the country when sending balloons aloft to measure wind directions and velocity. There remains the possibility that some super-

> secret upper-atmosphere balloon experiment had crashed near Corona, which would have accounted for all the confusion and secrecy involved in its recovery.
>
> Whether the pictured balloon equipment carried widely in the press was actually a photograph of the recovered fragments remained a question, but news editors should have been on their toes: other similar incidents had already been reported, like the discovery several days before of the weather device at Circleville, Ohio.[363] The New Mexico incident created an uproar in Washington, and high Army Air Force officials were reported to have delivered a blistering rebuke to Roswell Field spokesman for having fostered the confusion. But the damage had already been done and the next day "Another saucer Shot Down" was typical of the headlines found in American papers.[364]

That was the attitude of almost anyone who saw any of the newspaper articles or who wrote about the case. One of the exceptions was Frank Edwards who included a brief account in his 1966 book, *Flying Saucers – Serious Business*." Nearly everything that Edwards wrote about the case was wrong, other than it had happened near Roswell. Edwards wrote:

> There are such difficult cases as the rancher near Roswell, New Mexico, who phoned the Sheriff that a blazing disc-

---

[363] Those earlier stories published in various newspapers were written prior to the arrival the debris in Fort Worth and before any photographs were taken. According to the documentation available, the first photographs were transmitted over the newswire at 11:59 p.m. on July 8.

[364] Bloecher, Ted. *Report on the UFO Wave of 1947*. Washington, D.C.: The Author, 1967. P. I-13.

> shaped object had passed over his house at low altitude and had crashed and burned on a hillside within view of the house. The sheriff called the military; the military came on the double quick. Newsmen were not permitted in the area. A week later, however, the government released a photograph of a service man holding up a box kite with an aluminum disc about the size of a large pie pan dangling from the bottom of the kite. This, the official report explained, was device borne aloft on the kite and used to test radar gear by bouncing signals off the pie pan. And this, we were told, was the sort of thing that had so excited the rancher. We were NOT told why the military cordoned off the area while they inspected the wreckage of a burned-out box kite with a non-inflammable pie pan tied to it.[365]

These were two of the mentions of the Roswell case made in the 1960s, but no one paid attention to them. Neither contained any information that suggested there was more to the story and although Edwards' report suggested something strange had crashed, it didn't provide enough information to spark any sort of interest. They case had gone dormant.

This situation changed in 1978 with two events. First, Stanton Friedman, while in New Orleans, was told of a former soldier who had told his ham radio pals that he had picked up pieces of a flying saucer.[366] Friedman interviewed Jesse A. Marcel, Sr. and passed the information along to Leonard Stringfield who had been collecting tales of flying saucer crashes for a number of years. Stringfield presented the information at the 1978 MUFON Symposium though it didn't appear in the printed version of his paper.

---

[365] Edwards. *Flying Saucers*. pp. 41 – 42.

[366] Friedman, Stanton F and Berliner, Don. *Crash at Corona*. New York: Paragon House, 1992: 8 – 12.

Stringfield's paper created something of a stir in the UFO community but very few outside of it were aware of it. Stringfield was attempting to reverse a trend that had been around since 1952 when J. P. Cahn published an expose of the Aztec UFO crash tale.[367] Stringfield was advocating that it was time to reevaluate the stories of UFO crashes.

In 1980, Charles Berlitz and William Moore published *The Roswell Incident* which was an account of the Roswell UFO crash. It added additional witnesses to the events suggesting that something unusual had crashed. This marked the beginning of the real interest in the Roswell case, though it was derailed for a number of years by the appearance of the MJ-12 documents.[368]

Interest grew, first with the *UFO Cover-up – Live* in 1988 and then when *Unsolved Mysteries* aired a segment on the Roswell crash in 1989. It was also about this time that the J. Allen Hynek Center for UFOs became interested in the Roswell case and decided to investigate it further. Donald Schmitt and Kevin Randle were eventually teamed up to conduct the research. This began an expanded interest in the Roswell crash.

In 1991, *UFO Crash at Roswell* was published, followed by *Crash at Corona* and the *Truth about the UFO Crash at Roswell*. There were magazine articles and more television documentaries. Interest in the case was increasing and where, in 1990, most people had never heard of Roswell and the UFO crash, in a couple of years most people had.

---

[367] Cahn, J. P., "The Flying Saucers and the Mysterious Little Men." *True*. September, 1952; see also Steinman, William S. and Wendelle C. Stevens. *UFO Crash at Aztec*, Tucson, AZ: Wendelle Stevens, 1986: 538 – 549

[368] For the whole history of the MJ-12 saga, see Appendix A.

The GAO Roswell Document Search

It was in this environment that New Mexico Congressional Representative Steven Schiff, allegedly reacting to pressure from his constituents, asked the General Accounting Office to take a look.[369] The GAO investigation was limited to a search for documentation of various agencies including the CIA, FBI, Department of Energy and the Air Force, among others. They wrote, "We conducted an extensive search for government records related to the crash near Roswell. We examined a wide range of classified and unclassified documents dating from July 1947 through the 1950s."[370]

The GAO report did mention that "Our search was complicated by the fact that some records we wanted to review were missing and there was not always an explanation. Further, the records management regulations for the retention and disposition of records were unclear or changing during the period we reviewed."[371]

This meant, simply, that some records had either been lost or improperly destroyed, but the GAO was unable to determine what happened to them. It would seem that the disappearance of classified material, regardless of subject matter would require some sort of investigation but that didn't happen. Instead it was simply noted without real comment.

One of the few documents that was found was an FBI Telex from July 8, 1947, that had been created by the Dallas office. Although this document had been in the hands of UFO researchers for years in an uncensored form, the copy reproduced in the GAO report had been redacted, removing the names of both Army Air Forces officers and FBI agents who were mentioned in it.[372]

---

[369] Bitzer, J. Barry. "Schiff Receives, Releases Roswell Report." Press Release, July 28, 1995. See also, "Results of a Search for Records Concerning the 1947 Crash Near Roswell, New Mexico." General Accounting Office, July 1995, p. 1.

[370] GAO, "Records Concerning." p. 1

[371] Ibid. p. 7

[372] Ibid. p. 14

According to records, "In February 1994, the Air Force was informed that the General Accounting Office (GAO), an investigative agency of Congress, planned a formal audit to ascertain 'the facts regarding the reported crash of a UFO in 1949 [1947] at Roswell, New Mexico…' Thereupon, the Secretary of the Air Force directed that a complete records search identify, locate, and examine any and all information on this subject. From the outset there was no predisposition to refute or overlook any information… In short, the objective was to tell Congress, and the American people, *everything* [emphasis in the original] the Air Force knew about the Roswell claims."[373]

While the GAO searched for documentation, the Air Force searched for an answer that would satisfy most people. In their report, they did not mention the various members of the Air Force who had served in Roswell who could have offered information, nor did they mention information offered to them including taped interviews with various witnesses, nor the names and transcripts of those who had additional information. They interviewed only Sheridan Cavitt, who had been the Counterintelligence Corps officer assigned to Roswell in 1947.

Cavitt, who originally denied having been in Roswell in 1947,[374] was interviewed by Colonel Weaver on May 24, 1994. During that interview, Cavitt said, "So, I went out and I do not recall whether Marcel went with Rickett and me; I had Rickett with me. We went out to the site. There were no, as I understand, check points or anything like that (going through guards and that sort of garbage) we went out there and we found it. It was a small amount, as I recall, bamboo sticks, reflective sort of material… you would probably think it was aluminum foil."[375]

---

[373] Weaver and McAndrew. *Fact vs. Fiction*. p. 1.

[374] Sheridan Cavitt, personal interview by Kevin Randle and Don Schmitt, January 29, 1990.

[375] Weaver and McAndrew. *Fact vs. Fiction*. Section 18: Interview. Col. Richard L. Weaver with Lt. Col. Sheridan D. Cavitt (Ret). May 24, 1994.

Weaver asked, "What did you think it was when you recovered it?" and Cavitt answered, "I thought it was a weather balloon."[376]

This becomes important because of a skeptical argument that has been offered to underscore the Mogul balloon explanation. It is suggested that in the context of the times that Marcel and Haut were swept along with all the articles in the newspapers about the flying saucers so that when presented with unusual metallic debris, their thoughts took them to flying saucers as opposed to weather balloons. But, Cavitt, who walked the same field as Marcel, was asked what he thought it was, he said he thought it a weather balloon. No one asked him why he hadn't given this rather important piece of intelligence to Marcel, or later to the 509th Bomb Group Commanding Officer, Colonel William Blanchard. Instead, he said nothing, allowed the rumors to spread about the capture of a flying saucer and special flights take off for Fort Worth.

Although Cavitt returned to the notion that he had only been accompanied out to the site by Bill Rickett, when Weaver asked how they had learned that the debris was out there, Cavitt said, "That I don't recall. Looking back on it, I imagine somebody called the 509th. The 509th called Marcel and said there was something over here, whatever, and then… more and more thinking back on it now, he must have been… I must have been with him…"[377]

However, when asked why he thought Marcel would identify him as having gone out with him and told of the transcript made by Bob Pratt of an interview in which Marcel said that he had gone out there with a good West Texas boy from San Angelo, Cavitt said, "Sort of nails me, doesn't it."[378]

Cavitt also said that he had been too busy to be involved in any weather balloon recoveries during the summer of 1947. When asked specifically about

---

376 Ibid.

377 Ibid.

378 Sheridan Cavitt audiotaped interview by Kevin Randle and Don Schmitt, June 25, 1994.

aiding in the recovery of any rockets, prototypes or experiments, he said, "Never."[379]

But Cavitt continued with the different story when interviewed by Weaver. When Weaver asked if there was any doubt in his mind that it was a balloon and when identified it as some sort of a balloon, Cavitt said, "When I first saw it."[380]

With that single interview of one of those who had been involved with the 509th Bomb Group in 1947 (though Cavitt was assigned to the CIC rather than the 509th) the rest of the interviews were with those who had been assigned to Project Mogul, either as civilians in New Mexico or with the military at Wright Field. These interviews included those with Charles Moore, Athelstan F. Spilhaus, and Albert C. Trakowski, among others.[381]

Of interest was Trakowski's statement. He said, "I became aware of this [the recovery of material in Roswell] only after Colonel Duffy [Trakowski's predecessor at Project Mogul] called me from Wright Field from his home. This was just an informational call, he just wanted to let me know that someone had come to him with some debris from New Mexico and he said, 'this sure looked like some of the stuff that you launched from Alamogordo.' Duffy was familiar with the various apparatus and materials for the project, so if he said that it was debris from the project, I'm sure that's what it was. He was not concerned with a breach of security for the project."[382]

If we go back to Ramey's announcements in Fort Worth, he told the press that the special flight had been canceled when Newton identified the wreckage

---

[379] Sheridan Cavitt unrecorded interview by Kevin Randle and Don Schmitt, January 29, 1990.

[380] Weaver and McAndrew. *Fact vs. Fiction.* Section 18: Interview. Col. Richard L. Weaver with Lt. Col. Sheridan D. Cavitt (Ret). May 24, 1994.

[381] Weaver and McAndrew. *Fact vs. Fiction.* Table of Contents.

[382] Weaver and McAndrew. *Fact vs. Fiction.* Section 22, Statement of Albert C. Trakowski, June 29, 1994.

as that of a weather balloon and radar target.[383] Ramey had ordered his aide, Captain Roy Showalter to have the flight canceled.

The purpose of the Air Force investigation, however, was not to interview the witnesses from Roswell, but to search for documentation from July 1947. What they did was review the documentation that concerned Project Mogul and the majority of the documentation in the report related specifically to Mogul. For them, that satisfied the requirements of the mandate under which they operated, it answered the questions that had been posed by Congressman Schiff and suggested that nothing extraordinary was being hidden from the public.

This, it would seem, would have answered the questions that had been posed, provided what the Air Force and many others believed was an explanation for the material recovered near Roswell, and should have ended the discussion. However, according to McAndrew, "Although Mogul components clearly accounted for the claims of 'flying saucer' debris recovered in 1947, lingering questions remained concerning anecdotal accounts that included descriptions of 'alien' bodies. The issue of 'bodies' was not discussed extensively in the 1994 report because there were not any bodies connected with the events that occurred in 1947."[384]

The Air Force Explains the Bodies

McAndrew then suggests that further investigation by the "Air Force researchers discovered information that provided a rational explanation for the alleged observations of alien bodies associated with the 'Roswell Incident.'"[385]

---

[383] Carey and Schmitt, *Witness to Roswell*, 42; Randle and Schmitt, *UFO Crash at Roswell*, 74: Berlitz and Moore, *The Roswell Incident*, 29.

[384] McAndrew, James. *The Roswell Report: Case Closed.* New York: Barnes & Noble, Inc. 1997: p. 1.

[385] Ibid. p. 2

In the introduction of that second publication, *The Roswell Report: Case Closed,* McAndrew provided the conclusions of the research. He wrote that activities conducted by the Air Force, including various types of high altitude research that had been conducted over a period of years had been "consolidated… and represented to have occurred in two or three days in July 1947."

Among those activities were the result of "anthropomorphic test dummies" that were part of those high altitude tests, that bodies taken to the base hospital in Roswell were the result of an aircraft accident in 1956 and a "manned balloon mishap" in 1959 that injured two Air Force pilots, and the recovery operations of the anthropomorphic dummies. The fact that these events took place about a decade after the debris was found didn't bother McAndrew and as noted, he believed these events had been "consolidated" with those from July 1947. Or, in other words, witnesses were accurately reporting what they had seen; they just had the timing skewed.

The problem is that McAndrew's analysis is flawed by not using the best information available in the late 1990s when he wrote his report. To prove that what had been reported by witnesses, he used testimony that had been gathered in the early 1990s. To tie his theory together, McAndrew cited the testimony of Jim Ragsdale.

Using quotes lifted from the January 26, 1993, interview with Ragsdale and conducted by Don Schmitt, McAndrew wrote:

> Testimony attributed to Ragsdale, who is deceased, states that he and a friend [Trudy Truelove] were camping one evening and saw something fall from the sky. The next morning, when they went to investigate, they saw a crash site:[386]
>
>> "One part [of the craft] was kind of buried in the ground and one part of it was sticking our [out] of the ground." "I'm sure that [there] was bodies… either

[386] Ibid. p. 56

> bodies or dummies." "The federal government could have been doing something they didn't want anyone to know what this was. They was using dummies in those damned things... they could use remote control... but it was either dummies or bodies or something laying there. They looked like bodies. They were not very long... [not] over four or five feet long at the most." "We didn't see their faces or nothing like that... we had just gotten to the site and the Army... and all [was] coming and we got into a damned jeep and took off."[387]

This testimony then describes an assortment of military vehicles used to recover the "bodies":

> "It was two or three six by six Army trucks a wrecker and everything. Leading the pack was a '47 Ford car with guys in it... It was six or eight big trucks besides the pickup, weapons carriers and stuff like that." Ragsdale also said that before he left the area he observed the military personnel "gathering stuff up" and "they cleaned everything all up."[388]

McAndrew explained that the references that Ragsdale made to the procedures and equipment were "consistent with documented anthropomorphic dummy recoveries from projects High Dive and Excelsior." McAndrew wrote

---

[387] Ibid. p. 56. Ragsdale quotes come from the January 26, 1993 interview with Ragsdale conducted by Don Schmitt and recorded on audio tape. A complete transcript was shared with McAndrew, but only five pages are reprinted in McAndrew's report.

[388] Jim Ragsdale interview by Don Schmitt, January 26, 1993. See also, Randle and Schmitt. *The Truth about UFO Crash*, pp. 3, 7 – 8.

that the repeated use of the terms of "bodies or dummies" left little doubt that what Ragsdale had seen was one of the anthropomorphic dummy recoveries.[389]

McAndrew explained some of the technical details of the two projects that seemed to coincide with what Ragsdale described, though the number of vehicles seems to be too high for the recovery of a single anthropomorphic dummy. A single six by six or weapons carrier would have had sufficient cargo capacity for any of the equipment recovered.

But that isn't the real problem with the assessment made by McAndrew. He didn't use later information, much of it developed before he began his investigation. That information, which is in conflict with what Ragsdale had said in 1993, was available to McAndrew, had he wanted to compare it with the theories that he was developing. Had he done that, he would have realized that the Ragsdale testimony was badly flawed and did not lead in the direction that he, McAndrew, wished to take.

According to the information published in *The Jim Ragsdale Story* in 1996, he was camped out on the night of July 4, 1947, and saw the object fall. He told Max Littell, of the International UFO Museum and Research Center:

> From the northwest, there was a big flash, an intense, bright explosion, and then, shortly thereafter, with a noise like thunder, this thing came plowing through the trees, shearing off the tops, and then stopped between two huge rocks. It was about twenty feet around…
>
> The damned thing stopped about sixty yards from the pickup, and we thought at first it was going to hit us. After the impact, we were scared but curious. We went down to the crash of this disc-like thing. There was a hole in one side about four feet wide and two feet high. I looked inside the hole, and inside there was a chair that looked like a throne. It looked like it was made of rubies and diamonds…

---

[389] McAndrew. *The Roswell Report: Case Closed.* pp. 56 – 57.

> There were also little people, four of them. They looked like midgets, about four feet long. Their skin was gray and when I touched one of them, it felt like a wet snake.[390]

In his affidavit, Ragsdale would provide more description of the alien creatures he claimed to have seen. He said, "The bodies of the occupants were about four feet or less tall, which strange looking arms, legs and fingers. They were dressed in a silver type uniform and wearing a tight helmet of some type. This is positive because I tried to remove one of the helmets, but was unable to do so. Their eyes were large, oval in shape, and did not resemble anything of a human nature."[391]

Here is a much more detailed description of the event and it does not agree with what Ragsdale had said originally or with what McAndrew wrote. Here is the problem for McAndrew, if this information is accurate, then his assessment suggesting anthropomorphic dummies seen in the distance fails. Ragsdale was close enough to touch them, and if that is true, he would have recognized the dummies for what they were.

On the other hand, the changes in the story actually render the Ragsdale testimony as false. He contradicts himself on many of the key points, including the type of vehicle he was in, the location of the crash, and what he did immediately after he claimed to have seen the thing fall. If he is inventing the tale, then any assessment of those claims is false and McAndrew's theory fails again.[392]

---

[390] *The Jim Ragsdale Story: A Closer Look at the Roswell Incident*. Roswell, NM: Ragsdale Productions, Inc. 1996, pp. 2 – 3.

[391] Jim Ragsdale affidavit dated April 15, 1995; for the affidavit see also Pflock, Karl. *Roswell: Inconvenient Facts and the Will to Believe.* Amherst, NY: Prometheus Books, 2001. pp. 272 – 274.

[392] Barret, William P. "Now Where was it Those Aliens Crashed?" *Crosswinds*, August 1996, pp. 14 – 16, 34. See also, Randle, Kevin D. *The Randle Report: UFOs in the 1990s*, New York: M. Evans and Company, 1997. pp. 163 – 177; Ragsdale, Jim and Max Littell, *The Jim Ragsdale Story*, Roswell, NM: Ragsdale Publications, Inc. 1996.

There are similar problems with other aspects of McAndrew's attempt to shoe horn testimony into his theories. He mentioned the Glenn Dennis testimony. Dennis had been a mortician in Roswell in 1947, though the documentation suggests he was an embalmer. He told researchers that he had been called and asked about the availability of small caskets in early July.[393]

All this is important because it makes up some of the argument that the tales of bodies were caused by Air Force experiments that took place years later. At the end of his retelling of the Dennis tale, McAndrew wrote, "Air Force research revealed that the witness made serious errors in his recollection of events. When his account was compared with official records of the actual events he is believed to have described, extensive inaccuracies were indicated including a likely error in the date by as much as 12 years."[394]

The trouble with the analysis of the Dennis testimony is that it suffers from the same problems as that of Jim Ragsdale. It is clear from statements that Dennis made later, that what he said about these events were probably the result of his inventing the tale. That can be seen in a number of places. For example, though reluctant, he eventually did reveal the name of his friend who was a nurse. She was Naomi Self or a variation of that name given to various UFO researchers including Karl Pflock and Philip Klass.

The changing name of the nurse, along with the tale of what happened to her suggested that Dennis was not being candid with UFO researchers. This was played out by some of those who believed Dennis. Carey and Schmitt wrote:

---

[393] For various accounts of Dennis' testimony, see Randle, Kevin D. and Donald R. Schmitt, *The Truth about the UFO Crash at Roswell.* New York: M. Evans & Company, Inc. 1994: pp. 15 – 16; Randle, *The Randle Report.* pp. 186 – 189, 191 – 192; Friedman and Berliner, *Crash at Corona,* pp. 83, 112, 114 – 120; Carey and Schmitt, *Witness to Roswell,* 138, 143 – 148, 150 – 151; McAndrew, *Case Closed,* pp. 75 – 84. Also interview with Dennis conducted by Stan Friedman on August 5, 1989, by Kevin Randle and Don Schmitt in November 25, 1990 and videotaped interview for *The Roswell UFO Crash,* November 1990.

[394] McAndrew, *Case Closed.* p. 79; see also Pflock, Inconvenient Facts, Dennis Affidavit, pp. 254 – 256.

> After several years of searching for any record of a military nurse by the name of Naomi Selff, or any record of a plane crash in England involving American nurses during the pertinent time frame, Glenn Dennis was confronted by Roswell investigators with this information in the mid-1990s. His surprising and disappointing response was, "That wasn't her real name. I gave you a phony name, because I promised her that I would never reveal it to anyone." In any court of law, when someone is caught in a lie, that person is said to have been "impeached" as a witness, meaning that his or her testimony, as evidence, cannot be relied upon. Even though we know of witnesses who have told us that Dennis had told them about his run-in at the base long before Roswell became a household word – still, Dennis was found to have knowingly provided false information to investigators, and must technically stand impeached as a Roswell witness. There is no way to get around that fact without believable, clarifying information from Dennis himself. To date, no such information has been forthcoming, and Dennis's health no longer permits questioning him on the matter.[395]

McAndrew, in his investigation, uses the testimony from Dennis to help establish the validity of his theory that anthropomorphic dummies. The problem is the same for Dennis as it was with Ragsdale. If he was inventing his tale, and the story of the missing nurse, then any extrapolation from that point is flawed. In other words an argument that is based on the tales told by Dennis does not support the theory that those who saw bodies had misidentified the Air Force high altitude experiments.

---

[395] Carey and Schmitt. *Witness to Roswell*. p. 145; see also Randle. *The Randle Report*. pp. 189 – 192; Pflock. *Inconvenient Facts*. 129, 131 – 134.

According to the information, the Air Force's first investigation, in response to Representative Schiff, was a search for documents rather than interviews with the members of the 509th Bomb Group who would have been in the best position to provide information. Even when offered copies of audio and video taped interviews, they were uninterested, chasing, instead, the Project Mogul explanation. This satisfied them, and they concluded that the most likely answer for the strange debris found were the remains of a Project Mogul balloon train. Their report is hundreds of pages that relate to the constant level balloon flights, the notes and field diaries of those experiments, and other irrelevant documentation. In the end they fail at one point and that is if there was a Flight No. 4 and how it was configured.

McAndrew, worried about the tales of alien bodies, and with the approval of the Air Force, researched the idea, eventually concluding that high altitude research and anthropomorphic dummies were responsible for these stories. Unfortunately, much of the information he used to support his theory was flawed, and even at the time he was conducting his research, had been rejected by UFO researchers.

In the end, the Air Force didn't supply much in the way of enlightenment for the Roswell case. All they did was show that they had an interest in it, but that interest was only in explaining it away and not learning what had happened.

Chapter 8:

The Jesse Marcel Conundrum

The Roswell case remained relatively dormant until Jesse Marcel, Sr. began telling his ham radio buddies about picking up pieces of a flying saucer some three decades after the event.[396] According to Stan Friedman, who was in Baton Rouge on February 20, 1978, when one of the directors of a television station told Friedman he should talk to Jesse Marcel. Marcel, according to the man, had said that he had handled pieces of a flying saucer, and though Friedman said he was somewhat dubious, he took down Marcel's name and using directory assistance was given Marcel's telephone number. At the airport with some time to spare, he called Marcel. According to Friedman, Marcel related to him, during that telephone conversation, the details of the crash, though Marcel couldn't remember the exact date. He knew he was assigned to the base in Roswell where it happened.

According to Friedman's book, *Crash at Corona*, Marcel had given him the facts as have been reported over the years.[397] What seems to be relevant here is that Marcel had already formed the idea that the debris was left by something alien. Friedman quoted indirectly from Marcel but I don't know if

---

[396] There are been references to the Roswell case in the years that followed, but most of them had little in the way of evidence. Frank Edwards, in *Flying Saucers – Serious Business* mentioned that a rancher had reported a flying disk crash into a hillside near Roswell and that the Air Force had explained it as a large pie pan dangling from a kite. While nearly everything Edwards had written was wrong, he did mention Roswell and a UFO crash.

[397] Friedman, Stanton T. and Don Berliner. *Crash at Corona*. Paragon House: New York, 1992, pp. 9 – 13. Most of the quotes used in the chapter of that book are from Friedman, telling what he had heard from Marcel. Marcel's story of how he became involved is provided in Chapter One.

the quotes came from that first conversation or if in writing the book and knowing the full story in the 1990s that he recreated the quotes from other, later interviews. Since this was a telephone interview using an airport telephone, there is no tape. Friedman wrote, "That this sort of information is worthless until there is something or someone to back it up. One person's testimony simply isn't sufficient to make such a report worth more than scribbled notes."[398]

Friedman apparently alerted UFO researcher Len Stringfield about the Marcel story since Stringfield was writing a paper for the MUFON Symposium to be held that summer. Stringfield's topic was something he called crash/retrievals and the tale told by Marcel fit into that category. The paper had been submitted to MUFON before Marcel was discovered so there is no reference to his sighting in it. However, Stringfield was able to update the presentation so that the Marcel story was covered during the Symposium.

On April 7, 1978, according to Stringfield, he linked a Chicago NBC radio newsman, Steve Tom with Jesse Marcel at his home in Houma, LA.[399] Marcel again talked about the event, the strange material and suggesting that it had come from a craft that had exploded in the air. The debris was spread over a one-mile-square area. According to Stringfield, Marcel said that when the press learned about the retrieval operation and "To get them off my back, I told them we were recovering a downed weather balloon."[400]

Stringfield wrote, "Since the Major's story got publicity, it has been said by some researchers that the retrieved fragments were possibly part of the Skyhook balloon, at that time classified as Secret. On October 5, 1979, I called him and got this comment:

---

[398] Ibid. p. 11.

[399] Stringfield, Leonard H. *The UFO Crash/Retrieval Syndrome: Status Report II: New Data*. Seguin, TX: MUFON, 1980, p. 16 – 17. Stringfield did not record this conversation but did take notes. Members of CUFOS have attempted to locate a copy of the interview from the NBC side but have been unable to do so.

[400] Ibid. p. 16

> The material I gathered did not resemble anything off a balloon. A balloon, of any kind could not have exploded and spread its debris over a broad area… I was later told that a military team from my base was sent to rake the entire area.[401]

There are no indications that Stringfield recorded any of these additional conversations with Marcel. After Stringfield died, his files were donated to MUFON which restricts access to them (and most of their other material) to members.[402] Bob Pratt interviewed Marcel on December 8, 1979 and did record his interview.[403] A transcript of it was shared with UFO researchers.

Karl Pflock published a cleaned up version of this in his book, *Roswell: Inconvenient Facts and the Will to Believe*. As I have noted elsewhere, in his cleaned up version of that conversation, some of the problems with the Pratt interview were cleaned up. In one place the insertion of a comma changes the meaning of the sentence.

This has caused a bit of a dispute among UFO researchers who said that Marcel was claiming to have been the sole survivor on a combat mission but the records were not clear on that point. The Pratt interview was sometimes difficult to follow as it is almost one long paragraph with his questions sort of thrown in. Toward the end, Marcel was saying, "I got shot down one time, my third mission, out of Port Moresby."

Pratt asked, "Everyone survive?

---

401 Ibid. p. 17.

402 MUFON has reported that the Stringfield material donated to them was in 64 binders and there were no audio files with that material. He might have made audio recordings and those would have remained with the family.

403 For the uncorrected transcript of Pratt's interview, see: http://kevinrandle.blogspot.com/2013/09/jesse-marcel-sr-bob-pratt-and-interview.html.

Marcel said, "All but one crashed into a mountain."[404]

Reading Marcel's response here it can be seen that Marcel said that, at the very least, one other man survived that mission. The criticism that he claimed to be the sole survivor is unfair, given the structure of the transcript.

Pratt's transcript said, "I got shot down one time, my third mission, out of Port Moresby (everyone survive) all but one crashed into a mountain."

That shows that it is not quite as clear as has been suggested. But then, insert a comma so that it says, "All, but one crashed into a mountain."

Now it suggests that everyone survived except for the one who crashed into the mountain. Either way, there was no claim by Marcel that he was the sole survivor. However, in the interview by Linda Corley, Marcel talked about another one of the crewmembers he saw on the ground, alive.[405]

There is another point in the Pratt interview that raises questions about Marcel's military service and it surfaces in both the transcript and the article that Pratt wrote. According to that article:

> Wreckage from the UFO "was scattered as far as you could see," revealed Marcel who was awarded five air medals for shooting down five enemy aircraft on bombers in World War 2.[406]

According to military records, Marcel was awarded two Air Medals for combat missions during that war. There are no indications that he shot down

---

[404] Pflock, Karl. *Roswell: Inconvenient Facts and the Will to Believe*. Amherst, NY: Prometheus Books, 2001, p.233.

[405] Corley, Linda G. *For the Sake of My Country: An Intimate Conversation with Lt. Col. Jesse A. Marcel, Sr. May 5, 1981*. Bloomington, IN: Author House, 2007. This is an ebook and there are no page numbers associated with the book.

[406] Pratt, Bob. "I Picked Up Wreckage of UFO That Exploded Over U.S." *National Enquirer*, p. 8. (No date on the copy in my files.)

any enemy aircraft but it isn't clear if the error was one of Pratt's misunderstanding what Marcel said, or if it was a claim made by Marcel.[407] When these questions arose in the 1990s, I did ask Pratt if he had retained the tape recording but he told me that once the story had been published, they reused the tapes. There is no way to resolve this problem.

Audio Tape and the Marcel Interviews

Here's what is known at this point. The first Friedman interview was by telephone while Friedman was at the airport. Marcel didn't say anything about the photographs taken in General Ramey's office and while in today's world, if Friedman had photographs, he could have shown them to Marcel using the Internet, in 1978, that capability just didn't exist. Besides, Marcel wasn't sure of the date and it was Bill Moore who finally found the newspaper articles that included pictures of Marcel about a year later. According to Friedman, it was on February 10, 1979, "while [Bill] Moore was rooting through the newspaper files in the main library of the University of Minnesota, he came across several clippings describing much the same thing that Jesse Marcel had outlined to Stan Friedman a year earlier." This proves that Friedman didn't have the pictures taken in Ramey's office when he first interviewed Marcel and didn't know they existed.

The subsequent Moore/Friedman interviews were apparently also conducted over the telephone, according to a quote attributed to Marcel in Linda Corley's book about Marcel and Roswell. According to her taped interview, Marcel said, "Now those guys, Charles Berlitz and William Moore, I bet I have spent as much as ten hours along these telephones with these two guys. I've never met either one of them."[408]

---

[407] Marcel, Jesse A. Jr. and Linda Marcel. *The Roswell Legacy*. Franklin Lakes, NJ: New Page Books, 2009, pp. 42 – 43. Unit History of 43 Bomb Group (Marcel assigned to 65th Bomb Squadron) and Military Records of Jesse A. Marcel, Sr.

[408] Friedman and Berliner. *Crash at Corona*. p. 15

Corley questioned Marcel about the interviews and was surprised when he said, "I never met them. They interviewed me on the phone."[409]

At that point, according to Friedman, they had found copies of the newspaper clippings some of which contained a picture of Marcel and the balloon wreckage photographed in General Ramey's office. Moore might have even gotten prints of the pictures from the *Fort Worth Star - Telegram*, but they couldn't show these pictures to Marcel over the telephone. According to Berlitz and Moore, Marcel told them, again, over the telephone:

> General Ramey allowed some members of the press in to take a picture of this stuff. They took one picture of me on the floor holding up some of the less-interesting metallic debris. The press was allowed to photograph this, but were not allowed far enough into the room to touch it. The stuff in that one photo was pieces of the actual stuff we found. It was not a staged photo. Then they allowed photos. Those photos were taken while the actual wreckage was already on its way to Wright Field. I believe these were taken with the general and one of his aides…[410]

Looking at the captions on the two photographs that appear in Berlitz and Moore's book, it is apparent that there was some manipulation going on. They got a quote from Marcel that endorsed the picture of him with some of the real debris without having him examine the pictures. What they then printed was a highly cropped version of the picture of Marcel, but the other picture, of Ramey and DuBose was not cropped that heavily. It appears that the debris held up by Marcel is significantly different that that seen with Ramey and

---

[409] Corley, *Country*.

[410] Berlitz, Charles and William L. Moore. *The Roswell Incident*. (Hardback edition) New York; Grosset & Dunlap, 1980, p. 68.

DuBose. These pictures, as printed, give the impression of two separate sessions when it is clear when looking at full prints of the pictures, they were taken at the same time by the same photographer, showing the same mundane debris.

This is another point where it is claimed by skeptics that Marcel changed his testimony. Moore, after learning that there were two pictures of Marcel that were taken at the same time holding the same debris, sent a new version of the interview transcript to various UFO researchers. This one said:

> General Ramey allowed some members of the press in to take a picture of this stuff. They took **two** pictures of me on the floor holding up some of the less-interesting metallic debris. The press was allowed to photograph this, but were not allowed far enough into the room to touch it. The stuff in **those two** photos was pieces of the actual stuff we found.

This wasn't a case of Marcel changing his testimony, but of a researcher altering the transcript to reflect new evidence. At the time that Marcel was originally interviewed by Moore, Marcel only remembered the single photograph being taken. In his research Moore might have only found a single photograph. There was only one picture of Marcel in the negative file when the *Fort Worth Star - Telegram* donated their photo archives to the Special Collections at the University of Texas at Arlington.[411]

It was alleged that the photograph of Marcel had been taken with the real debris, but the whole negative showed that it was the same balloon debris that was held by Ramey. Marcel had not been shown those pictures. When he was, he told a different story.

---

[411] Information about the negatives from the University of Texas at Arlington Library, Special Collections. According to them, they held one negative of General Ramey and one of Jesse Marcel. The Bettman Photo Archives held the photograph of Irving Newton.

It is entirely possible that Marcel, not having seen the heavily cropped pictures that would be published in *The Roswell Incident* because the interviews were conducted over the telephone, would remember pictures being taken at the time but might have not remembered the sequence. His statement about him being with the real debris when the pictures were taken could easily refer to another event and not what had transpired in Ramey's office. So, when shown copies of the pictures from Ramey's office, he rightly said, "That's not the stuff I picked up."

This seems to be a plausible explanation for this dilemma. He was photographed with some of the real debris, but those pictures were not taken in Ramey's office. It wasn't Marcel getting coached into what to say about it, but how he actually remembered the events. Though there is another problem with this analysis and that is Marcel told Corley that on orders from Ramey, he had covered the real debris with the brown paper that is seen in the uncropped versions of all the pictures taken in Ramey's office.[412] This suggests that Marcel's memories of those events might not be completely accurate.

Then it gets worse, if possible. Marcel is quoted in the transcript, apparently paraphrasing General Ramey, "You can go ahead and scatter some of those pieces on the floor for the photographers and press but make sure they don't get any details about anything."[413]

There was a question, "Was that the actual material you had found?"

"I prepared that for the press. (That big piece was not part of it). [parens in original document.]"[414]

---

[412] Corley, *Country*.

[413] *UFOs Are Real*, documentary shot script, May 1979, Tape Two, page 1.

[414] Ibid.

And then it gets better. Marcel said, "Let me show you something. There's a picture of the same room [Ramey's office, I will assume here] It's not the material I brought there."[415]

Then on "Tape 3" comes more information. The director or the interviewer or the one asking questions said, "I talk about book I'm showing him [Which has to be something with the various pictures from Ramey's office.] Book in Jesse's lap showing warrant officer [has to be Irving Newton and that rules out *The Roswell Incident*]. The director asked, "This is not the material you found?" Marcel said, "Definitely not."[416]

The trouble here is that Marcel in the same film said that if he was in the photographs it was the real stuff and if it was anyone else, it was the fake. But now we learn that he was shown those photographs during that filming and said that it wasn't the material he had brought from Roswell. It seems that editing has created the contradiction in Marcel's statements.

There is additional corroboration for this identification of the photographs in Ramey's office. Johnny Mann, then a reporter for WWL-TV in New Orleans was filming a news series on UFOs for the station. The Roswell case with Jesse Marcel was one of the segments. As they were finishing the filming, Mann showed Marcel the photographs published in *The Roswell Incident*, and told him, "That looks like a balloon to me."

Marcel looked at them and said, "That's not the stuff I found."[417]

---

[415] This is a reference to the pictures taken of Marcel in Ramey's office. He was telling the interviewer, who showed him those pictures that it is not the debris he had collected on the Brazel ranch. Quote is from *UFOs Are Real* shot script.

[416] *UFOs Are Real*, documentary shot script, Tape Three.

[417] Johnny Mann, telephone interviews in October 1989, December 1989, in person, May 1994, by Kevin Randle.

The Other Problems for Marcel

During the interviews that Marcel granted, he apparently said things that simply could not be verified. Pratt wrote that Marcel was awarded five Air Medals for shooting down five enemy aircraft, but given Pratt's transcript and other information, this mistake might be Pratt's rather than Marcel's meaning that Marcel said that he had five Air Medals and he had shot down an enemy aircraft. Pratt assumed he got one Air Medal for each plane he shot down.

In the transcripts of the raw footage of *UFOs Are Real,* it said, "10 April 1942 volunteered for active service... 1st assignment Washington five Air Medals, the Bronze Star [Medal] and several comendations [exactly as written out in the transcript]."[418]

Since there is no mention of the five enemy aircraft in this interview, it would seem that the mistake was Pratt's. The real problem here is that there is only documentation for two Air Medals in his files and in the Unit History of his bomb group created during World War II. The lack of documentation for those additional Air Medals is a problem, but not an overly large one. Military records are not always accurate on these sorts of points.

What is more disturbing is his claim to have received a commendation for performing an appendectomy while in the service. This is something that should be recorded in his military record but there nothing about it in the file. There is a mention of another commendation for Marcel, but it was for service during Operation Crossroads. Maybe the commendation for the emergency surgery slipped through the cracks, but there comes a point when you can no longer blame Army inefficiency for the errors in the record.[419]

Marcel also claimed that he had received an Army Commendation Ribbon for his service during Operation Crossroads, but, again, the record does not

---

[418] *UFOs Are Real*, documentary shot script, Tape One, May 1979.

[419] During Linda Corley's interview with Marcel, he begins to the story of the appendectomy and his wife, said, "Oh, here it comes." She suggested throughout that interview that her husband had a habit of embellishing his stories of military service which does not bode well for Marcel.

verify this. In fact, there is a letter from the Army Decorations Board that is dated 12 March 1947 that reports, "By decision of a majority of the Board, the above-named individual [Major Jesse A. Marcel] is not recommended for the award of the ARMY COMMENDATION RIBBON for services as Intelligence Officer of the 509th Composite Group and the Air Attack Unit 1.5.1 from February to 16 August 1946, Operation Crossroads."

In Marcel's personal file there is the letter of commendation, signed by Ramey, for Marcel's activities for Operation Crossroads. Ramey wrote the letter of commendation after Marcel's nomination for the Army Commendation Ribbon was denied.

During the Pratt interview, Marcel said that he had flown as a pilot as well as other positions on the bombers during his service during the war. He then said that he had 3000 hours as a pilot and 8000 hours of total flight time.

While some have condemned Marcel for claiming to be a pilot, the wording seems to suggest that he wasn't making the claim of being a pilot, only that he had flown as one which is not the same thing. Given his assignment in an aviation unit, it is not beyond the realm of possibility that he had flown as a pilot, but mentioning it also suggests a bit of resume inflation.

Worse is the claim of so many flight hours. The 3000 hours of pilot time for someone who wasn't rated, meaning aircrew qualified and in this case trained as a pilot, seems excessive and there is no documentation to back up the claim. He doesn't even have a pilot's log book which many military pilots kept to record their time even though that record was kept officially by the military.

And 8000 hours of flight time again sounds excessive. While he was assigned to a bomber unit during the war and those missions often lasted twelve to eighteen hours at a time, his squadron records show a mere 485 hours of flight time. It might be that he included flight time in which he was a passenger in the rear of a transport, but again there is no record to verify any of that.

In keeping with these claims, although the transcript is less than clear on the point, Marcel said that he had attended several colleges and universities. The only claim that is documented by his records is a year and a half at LSU. In the Pratt interview, Marcel said, "...been in 68 countries... degree in nuclear physics (bachelors) at completed work at GW Univ in Wash... attended (LSU, Houston, U of Wis, NY Univ, Ohio State, Docotr [sic] pool? And GW..."[420]

No record has ever been found that he was awarded a college degree and no record was found that he had attended the universities claimed here. Even classes that were taught as military extensions off the university campus would have been recorded, so the record does not confirm any college education except for the year and a half of apparently noncredit work at LSU.

While it might be said that one or two of these sorts of embellishments might be ignored, there does come a point where it is problematic. There are clear areas of resume inflation but none that is particularly egregious by itself. It is only in the aggregate that it suggests that Marcel had a habit of stretching the truth. When dealing with something like UFOs, these sorts of problems are magnified.

Marcel and the Bodies

There are those who have suggested that it is strange that Jesse Marcel never mentioned alien bodies. They believe that as the base intelligence officer he would have been involved in the entire story once he returned from the debris field with samples of the strange metal. Since he was involved from the very beginning, it only makes sense that he was involved in the whole story, start to finish.

While that assumption might be in error given the nature of military secrecy, it is also true that Marcel didn't say anything about bodies to any of the investigators of the case. None of them including Linda Corley mentioned bodies in their discussions with Marcel. Corley spent hours with Marcel talking about the UFO crash, his theories on alien life and interstellar flight and provided the most comprehensive transcript of any those interviews with Marcel.

---

[420] Pratt interview with Marcel on December 8, 1979. This is how it appears on the Pratt transcript which is somewhat garbled. Nothing it left out and the ellipses are in the original. Pflock cleaned this up for his book.

In fact, when Corley brought up the idea of bodies, Viaud Marcel (Jesse's wife) said, "No dead bodies."

And Marcel said, "I would have picked them up and brought them in."[421]

But according to a few relatives, Marcel did see bodies. The first hint of this came from a newspaper article about the case written in 1996. Nelson Marcel, a cousin of Jesse Marcel, claimed that he, Nelson, had been told about bodies by his cousin. He said, "He [Jesse Marcel] said himself, 'It was nothing of this world."

According to the article, "At some point, Jesse Marcel claimed to have seen one of several pygmy alien bodies that were recovered from the wreckage. A picture he later drew of the alien's big, round, hairless head and slit-like eyes resembles the image described by other UFO witnesses. The picture has since been lost."[422]

Nelson Marcel was not the only person to make such claims. Sue Marcel Methane said that she had a chance to talk with Jesse shortly before he died. She provided additional information about the bodies, saying that she was told they were "white, powdery figures.[423]

To stretch this out even farther, Hayes Marcel told a reporter in Houma, LA, strangely that, "Those must have been aliens in that spaceship that crashed. Jesse said that he saw them."[424]

According to Carey and Schmitt, Hayes Marcel had heard none of that directly from Jesse Marcel but from his uncle, Nelson, which makes this third hand at best. According to Hayes, Nelson had discussed the crash in depth with

---

[421] Corley. *Country*. Section called "National Security."

[422] Reineck, Frederic. "Not of This World." *The* [Houma, LA) *Courier,* February 25, 1996, p. 6D.

[423] Carey, Tom and Donald Schmitt. *Witness to Roswell*. Pompton Plains, NJ: New Page Books, 2009, p. 84. It should be noted here that the story is second hand and comes from Nelson Marcel.

[424] Ibid. p. 84.

Jesse and that in addition to seeing the bodies, Jesse had kept a piece of the strange metal.

There is another source for the story that Marcel had seen bodies. Herschel Grice said he was a ground maintenance chief with the 715th Bomb Squadron and claimed that he had been working as part of the squadron intelligence team with Marcel in 1947. Grice said that Marcel had talked to him about seeing the alien bodies. He couldn't remember much about it, other than Marcel had described them as "white, rubbery figures.[425]

Jesse Marcel, Jr. made it clear that he and his father never discussed alien bodies with him. In interviews with Jesse, Jr., he was clear that they had discussed between themselves the extraterrestrial nature of the crash debris prior to Friedman's first interview in 1978 but there had been no mention of alien bodies.

It seems odd that Marcel, Sr. would have withheld that information from his son. He shared with him other material that was classified at the time. Marcel, Jr. had asked his father what the atomic bomb looked like and his father drew a picture of what I thought of as Fat Man, the bomb dropped on Nagasaki in 1945. In 1947, the atomic bomb shape and size was classified information but they seem to have resemble Fat Man. It would seem that if Marcel, Sr. was going to confide in anyone about alien bodies, it would have been his son, especially after Marcel, Jr. had finished his medical training.[426]

---

[425] Carey and Schmitt. *Witness to Roswell*, p. 83. It should be noted that Grice does not appear in the Yearbook, which is not necessarily proof he wasn't there in 1947. A search of the Unit Histories and *The Atomic Blast*, the base newspaper did not show his name. Further, documentation recovered does show him in Roswell but not in 1947 and not in the positions he claimed. Given all this, it is not necessarily evidence that he was not there doing what he claimed. "Information from Grice's DD 214, which is somewhat confusing and seems to be incomplete."

[426] Jesse Marcel, Jr. told the story of his father drawing the outline of the atomic bomb for him at the Citizen Hearing held in May, 2013 and information available at the Paradigm Research Group web site.

Add this to the statements that were given to Linda Corley by both Viaud Marcel and Jesse Sr., it seems unlikely that he would have shared this information with other members of the family. If he was going to talk about alien bodies, it seems that it would be most likely he would have said something to his wife and to his son rather than others. At this point, the best conclusion to be drawn is that Marcel, after his initial involvement, after he had returned from the debris field, and after he had been sent to Fort Worth to brief General Ramey, was taken out of the loop. He would have nothing to contribute beyond that point and others, such as the Provost Marshal, the Adjutant, would be brought into the secret.

The Roswell Timeline of Jesse Marcel

Although most Roswell researchers have established a timeline that flows nicely from Mack Brazel to the trip to Fort Worth, but Marcel, in some of the interviews conducted with him, turned all that upside down. Marcel said, for example, that after loading his car with the debris, "I stopped by the house [I] had left the day before and son and wife were waiting for me."[427]

This is curious because the story had always seemed to be that he had awakened them when he got back to Roswell in the middle of the night to show them strange debris. This suggests that he had stopped by so they wouldn't worry and that it wasn't very early in the morning but late evening. A minor problem at best but a problem nonetheless.

In the transcript (which, by the way, is strange) it is noted that on "Tape 2" at the 24:16 minute mark, Marcel apparently told them that Haut had issued the press release before he had even returned from the field. The exact quote on this transcript is, "He (Walt Haut had released statement to press before

---

[427] *UFOs Are Real*, documentary shot script, Tape One, May 1979.

Jesse had even returned home that night."[428] (Yes, this is how it was written out, so it is not exactly Marcel's words.)

During that interview, Marcel said that his wife had been "pestered by the news media."

In Linda Corley's book, Marcel makes a similar statement. He said, "She [wife] didn't even know where I was. By the time I got home, she had already faced the press that was out there."[429]

The script then noted, "Loaded it up next day in B-29 and was ordered to fly it to Wright-Patterson Airfield in Ohio [actually Wright Field in July 1947] (analysis laboratory) but Jesse suggested it be taken to Fort Worth 8th AF commanding general."[430]

There has always been something of a controversy about when the call came into the Officer's Club for Marcel. The timelines as developed all seemed to suggest that it was Sunday, July 6 while originally Marcel had said it was on a Monday, July 7. He left for the sheriff's office, then returned to the airbase, and then back to the sheriff. Then he, along with Sheridan Cavitt, who was a newly assigned counterintelligence officer, went followed Mack Brazel back to the ranch.

In the *Fort Worth – Star Telegram*, Marcel is quoted as saying, "Brazell [sic] then hurried home, and bright and early Sunday, dug up the remnants of the kite balloon… and on Monday headed for Roswell."[431]

---

[428] Ibid. This is the first real problem because if Marcel had yet to return to the base with the debris, how would Walter Haut know to write the press release. Cavitt, who Marcel sent back earlier in the day, hadn't arrived at the base in time to provide the information for the press release and he wouldn't have talked with Haut about it anyway.

[429] Corley, *Country.*

[430] *UFOs Are Real*, documentary shot script, Tape Two, page one, May 1979.

[431] "New Mexico Rancher's 'Flying Disk' Proves to Be Weather Balloon-Kite." *Fort Worth Star-Telegram*, July 9, 1947, p. 1.

Given the distances, given that Marcel had been eating lunch, and given all the other running around that has been suggested, it doesn't seem as if they would have left much before four in the afternoon. That would have put them at the ranch about seven or later that evening. With the light fading, there wasn't much that they could do that night. Marcel is quoted as saying that they arrived "at sundown and decided it was too late to find anything… slept at that old house."[432]

The next morning, according to what Marcel said, Brazel saddled a couple of horses to ride out to the debris field. Marcel apparently followed in the truck. Marcel said, "He took us to that place, and we started picking up fragments, which was foreign to me. I'd never seen anything like that. I didn't know what we were picking up."[433]

In a quote that has been circulated because it was on video tape, Marcel said, "I was amazed as to what I saw… It took me a little while to realize there was something strange about it."[434]

Marcel said that they picked up what they could but most of it was left behind. He said he sent Cavitt back early and he followed later. In the shooting script, it is noted that Haut issued the press release before Marcel even returned home that evening.

This is where the information becomes confusing. There is no way that Haut could have issued the press release before Marcel returned home from the debris field because Haut wouldn't know what was going on. There couldn't have been reporters calling the house if the press release had not been written at that point so it is clear that Marcel was confused about that.

In a point that is not relevant to the timeline, it must be noted here that Marcel said that Walter Haut had been ordered to issue the press release.[435]

---

[432] Ibid. Tape One, page one, May 1979.

[433] Marcel interview by Bob Pratt. See Pflock, *Inconvenient Facts*, pp. 227 – 228.

[434] *UFOs Are Real*, documentary shot script, Tape One, page one.

[435] Ibid. Tape Two, page one.

This is somewhat different than his claim that their "eager beaver" PIO had taken it upon himself to issue the press release. Marcel said, "…he called the AP about it. Then that's when it really hit the fan… I probably got telephone calls from everywhere."[436]

According to Marcel, the next day some of the material that he had recovered with Cavitt was loaded into a B-29 and ordered to fly to Wright Field, but, of course, it was diverted to Fort Worth. Marcel said that he had suggested that they stop at Eight Air Force Headquarters to inform their immediate higher headquarters about the discovery. Ramey ordered them to leave it and go back to Roswell.[437]

As Marcel had said to a number of researchers, when he got to Ramey's office, Ramey told him to keep his mouth shut. He wasn't to talk to reporters. He was told to scatter some of the pieces around, on brown paper and later he was photographed with that debris. The entire, uncropped picture makes it clear that Marcel is crouched with the remnants of a balloon and a badly degraded rawin radar target that showed no signs of having been outside in the desert for several weeks.

Although it is not clear, it seems that Marcel remained in Fort Worth overnight. The first *Fort Worth Star-Telegram* article has no quotes from Marcel, but an updated version of the story published later in the day had him saying, "The ranch is out in the middle of nowhere… and we spent a couple of hours Monday afternoon looking for any more parts of the weather device. We found a few more patches of tinfoil and rubber."[438]

---

[436] Marcel interview by Bob Pratt. See Pflock, *Inconvenient Facts*, p. 228.

[437] *UFOs Are Real*, documentary shot script, Tape Two, page one

[438] "New Mexico Rancher's 'Flying Disk' Proves to Be Weather Balloon-Kite." *Fort Worth Star-Telegram*, July 9, 1947, p. 1.

According to the article, "Marcel brought back the discovery to Roswell Army Air field [sic] early Tuesday morning and at 8 a.m. reported to his commanding officer, Col. William H. Blanchard, 509th Bomb Group chief. Blanchard, in turn, reported to General Ramey, who ordered the find flown to Fort Worth immediately."[439]

Marcel reporting to Blanchard on July 8 puts the timeline back on track. Haut issued the press release later in the day and although he suggested he made the rounds about noon, it seems that he might have used the telephone and it wasn't until after lunch that he made the telephone calls.

The True Timeline

Can all this be straightened out without suggesting that Marcel was confabulating at best and lying at worst? As everyone has said, memories that are more than three decades old are often unreliable. Can we make any sense of all this?

First, it should be noted that according to the newspapers, Marcel reported the find to Blanchard on July 8, and further that he met with Blanchard at eight in the morning.[440] The actual time of the staff meeting isn't of vital importance. It is the day on which it was held, which was July 8. Marcel was advising his commander that he had recovered some of the metallic debris. From that point, the timeline flows. The problem is the date that Brazel brought the debris into Roswell and started the ball rolling.

For the timing to work, given the distances involved, the state of the road system in 1947, and the testimony of several of the participants, Brazel had to come into Roswell on Sunday. There simply wasn't time on Monday for him to take his two kids to Tularosa, drop them off with his wife, then drive over

---

[439] Ibid.

[440] According to the timelines created by others, the regular staff meeting was moved to 7:30 a.m. which means it began before Marcel returned with the debris. The meeting time for the meeting was based on the memory of Walter Haut. See Carey, Tom and Donald Schmitt. *Witness to Roswell.* p. 285

to Roswell. If Marcel was eating lunch, it can be assumed that it was sometime after noon, though, in the military, lunch sometimes begins an hour earlier.

When he got the word, it is doubtful that he left his lunch but probably finished it because the problem was not one that demanded immediate attention. According to Marcel, he had gone to see the sheriff, then returned to the base, conferred with Blanchard, probably over the telephone, had to contact Cavitt to alert him, and then both returned to the sheriff's office.

Brazel had said he had some errands to run, so he went into town to complete those. Once everyone joined up at the sheriff's office it could easily have been four or later. The drive would take hours and Marcel said that they arrived too late to do anything. They stayed overnight, which puts them onto the field early on Monday morning.

Marcel said, "We collected all the debris we could handle. When we had filled the carry-all, I began to fill the truck and back seat of the Buick. That afternoon we headed back to Roswell and arrived there in early evening."[441]

Everything, all the testimony, agrees that this was Monday evening. Jesse Marcel, Jr. talked about being awakened in the middle of the night to look at the debris, but it seems that it would have been in the early evening. According to Linda Corley, the debris remained at Marcel's house overnight.

Then, the next morning, July 8, he took it into the base. Cavitt, it is assumed, having arrived earlier on July 7, had briefed Blanchard on what had been found and showed him some of the debris that he had collected. This does, of course, raise another question which is what was Blanchard's reaction to Marcel keeping some of this at his house overnight? Did Marcel check in over the telephone and was told there was no rush to get out to the base?

And what of the report that reporters have been calling the house prior to Marcel even returning with the debris? The answer is simple. When the press release hit the news wires with Marcel's name in it, the reporters would have attempted to find him for a comment. The only reporters who might have gone by the Marcel house would have been those working for the two daily newspapers in Roswell and Jason Kellahin and Robin Adair who were sent down

---

[441] Berlitz and William. *The Roswell Incident*. p. 67.

from Albuquerque to cover the story. The others would have called. But Marcel was no longer in Roswell, he was on his way to Fort Worth. By the time he returned, the story was dead. The weather balloon explanation had been published in newspapers around the country.

Marcel had simply confused coming home from the debris field with his coming home from Fort Worth after his quick trip there. He seems to have mixed up elements of the two events, confusing the timing, but then, that is just my speculation.

The questions then, have been answered, sort of. The first interview was not recorded according to Friedman and we are treated to his recollections of what was said. The interviews of Marcel by Berlitz and Moore, though it is likely that they were all conducted by Moore, were conducted over the telephone. There are quotes from Marcel that seem to contradict his story as it has been told for all these years though much of the problem is an outgrowth of the agendas of the researchers conducting those interviews rather than what Marcel actually said. The transcripts of some of those early interviews were altered so that they reflected the state of the investigation as new information was learned. That was not the fault of Marcel and the trouble between what he had said in those earlier interviews and what was reported later should not be used to discredit him.

At the other end of the spectrum, there are other quotes that seem to suggest that Marcel was embellishing his record, his wartime activities, and his overall importance. He said, for example, that he had two counterintelligence corps agents working for him when the truth was that he had no command authority over them. Their chain of command did not go through his office and while he outranked all of them, we move into one of those funny areas where rank is not the final arbitrator in who is in charge of whom. Cavitt's chain of command ran through the CIC and not the 509th. While Marcel, and Blanchard, could issue orders and assign details, in the end, Cavitt could ignore them by suggesting it was in conflict with his duties in the CIC.

That isn't, of course, a big embellishment by Marcel, and when interviewed three decades later he could have forgotten the nuances of the situation. It is more difficult to believe that he didn't know how many Air Medals he had been awarded, or if he had a college degree and there is a great deal of ambiguity surrounding the alleged appendectomy he performed. We can also look at the claimed hours as a pilot as well as the claimed number of flying hours. These seem to be excessive for someone who was not rated. Is it enough to reject all of what Marcel said? Depends on your point of view, but we do have more information that suggests Marcel's memories are not completely reliable on many of these points which calls into question much of what he said and at this time, that might be the best that can be said.

Chapter 9:

The Roswell Slides

The first public hint that there was something new in the Roswell investigation was when Rich Reynolds at his *UFO Iconoclasts* (now called *UFO Conjectures*) blog reported that the group that had stumbled on the UFO crash site was not archaeologists as had been suggested in the past, but were actually geologists.[442] His post didn't provide much new information other than to suggest that certain investigators had additional evidence that might be released in the near future to clarify all of this. But he stressed that everyone should have been looking for geologists rather than archaeologists.

At about the same time, I received an email from someone hinting that a number of us, who allegedly knew the truth about this revelation, had signed nondisclosure agreements that prevented a premature release of the information which was why we had said nothing about it. Nick Redfern's name had been mentioned as a source for some of the new revelations and it was suggested that he might be able to clarify the situation. I sent an email to ask him about this.

He emailed back, "No, I have no info for you... [about] being asked to enter a non-disclosure agreement. All I heard was of one person who signed a non-disclosure which seemed to cover the same subject I was told about."

From the email it seemed that he had heard something more, though there wasn't any corroboration for any of his information. He wrote:

> Basically, I got a "withheld number" call from what was without a doubt an elderly man with a southern accent who was looking for guidance on how to sell a picture/pictures to the

[442] Rich Reynolds, *UFO Iconoclasts* blog posting February 9, 2013.

> world of TV. The data was extremely detailed, with names, places etc. and was related to the "archaeologists" story from Roswell…"[443]

What should have been the first red flag was the caller, who wished to remain anonymous and had worked to hide his identity from the very beginning, was talking about selling the pictures that might be related to the Roswell case. This was not the first time that someone claimed to have photographs of either a crashed object or the alien occupants from inside it. As early as 1950 there were stories of pictures taken at the Aztec, New Mexico, UFO crash circulating.[444] The pictures never surfaced, though they rumor had come to the attention of the military and resulted in the creation of some authentic documentation about this.

More recently there was the Alien Autopsy which was supposed to show the medical examination of the body of one of the extraterrestrial creatures that had been recovered after the Roswell crash. There was an unidentified cameraman, supposedly raw footage made at an undisclosed military installation, and no identifiable people in the film. Eventually it was an admitted hoax though not before a great deal of time, energy and money was wasted on an investigation.[445]

Redfern supplied his telephone number in his email and suggested that he could provide additional information if I was interested. During the follow up

---

[443] Nick Redfern, email to Kevin Randle, February 11, 2013.

[444] For more information, see Ramsay, Scott and Suzanne Ramsey. *The Aztec Incident: Recovery at Hart Canyon.* Mooresville, NC: Aztec.48 Productions. 2011. pp. 175 – 179. Letter from the Department of the Air Force, October 9, 1950 describing the attempt to sell pictures of the "flying saucer" that had crashed. The letter was signed by Lieutenant Colonel Wallace B. Scholes.

[445] For more detailed information see Mantle, Philip. *Roswell Alien Autopsy: The Truth Behind the Film that Shocked the World.* Edinburg, TX: RoswellBooks.com: 2012. See also, Randle, Kevin D. *Alien, Mysteries, Conspiracies and Cover-ups.* Canton, MI: Visible Ink Press, 2013.

conversation, he told me that he had taken the telephone call from a man who either lived in the Midland, Texas, area, or had been in Midland, and had either been part of an inquiry or overheard parts of a conversation about a search for a man named Bernerd Ray.[446] Nick said that he had been told by his informant that a man held a couple of slides that apparently showed an alien body on a gurney. The man's sister had been in Sedona, Arizona, assisting her boyfriend cleaning out a house prior to an estate sale of Hilda Ray, who had been married to Bernerd. In the attic they found a box (or boxes) of Kodak Kodachrome slides and taped to the lid of one of those boxes were two slides.[447] They had been separated from the rest of the slides in the box for some unknown reason.

Redfern said that he didn't know when the slides were taken or who had taken them, but it seemed that the working assumption was that either Bernerd or Hilda was the photographer and one of them had hidden the slides for some unknown reason. Redfern didn't know much more about the slides, but later (much later) said that he did know that the genital area was covered (which raised all sorts of other speculations about why the Army would do something like that).[448]

The man, who had physical possession of the slides now, later identified as Joe Beason, was attempting to find out something about the photographer, who had been an important man in the Permian Basin geology and lived in

---

[446] Although the caller has never been identified, it seems more logical that either Joe Beason, the owner of the slides or Adam Dew who would eventually create a video trailer that was available on YouTube called Redfern. It seems just too unlikely that someone in Midland, TX overheard the conversation with enough detail that he would be inspired to call Nick Redfern. There are a few others who had the information prior to Redfern and it is possible that one of them was responsible.

[447] Bragalia, Anthony, "Do Slides from 1947 Show a Roswell Humanoid?" *UFO Today* (Internet Magazine). Issue 2, pp. 67 – 68 Reprinted with permission from the UFO Iconoclasts [now UFO Conjectures] website.

[448] It turned out that the genital area was not covered but it was the placement of a placard that gave that impression to some. At that point it seemed that Redfern had not seen the slides himself but was relying on information given to him by others.

Midland in the late 1940s. It was during this search that the man who called Redfern heard the tale or became involved in the tale and this was the source of Reynolds information about geologists as opposed to archaeologists.[449]

During this conversation, it came out that either Tom Carey or Don Schmitt or both had signed non-disclosure agreements so that they'd be allowed to see the slides and assist in an attempt to authenticate them. There wasn't much additional information available to Redfern and Carey and Schmitt were remaining quiet about the photographs because of those non-disclosure agreements. They answered few questions and shared almost no information about them.[450]

The situation remained fairly static for the next several months. None of those involved in the research were saying much publicly though Reynolds was periodically making references to some incredible evidence that would be revealed soon. There were attempts by others, outside that small group, to learn more about the slides and according to those inside the investigation, the pressure for information was increasing on a daily basis.

Early in the investigation, the slides were taken to Rochester, New York, home of Kodak for analysis. Robert Shanebrook, who literally wrote the book on Kodak history, was said to have performed photomicroscopy and spectral density testing. According to Anthony Bragalia, this confirmed that the film had been manufactured in 1947, and that it had been exposed and developed in that time frame.[451]

Although this was the first positive information from the investigation, it couldn't prove that the slides were of alien creatures killed in the Roswell UFO

---

[449] Reynolds made a big deal out of the information about the geologists assuming that this would help identify those who had stumbled onto the scene in 1947. It turned out to be a dead end especially when the truth about the slides was revealed.

[450] One of the exceptions was David Rudiak who had seen a poor quality, ink jet printer copy of one of the slides in July 2012.

[451] Anthony Bragalia, email to Kevin Randle, July 22, 2014. He also made similar suggestions to others about the authenticity of the slide stock and the date of exposure.

crash, but it did, at least, place the exposure of the slides in the right time frame. At that time, the late 1940s, the Rays lived in Midland, Texas, and Bernerd Ray's area of operations was the Permian Basin, a huge oil field that included Roswell. That suggested that he could have seen the bodies at the retrieval site or had gained access at some other time because of his position in the geological community and his area of responsibility.

To those investigating the slides that added to the credibility, suggesting that the being in the slide was one of the aliens from Roswell though they would periodically deny that the slides were of a Roswell creature. In an effort to validate this theory, the background of the Rays became one of the areas of investigation. Their apparent connection to Eisenhower, as shown in some of the other slides recovered at the same time suggested that the Rays might have been allowed to view the alien remains even though they would have been classified.[452] These two points, the apparent age of the slides and the connection between the Rays and Eisenhower, seemed to suggest to some that the slides were of an alien entity.

They were also trying to find experts in human anatomy to review the slides and attempt to determine if they showed something alien. On July 14, 2014, a man who was said to be a former NASA "Aerospace Professional" and only identified as Larry, provided some information about the creature in the slides. He gave the first detailed analysis of those slides providing those outside the investigation with some interesting information when he wrote:

> 1. There are 2 photos, taken in an indoor setting.
>
> 2. The photos are of poor quality (focus, exposure) compared to virtually all the other photos in the same collection. For this

[452] Several of the photographs in the box were of Eisenhower in his general's uniform. None of the pictures showed either of the Rays near Eisenhower, but the fact they had these pictures suggested to the slide researchers that some sort of intimate relationship existed.

reason, edge detection, contrast enhancement and other photoanalysis techniques are warranted and are being used.

3. The photos appear to have been taken about 4 or 5 feet from the humanoid, from a position slightly above it.

4. To my eye, the humanoid is lying on a clear glass shelf and is surrounded by either clear glass walls and/or a full glass enclosure. The enclosure appears to be more like a rectangular box than like a bottle.

5. In one of the photos, a woman is standing behind the glass case (visible from approximately the waist down). In the other photo a man is visible in the same location, leading to speculation that the man and the woman traded places and took turns taking pictures.

6. The humanoid is not immersed in a fluid; it appears to be open to the air (at least if the lid were off).

7. The glass shelf/ box that the humanoid is on/in appears to be supported on shelf brackets that are connected to vertical, metal supports. The vertical supports are perforated at regular intervals (nominally, 1 inch spacing) by drilled holes. The shelf arrangement gives the appearance of a laboratory apparatus rack.

8. The humanoid is lying on its back, with its head to the camera's left and feet to the right.

9. There is some type of placard on the front of the glass case, with (currently undecipherable, out-of-focus) writing on it.

(Shades of the Ramey memo!) It is my suspicion that this placard is the source of the idea that the genital area of the humanoid was deliberately covered up in order to escape the wrath of the censors when the slides were developed. I don't think that is the case. From what I could see, the genital area was not visible to the camera due to the view angle of the camera. I suspect that the placard simply serves to identify the contents of the glass case.

10. The proportions of the humanoid appear to be slightly different than a "normal" human, but probably, no single dimension of the body is outside the range of naturally occurring sizes. The length of the head (crown to chin) is approximately the same length as the torso (neck to crotch). The arm length (shoulder to wrist) is approximately the length of the torso (i.e., the wrist joint is approximately aligned with the hip bone). The leg bones are long, compared to the arms.

11. Given that the body is about 3 feet long, if it is human, then it must be either a child or an adult with a developmental disorder. (Human Trisomy 17 has been suggested as a candidate.)

12. I could not see digits on either the hands or feet, and so could not count them.

13. The mouth is open and no teeth are visible.

14. The skin of the humanoid is smooth and appears to have shrunk taut against the bones (ribs, legs, arms, cranium). Whether this is due to natural effects of death (saponification, dessication [sic], etc.) or is the result of some post-mortem treatment (embalming, freezing, etc.) is not clear.

15. The head appears to have been severed from the top of the spinal column and then replaced, lying at an unnatural angle relative to the torso.

16. There is nothing in the photo that would either definitively connect this to the Roswell event or definitively disconnect it. Any connection is coincidental (it appears to have been taken at about the right time).

The information was published with only the first name of the individual attached. He was also called an engineer and an "exo-biologist" who works at or who had worked at the Ames Research Center who was also in touch with a medical doctor and a Forensic Pathologist. This analysis, however, was based on seeing scans of the slides as opposed to seeing the actual slides themselves. It is a description of those slides with little formal analysis presented. These American scientists didn't provide anything to suggest the being in the slides was alien, but then, they didn't rule it out either.

Some three days after this information was published, there were suggestions that the creature was seen as matching, to some degree, the alien being in the drawing attributed to Glenn Dennis but drawn by Walter Henn. The alien face was angular with a high domed head, somewhat flat features and large eyes. The important feature was the hand, which had only four fingers. The body in the slide had only four fingers visible.

There was another important observation. According to Bragalia, Robert Sarbacher, who had revealed that he knew something about the crash retrievals, suggested that the alien creatures had an "insect-like feature … The lower face is like that but the face is more humanoid, like strange little people."[453] The slides seemed to confirm that point of view.

---

[453] Anthony Bragalia, email to Kevin Randle, July 18, 2014.

The trouble here is that quote, in a letter from Sarbacher to William Steinman isn't about the appearance of the creatures. Sarbacher wrote, "I remember in talking with some of the people at the office that I got the impression these 'aliens' were constructed like certain insects we have observed on earth, wherein because of the low mass the inertial forces involved in operation of these instruments would be quite low."[454]

Instead of talking about the appearance of the alien creatures, he was talking about the internal structure and the exoskeleton which might explain how they would be able to survive the high-speed maneuvers and right-angle turns that had been observed in many UFO encounters and observations.

Tom Carey, in fact, took the insect features further when he said that he had received an email from the slides owner after signing a nondisclosure agreement. Carey said, "He emailed the image. When I first looked at them… a chill went down my spine…to me they rung true because there were things on them that had been described by other witnesses…"[455]

Later in the same interview, Carey said that there is a woman standing by a glass slab. The body was about three and a half to four feet tall with large inverted pear-shaped head. He then added:

> There is one item on it, on top of the head that was described by one of the first hand witnesses, one of the first ones to the crash site… the local fireman named Dan Dwyer who described when he got home that night and he told his family about it… they asked him what did it look like. Instead of giving a detailed description he just said, "Child of the Earth."…

---

[454] Robert I. Sarbacher, letter to William S. Steinman, November 29, 1983; See also Steinman, William S. and Wendelle C. Stevens. *UFO Crash at Aztec*. Boulder, CO: America West Distributors, 1985; pp. 324 – 325; Clark, Jerome. *The UFO Encyclopedia: Second Edition*. Detroit, MI: Omnigraphics, 1998, pp. 814 – 815

[455] Richard Syrett, *The Conspiracy Show*, aired on April 12, 2015. The Carey interview begins at 1:04:20.

> It's something on the head, I don't want to give everything away here… there is something on the top of the head that one of the eyewitnesses described and it's on this particular creature on the slide… Oh my goodness, that's why Dan Dwyer called it the Child of the Earth."[456]

The trouble here is that no UFO investigator ever talked to Dan Dwyer. The tale is told second hand by one of Dwyer's daughters.[457] Carey, in this interview seemed to suggest that he had talked to Dwyer by calling him a first-hand witness. While Dwyer might well have seen something himself, it is misleading to call him a first-hand witness, giving the impression that the information was heard directly from Dwyer.

Much of the discussion about the slides, at this point, had been speculation by those who had not actually seen them. It was based on what they had heard from others who had seen them but those who had seen the slides were not saying much about them in public.

There was also discussion about the lack of provenance for the slides. While it was claimed that one of the Rays, probably Bernerd had taken the pictures, that wasn't established. If it was one of the Rays, both were dead so they couldn't provide details of how the pictures were taken, when they were taken or where.

To make it worse, the story about the discovery changed as well. It was first said that Beason's sister, Catherine, had been cleaning either the attic or garage for an estate sale in Sedona, Arizona in 1988. She had looked in the box and found somewhere between one hundred and four hundred color slides that were apparently taken in the 1940s and the 1950s in a variety of locations around the world. Many of them were stamped with Hilda Blair Ray which did establish a connection between the box and the Rays.

---

[456] Ibid.

[457] Frankie Rowe, daughter of Dan Dwyer, interviewed by Kevin Randle, January 2, 1993.

The two slides were in an envelope separated from the rest of them and taped to the inside of the top of the case holding the slides. She saw value in the other slides and rather than tossing them out, took them home. She didn't look at them until sometime later and found the two slides. She did nothing with them for about ten years.

Although it was assumed that the house being cleaned had belonged to Hilda Ray, that wasn't established, and she had died two or three years before the estate sale. It was also suggested that the house had belonged to an attorney. Given that, the chain of custody was broken as well.

Even though these problems had been pointed out by several others, the investigation continued, much of it in the background with little being said by either Carey or Schmitt. The speculation about the slides also continued, but those who had seen them were saying little about them.

Then, on November 12, 2014, Carey, as a panelist in a UFO discussion held at the American University in Washington, D.C., publicly provided information about the slides. He again said that the slides had been owned by a "high-powered couple" in Midland. He said that the slides had been found in 1989 and that he had learned about them two years earlier. This was the first time that he had been allowed to talk about the slides.

Carey said that the slides showed a partially dissected humanoid that was enclosed in a glass case. He said that the head was removed from the body but it had been replaced. He said, "The head looks almost insect like and exactly what I thought the Roswell aliens would look like based on the information that had been given to me over the last twenty years. Exactly what I expected."[458]

In December, 2014, I tried to figure out what could be on the slide if it wasn't an alien creature. I searched through a number books and web sites that were dedicated to science fiction films, especially those of the 1940s, assuming that the dating of the slides was accurate. There was nothing that related to the descriptions alien that had been given by Carey. In most cases, the movies relied on aliens that were either human or humanoid. They didn't have the

[458] Tom Carey, American University, November 12, 2014.

technical skills to create aliens that were very exotic looking. The easiest and cheapest way to create aliens at that time was put humans into strange costumes, sometimes with make-up to make them more exotic and claim they were from another world.

An alternative explanation seemed to be a mummy. On December 15, 2014, I asked Tom Carey in an email, "… is it possible that the body shown is a mummy? Given the nature of the terrain in that region (meaning, of course, West Texas and New Mexico) and given that some of the native peoples did create mummies, and given the dry nature of the area, especially about a 1000 years ago, is it possible that this was a mummy found by an oil exploration team?"

On December 16, Carey answered briefly and to the point. He wrote, "No, it's not a mummy."

From that point, with the first description from one of those who had been working with the slides, the discussions exploded. Tony Bragalia published a short article about the slides in early February, 2015. After providing a slightly different version of the slides discovery, he then explained why the slides were genuine. He wrote:

> Professionals from a range of disciplines who have seen the slides agree that they depict a small humanoid creature – a formerly living thing – that is not a prop or a genetically defective human. And importantly, the being that is shown in the slides does not correlate whatsoever to the depictions of aliens extant in the popular culture of the 1940s (such as those that appeared in pulp magazines like *Amazing Stories* or movies like *Buck Rogers*). What the slides depict were not even part of the public psyche of that time. This is not how people envisioned things from outer space to look like back then. Instead, these 1947 slides reveal a being that looks like the beings found in the desert in 1947 as described by the witnesses to Roswell.

> I was asked by Tom Carey to find the best available talent to test and analyze these Kodak slides. An extensive search was conducted and I found that talent. A Photo Scientist employed by Kodak for decades who will be named at the event [but was not, though his name was widely known by then], this expert has led engineering, production and product management groups at the company's Rochester, NY headquarters. Now a consultant, he also published the definitive book on Kodak film processes. Highly acknowledged in his field, he conducted extensive testing on the slides and conclusively authenticated the slides of the creature as having been exposed in the year 1947. It was also concluded that the slides had not been tampered with nor manipulated in any way. What is depicted is really there, accurately reflected in the emulsion as an actual moment in time in 1947. Science has weighed in and has determined that these are real slides that are really from 1947…
>
> This humanoid is not a deformed person, mummy, dummy, simian or dead serviceman. It is not a creature that finds its origin on Earth. And given that the slides of this creature were taken the very same year as the Roswell UFO crash; that the appearance of the creature matches the reported appearance of the Roswell crash aliens; and given that the person who was in original possession of the slides was a geologist working the New Mexico desert throughout the 1940's, it is not a jump or stretch to then conclude that these slides indeed show the corpse of one of the creatures found fallen at Roswell.

Bragalia was about to be upstaged by the release of a short YouTube film by Adam Dew, who had been one of those involved from the very beginning. Called *Kodachrome*, it told a little of the story of the slides, some of the things that had been done to verify the information, some blurred shots of one of the slides and a brief interview with a soldier who had been in Roswell in 1947 and who had claimed to have seen the alien bodies.

In the documentary trailer, Dew said, "I am a graduate of Northwestern University (BSJ '98). If I wasn't on an actual paying job (as opposed to the slide doc project) I'd post the diploma for you," which didn't tell us all that much. A great many of us have college degrees and we still don't know what his area of expertise is.

He did tell us however, that the slides have been shown to various experts such as Professor Rod Slemmons who had been the director of the Chicago Museum of Contemporary Photography who suggested that to fake the slide holder, that cardboard sleeve they are mounted in, would be difficult to do. Later analysis and evidence would underscore this. The slides were on film that had been manufactured at the right time and were encased in the proper mount for that time.

Dew also presented footage of an unidentified man, other than to say that he was a lieutenant in Roswell in 1947 and who said that what is on the film looked "identical… if this is a copy, they did pretty good."

This was PFC Eleazar Benavides, who had been identified in Carey and Schmitt's *Witness to Roswell* as Eli Benjamin. The records proved that Benavides was stationed in Roswell in July 1947 so he would have been in a location to see the results of the crash and bodies but doesn't necessarily prove that he did.

Dew said, "Logic tells me it's probably nothing but I simply can't shake the thought that maybe, just maybe it's something."

In the video clip, Carey provided the most information about what is on the slide and what the alien looked like. He said, "What we see in the picture is what appears to be a three and a half to four-foot-tall alien. It's got a frail body and the oversized head just as it was described to us [Carey and Schmitt]. It appears bipedal… [It] has two arms. It looks like it has some analog to a ribcage but it has been partially dissected. The pleural cavity and the abdominal cavity have been removed. The head has been severed from the body."

He added, "…We don't know where the picture was taken. It's taken indoors [which suggested that Ray or whoever didn't stumble across the body in the desert]... We don't know exactly what circumstances the photographer was being a part of to take those pictures."

There was one other aspect that was important. Several researchers were able to find an apparent mistake in the video. Dew, who had attempted to blur the slide wherever it appeared, missed one spot. The researchers were able to grab that shot and using a variety of computer programs, clean it up, to a degree. It wasn't the best photograph, but it did give the rest of the world a glimpse, a digitally altered and somewhat obscured glimpse of what was seen on the slides.

Bragalia wrote on February 9, 2015, that "I can confirm that this is one of the slides... however, the original is far clearer... and is in color. And the other slide is far more revealing."

Dew wasn't through. He appeared on WGN-TV telling a little more about the slides. He told us that the Rays, Bernerd and Hilda, divorced in the 1960s. He remained in Midland and she went to Arizona. Dew said that people have been contacting him about the Rays, so he was learning a little more about them. He implied that the Rays had taken the pictures, probably Hilda, though it could have been Bernerd and now that we have seen the slides know that each of them probably took one of the two. At any rate, the slides had wound up in Hilda's possession and were only discovered long after her death. Although it was suggested at the beginning of the story that it was Dew's sister who had been cleaning the house for an estate sale, it was actually the sister of a friend eventually revealed to be Joe Beason's sister, Catherine.

Not long after that, Bragalia provided another strongly worded defense of the slides. He gave additional details based only on an examination of the scans when he wrote [with boldface and underlining left intact]:

> Since then, opinions have been proffered and amateur "analysis" has been conducted. Verdicts on just what the slides show have been rendered, often with impassioned, mean-spirited response and heated accusations. Inflammatory remarks, name-calling and near-libelous allegations have been made by people who have not been privy to a clear version of the slide

nor seen the other existing slide at all- and without the benefit of review of the professional, scientific study that has been conducted on… them. And this negative, knee-jerk reaction to the slides existence began far earlier, even before the release of any image at all!

The truth of the matter could not be more different from what the noisy naysayers maintain…

WHAT ARE THEY LOOKING AT?

If, as the saying goes, a picture is worth 1000 words, this attempted enhancement gives only 250 of them. The fact is this: **this is a video screen grab from a computer monitor** – it is **a picture of a picture of a picture**- which **has been taken at a distance** of **a slide in its frame**. It is not a photographic print made from the slide, nor does it show the slide's projected image on a screen.[459]

Importantly, **this poor-quality image is not even in color as are the original Kodachromes** (a sepia-tone was applied to the image in the video.) **The size and perspective of the being –and its texture and shape- is hugely distorted** and **important key details are unable to be seen.**

A reproduction of an image can only be as good as its source material – and that source material was *intentionally modified in the preview video* [emphasis added to prove the point of

[459] It seems necessary here to point out that neither Bragalia nor many others had seen the actual slides. They were looking at scans made by Dew or Beason of the slides which were often modified, cropped and otherwise altered to obscure much of the information available on the slides.

alteration]. Bear in mind too that this is only one of the two slides that exists. This slide is the least interesting of the two. **The other slide provides greater clarity and with far more detail revealed.**

None of the photo-scientists who analyzed the slides were working with such degraded material like a video screen grab - they were working with the 'raw' original slides and with high-definition enlargements of them. This is not so of the many who give ill-informed opinions about them.[460]

Finally, the image on the video was only offered as to give an idea or preview of the 'real deal.' It was not intended by any means whatsoever to be used to technically dissect the image or to offer the 'full view' of what the slides actually show. It is difficult to understand what some people do not understand about that...

To address the question of dating of the slides and the possibility of photographic deception, here is a summation of analysis done by experts from industry and academia:

-The film is manufacture coded (edge code dated) as 1927 or 1947 or 1967[461]

---

[460] The photo experts could only suggest that the original slide film had been manufactured in 1947 and that they had been developed in the late 1940s. While it is clear, based on that research that the slide film was exposed at the proper time, it did nothing to establish the alien nature of the subject of the slides.

[461] After the reveal in Mexico City on May 5, 2015, it turned out that this was not true. Only one of the slides had been opened and there was no code on it to reveal the manufacturing date. It also seems, based on information provided by Kodak that the edge codes that were allegedly there were actually from movie film and not slide film.

- The protective lacquer used on the film is from the 1930s to 1960, eliminating the year 1927

- The cardboard sleeve used is 1941-1949, eliminating the year 1967 and leaving 1949 as the latest date the film was exposed

By simple process of elimination using these findings, we are left with the year 1947…

- Clear versions of the slides depict a being whose anatomy does not correspond to a human being. **The limbs (legs and arms) are exceeding thin, frail and fragile, characteristics that are not associated with hydrocephalus**. In fact, the torso (which has been opened) and rest of the body look nothing like any known case of hydrocephalus in history. The skull too, is enlarged but not 'bulbous' which is characteristic of non-shunted hydrocephalic.

- **The being's head is severed from the body** (not evident in the screen grab) and **one eye is missing**. **The chest and the abdominal cavity are missing**. **Hydrocephalic corpses are kept intact in medical study and display**.

-The being has no teeth and has wide-set eyes. **Lack of teeth and wide set eyes are not known to be conditions associated with hydrocephaly**.

-In the actual slides it is evident that **the being has only four fingers**. To my knowledge, **mummies and hydrocephalic are not typically missing a fifth digit.**

-A detail not known or revealed to anyone but those who have seen the slides is that close-ups of **the being's face show a very 'pointed' chin, a chin that in no way resembles a human**, mummified or hydrocephalic. In fact, the facial features do not in any way match that of other known hydrocephalic or mummies.

-One commenter (Gilles Fernandes) has shown a side-by-side comparison of the video grabbed slide and an infant mummy. He circles the feet of both, making a comparison and implying that they are one and the same. **However, the image Mr. Fernandes offers is that of a specimen who is far, far shorter than 3.5-4.0 tall. And what is depicted in the slide is not a foot at all, but something else, perhaps a piece of debris lying on the surface. The being's feet actually end *behind* the placard.** In the actual slide there is even another similar, smaller such item which can be seen.

-This 'placard' is not very evident in the video grab image. However, it has been enlarged by experts and **the writing, in red ink, is handwritten, not typed, as would be found in a biological display in a museum.**

-Most importantly, the placard, as well as the support structure that the being rests upon, are clearly 'temporary.' The structure looks very make-shift, resembling a quickly-assembled 'erector set' type deal, with beams that have ratchet holes in them. **The set-up in no way whatsoever resembles that of a professional museum display**. It is not a well-crafted, pristine glass museum display box, but something not meant to be at all permanent. There is also a military-green blanket upon which the being rests, atypical of any such museum display of other biological specimens.

-Mummies are desiccated. **This being was obviously either recently alive before the fatal pictures were taken, or had been embalmed.**

-The Rays hid these two slides away and separate from the other slides found in a chest and were only discovered by the owners much later, **as if to indicate that these two slides held special importance and meaning.**

A CIRCUS- BUT WHO ARE THE REAL RINGMASTERS?

Some rabid skeptics have disparagingly termed the whole slide affair as 'a circus.' If it has in some way become one, it is not at all due to the actions of those who seek to study and present the slides. In fact it is outsiders who have tried to insert themselves into the saga who are the real ringmasters.

It began with a 'leak' of the story nearly three years ago. An anonymous individual apparently contacted researcher Nick Redfern and divulged what he knew. Nick then –understandably- began contacting researchers to gain more information. When word of the slides existence became public, very sick behavior ensued:

- This author had his computer system hacked in an attempt to gain more information about the slides, or perhaps to obtain the slides themselves.

- Other researchers including Nick Redfern and Tom Carey (who had his stored documents 'crypto-locked' with malware) were also hacked.

> - Information and names obtained from my stolen emails on the slides investigation was made public on a website (before being deleted.)
>
> - Some people began contacting -or threatened to contact- involved photo scientists and witnesses (including a 90 year old man) in an effort to either gain more information or to derail the investigation…
>
> - Accusations of hoax were made even before any release of any type of the slides. I was directly accused of being 'a liar' and other defamatory and legally-actionable comments were made against me and my reputation.
>
> - Phone calls were placed to me in the wee hours by blocked callers who threatened me with 'exposure' as a fraud and my family members have even been harassed.
>
> **WHAT THE SLIDES SAY ABOUT US**
>
> Perhaps as interesting as the remarkable story of the slides themselves is the remarkable story of how people have dealt with such news. Jealousy, a sense of exclusion, and an inability to accept the possibility of what the slides do represent have all been in evidence during the slides saga. The compulsion by some to insert themselves into the story and to offer their judgment even before the slides and study are presented is worthy of a psychology study. Indeed, what the slides say about life beyond Earth is as telling as those who live upon it.

For the next couple of months this is the sort of argument that raged. The allegations were tossed around freely but there was little actual information supplied. A consensus seemed to suggest that the image in the slide was a mummy with dozens of people using the Internet to search for a duplicate of the slide image. There were some that seemed to be quite close, but nothing that could be identified as exact. This might have been the problem for those on the inside. They were looking for an exact match rather than comparing the features of the mummies from various museums with that on the slides.

The arguments continued, but with the trickle of information having stopped, most of the discussion was built on speculation and a suggestion that everyone wait until the May 5, 2015 presentation. It was here, according to the proponents, that everything would be revealed and the evidence would be presented. It would end the discussion and Carey suggested that the skeptics and debunkers would have a difficult time dealing with the facts when they were presented.

In Mexico City, at the May 5 presentation, Schmitt reinforced the claim that the slides had been subjected to rigorous testing by experts in the field of photography. According to the newspaper accounts from Mexico City, "Exhaustive investigations by other photographic and medical experts have concluded that the photos are genuine. The experts list presented at the Mexico City event include Dr. David Rudiak, an expert in photographic analysis, Dr. Donald Burleson, a specialist in computer enhancement; Ray Downing, materials expert from the Studio MacBeth, New York; Col Jeffrey Thau associated with the Pentagon's Photo Interpretation Department, and Prof Rod Slemmons, a former Director of the Chicago Museum of Contemporary Photography.

Some of this, however, was not exactly accurate. David Rudiak is not an expert in photographic analysis, but has experience in attempting to read the Ramey Memo. Because of that, he was asked to look at the placard with the body but said he was unable to unscramble or deblur the image from the scan that he had been given.

Colonel Jeffrey Thau is a retired Air Force officer who once had offices at both Wright-Patterson Air Force Base and the Pentagon. The Photo Interpretation Department had been moved from the Pentagon to Fort Meade, Maryland. Their expertise was not in attempting to read messages that were obscured but in interpreting photo intelligence of various kinds including ground based military facilities and movements. It seems that this failed attempt to read the placard wasn't actually an attempt by the experts at the Pentagon or Fort Meade, but friends seeing if they could make out anything on the placard as a favor to Colonel Thau. To suggest the Pentagon had attempted to read the placard was, at best, hyperbole and more than a little misleading.

Three scientists, two Mexican and one Canadian also made presentations about the unusual nature of the body seen on the slide. The Canadian, Richard Doble, a professional anthropologist said:

> The creature looks very much like a human being… a deformed one you often see in anatomy labs or museums. But on closer examination it's not a human being at all. It is not even a mammal. We are mammals because we have mammary glands. There is no hint of nipples or mammary glands. There is no hint of hair that on this creature but we can't be absolutely certain. We don't know how it bears its young. We don't know how they look after themselves… This creature is in no way related to us.
>
> Now it gets more interesting as you go along because it starts to show a bunch of primate traits. This happen because this population has evolved in an environment somewhat similar to our own and it has in the course of interaction with this environment… selected for traits that are similar to ours… it's got a rather flat face. It's lacking… a snout. It has reduced now. It has binocular vision. It has an enlarged skull. It has what we call prehensile hands. It looks like it can grasp but we can't see because of the way the specimen is mounted…
>
> But again it's not quite a primate and it's badly damaged… But what's really interesting about it it has three classic traits that make it a hominid or a human-like creature. It has a large brain… presuming it came from outer space on a spaceship of

> some kind, who knows, it has technology or tools and it is bipedal but particularly the large brain and the bipedal nature of the creature says, hey, this looks like a hominid but it's not… it just doesn't add up to being a human being… It is not at all like a human being… what we are looking at is not a human child or anything like that… we know from the bones alone this is nothing like us…

His Mexican colleagues made similar claims, suggesting that their studies of the anatomy as seen on the slides, suggested that the creature, while humanoid, was not human. Jose de Jesus Zalce Benitez and Luis Antonio de Alba Galindo were both highly trained, capable, and convinced that the body was of something other than a human and was certainly not a mummy of a human.

Richard Dolan, who not only attended the event but was a participant in it, wrote on May 6:

> Such people [referring to those who believed that the creature in the slide was either a mummy or some sort of science fiction creation] - all of them English-speakers - obviously did not acquaint themselves with the detailed and technically proficient treatment of these very questions by the three scientists who were featured last night [May 5, 2015]: Jose Benitez, Dr. Luis Antonio de Alba Galindo, and Richard Doble. The first two spoke in Spanish, and I understand there may have been glitches at times with the translation on the livestream. However, Richard Doble's Skype interview [Doble had not been physically present in Mexico] was in English and extremely easy to follow. The Spanish speakers were simply outstanding, and I was able to listen via translation. All of these gentlemen spoke in detail and with deep analysis as to why that body was not a human being. I am not going to repeat their reasons here, but I have been assured that a website is being constructed right now that will feature full translations of their

> analysis in text form. Incidentally, in addition to these three individuals, there was some excellent video testimony presented from other technical professionals relating to the physical properties of the slides themselves.

This addressed the technical expertise of those presenting information at the Mexico City event and suggests that it was impressive testimony. The presentations were professionally done and quite dynamic providing reasons to believe that the being on the slides was not human. Those watching on-line were apparently having trouble with the translations and the on-line feeds cut out before Dolan made his presentation at the end of the program.

The next day, that is May 6, 2015, possibly because of the criticisms about the slides from the presentation, Dew put up a higher resolution scan on his web site and a number of different people around the world including a loose-knit organization called the Roswell Slides Research Group downloaded it. They began work to deblur the placard using a variety of software programs including Smart Deblur. Within forty-eight hours of the presentation, with the scientists explaining why the body on the slides couldn't be human and that it certainly wasn't a mummy of a child, those half-dozen independent researchers around the world had been able to read the placard. The first line, which almost had to be readable in the slide and could certainly be read in a process of deblurring, said, "Mummified body of a two year old boy."

Although some of the remaining words were obscured, the rest of the placard added details that confirmed the first line. According to Curt Collins, one of those working with the Roswell Slides Research Group, posted at his Blue Blurry Lines blog, what the placard said:

MUMMIFIED BODY OF TWO YEAR OLD BOY
At the time of burial the body was clothed in a xxx-xxx cotton shirt. Burial wrappings consisted of these small cotton blankets.
Loaned by the Mr. Xxxxxx, San Francisco, California

Given that the scan used to see read the placard had come from Adam Dew's web site, and those at the Roswell Slides Research Group used that scan, there was little real suspicion that this was anything other than what it said. Still, there was push back by several of those who had been on the stage on May 5. Dolan was convinced by those experts with him on stage and believed in their expertise. Jaime Maussan claimed that those who had deciphered the placard had faked their results. He remained convinced that what was on the slides was an alien creature.

Tony Bragalia, who for months, if not for years, raged against those who didn't understand that what was on the slides was alien, and who suggested that only those who had only seen a poor quality screen grab or nothing at all would be surprised when the actual slides were revealed quickly reversed his opinion. Within hours of learning what the placard said, conducted his own research, which not only confirmed what the placard said, but identified the location of the mummy. He wrote, [again leaving his underlining and boldface type intact]:

> Working with a colleague from Europe and with the text of the de-blurred placard, I discovered … that this interpretation of the text was correct. Found in the September 1938 Volume VIII, Number 1 *Mesa Verde Notes* that was published by the National Park Service was an article that definitively solves the mystery of the "Roswell Slides." In paragraph four of the section of the publication entitled *Around The Mesa* was found this:
>
> **"A splendid mummy was received by the Park Museum recently when Mr. S.L. Palmer Jr. of San Francisco returned one that his father had taken from the ruins in 1894. The mummy is that of a two year old boy and is in an excellent state of preservation. At the time of burial the body was clad in a slip-over cotton shirt and three small cotton blankets. Fragments of these are still on the**

> **mummy." The full text of the article can be found in this link**: http://npshistory.com/nature_notes/meve/vol8-1f.htm
>
> This paragraph corresponds directly to the slides placard: **the mummified body of a two year old boy, three small cotton blankets (the word "three" understandably seen by the de-blurring program as "these") and Mr. Xxxxx of San Francisco, California.**

It would seem that such confirmation would be the final blow in this saga. He had found, in a journal, an entry that nearly matched what was on the placard. That enabled him to complete the wording of the placard and to correct one or two minor errors. Nothing Bragalia found did anything to refute the work of the Roswell Slides Research Group. It underscored the accuracy of what they had released.

But the arguments continued about the slides. Tom Carey responded with an announcement that seemed to ignore the facts. He wrote, in part:

> We believe that the recently released "reading" of the placard by the so-called "Roswell Slides Research Group" is faked.[462] Ever since Don Schmitt and I became aware of the slides three years ago, our modus operandi has been four-fold: (1) to authenticate the age and integrity of the slides; (2) to obtain professional anthropological and forensic opinion as to what the body on the slides represented; (3) to find out as much as we could about Bernerd and Hilda Blair Ray, the long-deceased owners of the slides; and (4) to "read" the placard located at the foot of the body on the slides.

---

[462] In a posting to *A Different Perspective*, and at the request of Don Schmitt, I had toned down this line. However, Carey had already sent it to several different researchers and websites, so the allegation of fakery had been published.

We physically took the slides to Kodak's historian, who is an expert regarding Kodachrome, and, using several parameters of interrogation, he determined that the slides dated from the 1947-49 time period (manufacture to exposure).[463] For the most part, the American anthropologists we contacted did not want to even look at the slides when they learned that they might be "UFO-related." Those who did, however, did so "off the record." They all concluded that the body on the slides was **not** that of a mummy but possibly that of a congenitally deformed child.[464] Fortunately, we were able to secure Canadian and Mexican anthropologists and forensic anatomical experts who went "on the record" at our May 5th "beWitness" event in Mexico City. In short, their detailed presentations concluded that the body on the slides was: not a mammal, not a primate and not human. One, Richard Doble, after a detailed morphological examination, concluded that the creature on the slides did not evolve on earth. You already have Doble's report, and the report of the two Mexican authorities is still in translation…

Regarding the placard, we quickly determined that (1) its content would be key to interpreting the slides; and (2) we could

---

[463] This isn't exactly accurate. The slides were taken to Rochester but neither Carey nor Schmitt had been there. Dew was the one who took the slides for analysis. While the results of that analysis don't seem to be in dispute, and the fact is that a Kodak expert did examine them, the idea that Carey and Schmitt participated in this aspect of the investigation is misleading.

[464] This turns out not to be accurate either. It seems that the American anthropologists wanted additional information about the slides. This information, including the provenance, names of those who had taken the pictures and the chain of custody was not available. Some of them suggested that it would be ill-advised to render an opinion without more information and the examination of the body.

not read it. So, we sent copies to Dr. David Rudiak and Dr. Donald Burleson. Both had done exemplary work in trying to decipher the so-called "Ramey Memo" - a situation very similar to placard issue here. Both responded to us that the placard was "unreadable." Through a contact, we had the Photo Interpretation Unit at the Pentagon in Washington, DC take a look at it. They said that it was "unreadable."[465] A copy went to a company in New York now requesting anonymity that conducted the analysis on a major historical artifact. That company's response to us was that the placard was "unreadable." Another copy went to the people at Adobe, Inc. (manufacturers of Adobe Photoshop and the Adobe Reader on your computer). Their response? "It's unreadable." A copy also was also sent to aggressive Roswell researcher Anthony Bragalia who also reported to me that it was "unreadable." (Bragalia has now aggressively joined in with our critics). Our own computer guy says that he applied the "SmartDeblur" software to the placard over a year ago without any success. He did so again this week to an enhanced, sharper version of the placard with the latest edition of the "SmartDeBlur" program, again without success.

Now, we are told (not asked) to believe that a cast of characters, one of whom has clearly become unhinged and was himself party to a known UFO body hoax some years ago, has used the same program (SmartDeBlur) on a distorted, "screen-

[465] The Photo Interpretation Unit moved from the Pentagon long before the slides were provided and their job is not to "deblur" placards but to interpret what is seen in high altitude and satellite imagery. That they didn't decipher it is not surprising and irrelevant.

grab" of the placard[466] and is somehow able to "read" it when all of the above, some of whom had much more sophisticated equipment and techniques at their disposal, could not. I ask you, what's wrong with this picture?

Finally, lost in all of the vile invective being hurled our way by the members of the RSRG and their fellow travelers, is what the analysis of the physical body on the slides is saying. The RSRG has used a note from an obscure late 1800's journal to weave their tale that the slides show the "mummified body of a two year old boy" (the word "mummy" or "mummified" appears nowhere in their alleged de-blurred "reading" of the placard).[467] In their excitement to play "Gotcha!," it apparently has not crossed their thought processes (I'm being charitable here) that a mummy of a two year old boy several thousand years old would be less than half the size of the body shown on the slides!

So, what are we to make of all this? Jaime Maussan, Tom Carey and Don Schmitt, relied on all of the above to reach the conclusions that were reached. They were not our conclusions but those scientists we consulted. We have, at this point in the proceedings, have sent out additional copies of the placard image to third parties whose opinions we can trust to run the SmartDeBlur application on it and are prepared to abide by their findings, wherever the chips fall.

---

[466] The distorted screen grab was in fact a scan put up on as web site that was a better copy than those used in the past. The screen grab came from Adam Dew and was taken down within days of his posting it.

[467] The journal contains a complete history of the discovery, examination and display of the mummy. The record is quite clear and almost unambiguous.

Don Schmitt added a note to this. He wrote, “As I said to Tom this morning, if the independent analysis of the placard comes back in support of the opposition's read, then I will accept that read. I will remain a gentleman and concede that point.”

Within hours of writing this, Schmitt learned that half dozen others in other parts of the world and working off the scan posted by Dew had reached the same conclusions. Bragalia’s find at Mesa Verde only underscored the legitimacy of the translation. Two days later, Schmitt issued his own mea culpa. The most important line in that document was, “I now realize that the image in the slides is a mummy as specified by the display placard. At this time I consider the matter concluded and intend on moving forward.”[468]

Richard Dolan also retreated from the alien slides. He wrote, “My problem is this: when I simply look at the pictures, I see what many other people are seeing -- an interesting museum piece. One that looks like other examples that are on the web. With the acknowledgment that looks can be deceiving, I still keep coming back to that.”

He also acknowledged one of the problems with the analysis of the slides when he said, “Although I previously stated the slides would not easily be debunked, it seems more relevant to me that they don’t need to be debunked so much as to be proven to be something anomalous. That is where the burden of proof lies.”

He concluded his published remarks by writing, “But for me, until the proper analyses are published, and until we have had time to read critiques of those analyses, and until there are strong replies to some of the critiques that are currently published, I cannot consider these slides as evidence of extraterrestrials.”

But the matter of the slides was not concluded. Tom Carey and Jaime Maussan continued to support the idea that the image could not be a mummy. Maussan offered a reward for anyone who could produce the body itself or another photograph of it. Maussan was also insisting that the experts he had

---

[468] Don Schmitt has since backtracked on this statement as well.

consulted had not changed their opinions, even given the identification that had been offered.

Richard Doble was interviewed again by Jaime Maussan. It was a restatement of what he had said during the May 5th presentation in Mexico City. He explained why he continued to believe that the image showed an alien creature as opposed to a mummy.

He said that it was obvious that this creature is not originally from Earth. He said:

> It is definitely not human nor is it even Mammalian. Rather it has evolved convergently by interacting with an environment similar to our own. It only looks superficially close to us.
>
> There appears to be absolutely no hint of a nipple or mammary glands. The mammary line should run down the front on either side of the chest. That alone makes it not a member of the Class Mammalia. There appears to be no hair but the photo quality is poor. Perhaps it is singed off. This could make this assumption incorrect but I doubt it. These two phenomena make the creature absolutely not a mammal.
>
> The chest cavity is proportionally wider and flatter than any in known hominids again save for a few anomalous humans with rare traits and conditions. There is marked asymmetry between the two sides probably from the autopsy. It has the equivalent of ribs. There are fewer of them and they are broader—probably 2 pairs that go right across and 2 pairs of floating ribs beneath them. Its left side is not all that clear. Its right side is much clearer. Humans have a sternum or breastbone and a dozen ribs that mostly attach to the sternum and floating ribs beneath them. We have clavicles or collarbones on either side and a scapula or shoulder blade in the

> rear but none is visible in this slide. The top pair of ribs appears to act as a combination clavicle/scapula.[469]

Doble provided more analysis, examining in greater detail why this was not a human and not a mummy. But all the analysis was superseded when the placard was deblurred. That provided clues that lead to the discovery of the mummy. Maussan wanted to see another photograph of the mummy and that was discovered. Maussan, in fact, said that he was going to hold another presentation in late June with more scientific evidence.[470]

Not long after that, Bragalia, working with a colleague in Germany, found another picture of the mummy on line. The picture, taken in the late 1950s, had been posted to the Internet in 2008. To many, that was the very end of the saga.[471]

Shepherd Johnson, another researcher interested in this, had submitted a FOIA request to the National Park Service in Mesa Verde and received a response from Charis Wilson, the NPS FOIA Officer in Denver on June 12, 2015. It included a 186-page document that provided a history of the discovery of the mummy, the physical description of it, and other important information about it. Also included were a series of pictures including one showing the excavation of the mummy and one that showed the mummy on display in 1938 that matched the pictures taken at later dates.[472]

At this point, given the photographic evidence and the documentation, it is clear that the slides showed the mummy that had been found in the late 19th

---

[469] Richard Doble interviewed by Jaime Maussan and posted to the Internet.

[470] That presentation was postponed until later and has not been rescheduled.

[471] The picture had been taken by a couple and posted to their Picasa account. After it was discovered, along with the name of the couple, they were bombarded by telephone calls. They have since removed the photograph, but not before others had seen it and downloaded it.

[472] National Park Service, Mesa Verde National Park, File No. 740-02.2.6, referring to the S. L. Palmer Collection.

century. Unlike the Roswell Slides, there is a provenance, there is a chain of custody and there is documentation that proves the point.[473]

One of the experts who was defending the idea that the slides showed an alien creature was Dr. Richard O'Connor, who wrote to Linda Moulton Howe that he had been able to confirm the deblurring of the placard to his satisfaction but that the statement on the placard "cannot be correct."

O'Connor joined the alien body team after the May 5th program. Maussan interviewed O'Connor via Skype because he had solid medical credentials and he spoke English. It was used as part of an article that claimed, "Doctors Agree: Roswell Slides Show a Nonhuman Body."

This interview that was posted to YouTube would be of some value in supporting that idea of "two bodies" meaning as Maussan claims that the image in the Roswell Slides is not the same as the image found in the other documentation.[474] It seemed that Maussan had found another voice to support his point of view but all that changed. O'Connor, having seen the FOIA material recovered by Shepherd Johnson, said, "Yeah, I've just, over the past 48 hours more or less, been looking at that, and it seems to me like it's drawing us toward the conclusion that in fact is this photograph probably does represent a native American child. There were some, a couple of photographs in the last pages of that set of documents, one of them in particular on page 176, and in my opinion it really does show a different photograph of what is very likely the same child."[475]

---

[473] It should be reinforced that the 186 page document provides clear evidence for the source of the mummy, detailing its excavation and history once discovered. It places the mummy on display in the museum at Mesa Verde, photographs of it, and its journey to Montezuma's Castle in Arizona. The Roswell Slides show enough detail to confirm that the pictures were taken in Mesa Verde. Details on the floor, other placards in the museum, and the background displays all confirm this. This is one of the best documented cases in UFO history.

[474] YouTube interview can be seen here: https://www.youtube.com/watch?v=rX0Ehq94rOM.

[475] See: http://www.blueblurrylines.com/2015/06/dr-richard-oconnor-on-putting-away.html

Now one of those who had once suggested the body was alien though based solely on an examination of the slide, and whose words had been posted to the Internet, had now reversed himself. After seeing all the available documentation, he changed his mind.

Curt Collins who did most of the research on this aspect of the case, sent an email to O'Connor and was surprised to get a response and an invitation to give him a telephone call. According to Collins, at his Blue Blurry Lines website:

> He told me that looking at a photograph is fraught with pitfalls, and mentioned the fact that the quality of the Slides photograph was not very good, the details were not clear due to the blurry photograph, which was taken at an angle from the body (and possibly distorted by the glass in the case).
>
> There were some characteristics that he still didn't quite understand, like the condition of the chest cavity, but it occurred to him that the terraced cliffs of Montezuma Castle must have caused the deaths of a number of children from falling off the ledges. He wondered if that could have accounted for the injuries to the child's body, particularly the damage to the head and the fractured femur. I pointed out the shallow grave may have accounted for some of this, particularly the loss of the lower leg. (I [Curt Collins] thought later that the excavation by amateur archeologists could also be a factor.)[476]

---

for the complete information about O'Connor and his conclusions about the slides. Accessed June 25, 2015.

[476] Ibid.

Even with this array of evidence, one of the suggestions is that no other scientists have gone on the record about the slides. But this claim is misleading. It might mean that none would go on the record based solely on examination of the slides without additional information. This doesn't seem to be an unreasonable request. In fact, it sounds just like the question a scientist would ask when presented with something like the Roswell Slides.

There have been statements by recognized scientists concerning what is shown in the slides. Tim Printy has published information about this, much of it found by Philip Mantle.[477] Here are the germane points:

> Dr. Daniel Antoine, Institute for Bioarchaeology - Curator of Physical Anthropology: Based on the photograph, this appears to be the mummified remains of a very young child. The mummification process is likely to have been natural (i.e. buried in a very hot or arid environment) but it may also have been intentionally embalmed.

> François Gaudard, University of Chicago: To me it looks indeed like a mummy: the mummy of a child. The item on the other side of the mummy appears to be remnants of mummy bandages, but it is difficult to tell for sure. However, since some parts of the mummy look a little shiny, for example, the right hand and just below the ribs, it makes me wonder whether it could be varnished or made of plastic? And also why is the text on the label not visible as if someone was trying to hide something? [This is an accurate statement. Someone was trying to hide something.]

---

[477] See http://www.astronomyufo.com/UFO/SUNlite7_4.pdf. accessed August 15, 2015.

Frode Storaas, University Museum of Bergen: This seems to be a mummy, but not from old Egypt. Mummies are found many places. The photo indicates that this mummy is exhibited, or stored, somewhere and by someone who probably can tell more. [Should I point out here that this is right on point. That documentation exists.]

Dr. Suzanne Onstine, University of Memphis: It does appear to be human remains (and likely a child), although the photo is too blurry to tell if artificial mummification procedures were done. It is certainly possible the body was naturally mummified due to dry climate and soil. That kind of thing happened all the time in many cultures.

S.J. Wolfe, Director of the EMINA (Egyptian Mummies in North America) Project: Okay, it is a mummy, but very hard to tell if it Egyptian, South American or European. I see no wrappings of any kind, it appears to be a child or youth. Do you have a provenance on the slide??? That may help the determination.

Dr. Ronald Leprohon, University of Toronto: Where was this shot taken? It looks like a museum. What did the label say? Did you ask the folks there? I'm sure they'd have information on their displays. It certainly looks like a mummy but it's pretty blurry so it's difficult to see properly. Sorry I can't be more helpful, and good luck in your quest. [A really astute comment by someone who only had a scan of the slide to examine.]

Dr. Patricia Podzorski, University of Memphis: Based on the image you sent, it appears that what you saw is the preserved remains of a human body, or a good imitation thereof. Since

> no wrappings are clearly visible in the photo, I can not determine the culture (Egypt, Peru, Asia, North America, etc.) or the date/ period (ancient or recent) of origin. Given that the head is turned slightly to the side and the color, it might not be an unwrapped ancient Egyptian mummy, but I am not able to be certain based on the visual information.
>
> Salima Ikram, American University in Cairo: I confirm that the photo is of a mummy of a child, possibly Peruvian or even Egyptian. [Another scientist who was to accurately identify the remains from the slide without going off into the extraterrestrial.]
>
> Denise Doxey, Curator, Ancient Egyptian, Nubian and Near Eastern Art. Museum of fine arts, Boston: Yes, that would appear to be the mummy of a small child.[478]

Interestingly, Tom Carey was interviewed on June 2 on a KGRA show about all of this. According to what Collins reported, "Of the placard being read he says, 'a day or two later, this bombshell hits about it being a mummified two-year-old boy. Well, talk about a right cross, or a left hook. He also seems to feel betrayed by two of the people who he'd asked to help with the placard have since 'joined our critics.' Of the critics, he said he'd have worked with them, 'had they been civil.' In the opening statement in the interview, he mentioned having plenty to keep him busy, a new book coming out with Don Schmitt, and another one planned beyond that, but first up is their appearance at the annual Roswell Festival."[479]

---

[478] See https://docs.google.com/document/d/1YXp9c1ACGJnIlr2rIRlHf-CcqME3ZiRt68KO5nSgIwk/edit. Accessed August 15, 2015.

[479] Interestingly, there was no public mention of the slides during the 2015 Roswell Festival. Sources suggest that the ban was put in place by the hosts of the festival.

There are other things that are evident when examining the slides. One of those is the ease with which the placard had been deblurred and decrypted when a fairly good quality scan was offered by Adam Dew on his web site. Few people have seen the actual slides projected on a screen and that includes most of those involved in the years' long investigation. All of this suggests that someone who was involved in this either knew or suspected what the placard said and that would explain the various scanned images that did not reveal the nature of the placard. Someone had to intentionally blur that placard so that it couldn't be read.

This theory was reinforced when it was discovered that the Roswell Slides were not sequential. The Roswell Slides are numbers 9 and 11. Number 10 has not been shown and could have resolved this easily. It also suggests that there was something in number 10 that was not visible in either of the others.

It should also be noted that the story of Bernerd and Hilda Ray is probably nothing more than a red herring.[480] No evidence that they knew either of the Eisenhowers has been found other than the photographs of Eisenhower during what looks like some sort of stop in a railroad yard. The idea that the Rays had an intimate relation with one or both which explained how they had managed to photograph an alien creature is unproven.

At this point it is clear that no evidence that the image on the slides is anything other than a mummy is going to be offered. The scientists who offered opinions based on what they saw on the slides have been superseded by the documentation that has been located. This has turned into a case in which a solution has been offered that is accepted by nearly everyone who had looked into it. Skeptics and believers have agreed that the placard referred to a mummy, the image on the slides is a mummy, the documentation proves it was a mummy and that solution should have been available to those investigating the case earlier. This will, however turn into another Alien Autopsy which

---

[480] Inquiries made to the Eisenhower museum found no suggestion of an associate between Eisenhower that either of the Rays. No link to Eisenhower has been found other than the two slides that show him on a platform at the rear of a rail car sometime in the late 1940s.

nearly everyone agreed the autopsy was a hoax and those who had created the alien and the video footage have admitted it but there are still those who insist that it is real. The slides show a mummy, but there will be those holdouts no matter the evidence presented.[481]

---

[481] As an addendum to all of this, Tony Bragalia in an email to Kevin Randle on April 14, 2016 reported, "Adam Dew emailed me in March [2016] saying he was going to be in Florida where I live. He said he would like to interview me for his Kodachrome documentary. I did not reply. He then emailed me several days later saying, 'Tony, you are part of this whole thing whether you care to be or not.' He implied he was going to make me look like an ass."

Chapter 10:

Walter Haut – The True Father of Roswell

In the early 1990s, as Don Schmitt and I were investigating the Roswell crash, it seemed to be very real to us. We had found dozens of witnesses who claimed to have seen and handled the debris, seen the craft or helped recover the bodies,[482] or who had participated in some aspect of the cover up of those events. There were civilian witnesses, law enforcement officials and military witnesses, both enlisted soldiers and officers. The testimony seemed overwhelming and there were hints of physical evidence that could be recovered and even talk of photographs of the bodies taken by a civilian who had blundered onto the scene before the military arrived. The promise of all this additional evidence suggested that the case was very solid and would be the one to break the UFO phenomenon wide open.

One of the most important of those first witnesses was Walter Haut who was critical to understanding what had happened in Roswell and who helped find witnesses to the crash, the recovery and the cover up. Upon my request, he supplied me with a copy of the Yearbook created in 1947 of the soldiers who had served at the base, that proved so valuable.[483] We, that is Don and I,

---

482 I once had talked to eight people who claimed to have seen the bodies and provided information about them and as of today, not one of those eight turned out to be telling the truth about what they had done or seen or even where they were in July 1947.

483 Just so there is no misunderstanding, I knew that there were no copies available so I asked him to make a photocopy of the Yearbook. Since the copying center would require payment, I estimated the total cost and sent him a check. He made the copy deleting some of the material that did not have names and pictures of the 509th Bomb Group members on them. George Eberhart made an index for me, which I shared with other researchers and some of the members of the 509th Bomb Group. The problem is the page numbering is based on my copy with some of the pages left out so the index only works well for the copy I have and the numbering I did. This also demonstrates that some others, who claimed they made the index had actually received the copies that Eberhart had made.

interviewed him a number of times and during those interviews he said repeatedly that he had done nothing other than write the press release.[484] He said, "Col. Blanchard told me to write a news release about the operation and to deliver it to both newspapers and the two radio stations."[485]

In discussing the press release, he said, "I mentioned that Major Marcel flew the object to Fort Worth [and the question became] how did he fly it? That was a poor choice of words… Colonel Blanchard told [me] to put out [a] news release concerning [the] flying disc but that [I] couldn't see it… This was unusual [because] he generally took me into his confidence but not this time."[486]

Haut, in fact, after relating more than once that he had only written the press release and taken it to the various media outlets in Roswell, said to me, "I think I told you my whole story."[487]

Haut, however, became a resource for those studying the UFO crash tale because he knew many of the players from 1947 and he lived in Roswell since he had been sent there by the military. His house became a sort of an unofficial headquarters and he even kept a guest book for those who visited.[488] He also

---

[484] It is clear that Walter Haut is the man who passed the press release around Roswell. His name appeared in dozens of newspapers of the time, he was listed in the both the Unit History and *The Atomic Blast*, the base newspaper, as the Public Affairs Officer. Haut's original affidavit was published in Pflock, Karl. *Roswell: Inconvenient Facts and the Will to Believe.* Amherst, NY: Prometheus Books, 2001: 261. His latest affidavit is in Carey, Tom and Donald Schmitt. *Witness to Roswell*, Pompton Plains, NJ: New Page Books, 2009, pp. 251-254.

[485] Walter Haut affidavit in Pflock, Karl. Roswell: *Inconvenient Facts* and the Will to Believe. Amherst, NY: Prometheus Books, 2001, p. 261.

[486] *UFOs Are Real*, shot script, Tape Seven, Page One, May 1979.

[487] Walter Haut, telephone interview with Kevin Randle, April 20, 1989.

[488] I signed the book on one of the earlier trips to New Mexico and saw that there were quite a few pages that had been used. The Haut's hosted many of those who came in with various production companies, assisting them with their documentaries.

provided the names of others who had been involved in the case in 1947, suggesting that we might wish to speak with them. I had heard from another source, for example, there was a mortician involved and it was Walter Haut who told me who it was.[489]

It was during one of the visits to Walter Haut that he mentioned someone we might wish to interview. He said that Frank Kaufmann, who had lived in New Mexico for decades, might have some information about the crash.[490] When I asked Kaufmann if he had been involved in UFO crash case some fashion, he said, "Well, I don't know."[491]

The Frank Kaufmann Saga

Although cagey in that first interview, Kaufmann, in referring to the 1989 *Unsolved Mysteries* tale about the Roswell crash, said, "I guess it was all right. Some things, I mean, were stretched a little bit. Let me put it this way, it just depends on how much of it you want to believe. I knew some of the people involved in it."[492]

Over the next several months, Kaufmann would provide additional information, taking a number of researchers out to where he claimed the UFO had crashed, and even said that there was a secret organization that was responsible

---

[489] I believe that it was Cliff Stone who told me that there had been a mortician involved but Stone didn't have the name. While talking with Walter Haut, in his home, I was asking some questions about some of this. Haut said to me, "I know the name you're fishing for. It's Glenn Dennis." Records do indicate that Dennis worked at the Ballard Funeral Home in Roswell in 1947. His involvement was both first and second hand, though the tale of the bodies, as he related it was second hand. He had not seen them but did report that he had seen some of the metallic debris and that he had been threatened by the military if he talked about it.

[490] At the beginning of the January 4, 1990, interview with Frank Kaufmann, Kevin Randle said, "I've been talking with Walter Haut…"

[491] Frank Kaufmann, telephone interview by Kevin Randle, January 4, 1990.

[492] Ibid.

for the cover up. He claimed that the crash was of an alien craft and provided drawings of both that craft and the creatures from inside it.[493]

Kaufmann, based on what he said, had been assigned to intelligence during the Second World War and when the war ended, he was discharged in Roswell, but hired on as a civilian doing the same job as he had done for the Army. He said that he was assigned to personnel, but that was merely a cover for his real duties that had something to do with Soviet spies trying to sneak close to the base.[494] Given that somewhat covert assignment, Kaufmann was brought into the recovery operation because he held the proper clearances and was already stationed at the Roswell base when the crash happened. His presence there would cause no questions to be asked because he was known locally.

Much of what he said seemed to be too good to be true. He was a witness who said he was there for it all, had seen it all, had participated in the cover up and certainly could supply information about all of that and might be able to give hints about the direction our investigation should go. And there was always the hope that he would give us the names of other participants to help verify his tale and provide additional witnesses.

Fearing that we might be lead astray here, both Don Schmitt and I asked Walter Haut what he knew about Kaufmann. After all, it was Haut who had originally supplied the lead for us. Haut, without qualification and without hesitation said, "Everything Frank tells you is golden."[495]

---

[493] For a variety of information about Kaufmann see: Randle, Kevin D. and Donald R. Schmitt. *The Truth about the UFO Crash at Roswell.* New York: M. Evans and Company, Inc., 1994, pp. 12, 74 (as Steve MacKenzie) pp. 4 – 5, 9 – 14, 137, 141 – 142, 144, 160; Pflock, *Inconvenience Facts*; for illustrations by Kaufmann, see, Randle, Kevin D. "Frank Kaufmann: Roswell Eyewitness?" *Fate* 54,12 Issue 621 (December 2001), pp.24 – 29.

[494] Apparently the mere size and shape of the atomic weapons would provide the Soviets with clues about their nature and their destructive force so even that information was classified.

[495] Walter Haut, unrecorded personal interview by Kevin Randle and Don Schmitt in Roswell.

After Kaufmann's death due to cancer, Mark Rodeghier, Mark Chesney and Don Schmitt were in Roswell on a research trip. They were asked by Kaufmann's widow to look through his papers and files to ensure that he had met all his financial obligations to various researchers and documentary producers. Those papers provided the information necessary to reevaluate Kaufmann's testimony. Kaufmann was not who he claimed to have been, did not have the military training that he claimed to have had, did not obtain the military rank that he suggested he had, and turned out to be little more than a clerk in the military during World War II.[496] Anything he knew about the UFO case was what he had seen on television or had read in books and magazines or possibly picked up from Walter Haut and if those sources were of no help, he just made it up. The documentation available proved the Kaufmann had no role in the investigation in 1947 and we were given his name in an attempt to keep the Roswell UFO crash ball rolling. Nearly everything that Kaufmann told us was a lie.

There is one more thing that should be mentioned here. The last time that I saw Kaufmann, he pulled me aside as we left a restaurant, and told me that I couldn't trust one of the Roswell witnesses. This man, said Kaufmann, was making up his involvement in the case. Then he added, "I've tried to give you hints about all of this." The implication, of course, was that he was telling the truth and some of these others were not.

More Sources Supplied by Walter Haut

Walter Haut was the source of another witness as well who had a fabulous inside story. Haut suggested that a fellow named Rick Tungate had been involved sometime after the events of 1947 in cleaning up the paper trail. According to the *Roswell Telephone Directory* Tungate had been a colonel

[496] For the complete story, see Randle, Kevin D. and Mark Rodeghier. "Frank Kaufmann Reconsidered." *International UFO Reporter*, 27,3 (Winter 2002) pp. 8 -11, 17 – 19, 26.

(which means nothing other than Tungate had suggested that for his listing), and Tungate, when I met him, said that he had been an O-7, which was a brigadier general. When I asked about this discrepancy, he said that the retirement benefits worked out better that way so he had retired in the lower grade.[497]

According to the story, Tungate, as a young intelligence officer had been assigned to the base in Roswell to review the records about the crash to determine what should be saved, what should be destroyed and what should be moved to another location. He made it clear that was his only connection to the Roswell case. He had not been there in 1947 and he hadn't talked to any of those who had been. He had merely gone through the classified documents there and removed or destroyed those that could lead to the truth.

While interviewing him one afternoon, he introduced me to a friend who had been a commanding officer of an Air Force unit during the war in Korea. He said that the man was a retired general. I found a book that listed all the commanding officers for the Air Force in Korea down to the squadron level and the man's name did not appear in it at any level.

I also attempted to find independent corroboration for Tungate's military service but failed. Had he actually been attached to one of the intelligence functions in the Air Force, his records might not be as easily accessible as those of officers who followed more traditional career paths. Still, it was a discrepancy that was somewhat worrisome.

Here was another source provided by Walter Haut whose story could not be verified in any way. Although Haut didn't vouch for him in the same way that he had for both Kaufmann and Glenn Dennis, he was the one who suggested that Tungate had something important to say.

---

[497] In the military, a colonel is an O-6 while a brigadier general is an O-7. I know of no circumstance in which the retirement benefits for a colonel would be better than those of a brigadier general. There is one circumstance where accepting retirement at a lower rank is beneficial. A CW-5 (Chief Warrant Officer 5) receives a higher retirement pay than a major with similar years of service. This is nearly the only place where the lower ranking officer retirement is better than that of the high ranking officer and is one of the bizarre situations in the pay structure of the military and the position as a warrant officer add to the confusion.

Glenn Dennis, Roswell Mortician

Once Haut had identified Dennis as the mortician for me, and for others including Stan Friedman, it wasn't long before Dennis reluctantly agreed to an interview. I interviewed him in November 1990 in Roswell and the story he told is basically the same that he told throughout the later years.

He said that he had received a number of telephone calls from the base mortuary officer asking questions about body sizes and small coffins. He told the officer they had nothing in stock but he could order it from Amarillo and have it to them in a day or so.

As was the case in many small towns of the era, the funeral home also ran the ambulance service. He received a call to take an injured airman out to the base. According to Dennis, they parked around the rear of the base hospital and he saw a number of military ambulances parked there.[498]

As he walked by the ambulances, he noticed that some wreckage had been loaded into one of them. Inside was what he thought of as the front of a canoe that had some strange writing on it. He said, "What was so curious about it, was that in those two ambulances was a deal that looked like half a canoe. It didn't look like aluminum … I glanced in and kept going."[499]

Inside the hospital he was surprised to find unusual activity. He thought that he would find a nurse he knew and buy a Coke. She saw him in the hallway and told him to get out of there before he got himself into big trouble. Before he could say much a nasty officer saw him and asked him what he thought he

---

[498] This is a minor problem with Dennis' tale. In 1947, the base hospital was not a single building but a cluster of narrow, single story buildings with each having its own function. Dennis indicated during the interviews that he had parked behind what was the hospital built in 1952.

[499] Friedman, Stanton T. and Don Berliner. *Crash at Corona*. New York: Paragon House, 1992, pp. 116 – 117. See also Carey and Schmitt. *Witness to Roswell*, pp. 143 – 145.

was doing there. He was ordered out of the hospital when another, red-haired officer spotted him and called to him. He wanted a few words with him.

According Dennis, the officer was accompanied by an African-American NCO and both of them were angry. Dennis was told there had been no aircraft accident, there was nothing going on, and he should get back to town. He would tell no one what he had seen and if he did, they're be picking his bones out of the sand.

Dennis left the hospital a little annoyed and more than slightly puzzled but a day or two later he spoke to a nurse he identified as Naomi Self. He met her at the Officer's Club where she gave him all the inside, classified information. She told him of a preliminary autopsy of little creatures from another star system. She told him not to tell a soul about that but she did draw for him a representation of what they looked like. Dennis later had an artist friend recreate that illustration.

Within days, according to Dennis, Self was transferred to England. He said the he had received a note from her telling him her new address. He wrote once but the letter came back marked "deceased." Dennis said that she had been killed in an aircraft accident that had killed five nurses.

Unfortunately for Dennis, the name, Naomi Self, was not in the Yearbook, and there were no indications that anyone by that name had ever served with the 509th Bomb Group. As mentioned some ten to twenty percent of the personnel assigned to the base were not featured in the book.

I checked the *New York Times Index* in the world before the Internet and the like. The Index contained a section on aircraft accidents that were broken down by date, type of aircraft, location, and when and where it appeared in the newspaper. I didn't have to search each edition of the newspaper and I searched everything in the Index from July 1947 through 1955 and found no aircraft accidents in England, Europe or the United States in which five Army nurses had been killed.

Don Berliner did much the same thing looking through issues of *Stars and Stripes*, which is a newspaper printed for the military service members overseas. Berliner's search failed to find anything that remotely sounded like the story told by Dennis of the Army nurses.

There were other avenues to be explored. I attempted to get the morning reports for the medical unit in Roswell but was sent the morning reports for the Headquarters Company instead. Repeated efforts to get at the proper morning reports by me failed. I was told they didn't exist anymore.

There were other hints about this. Rosemary Brown, a nurse in Roswell at the proper time was interviewed. She said that she had no memory of a nurse who fit the description offered by Dennis. David Wagnon, who was also in Roswell at the proper time said that he did recognize the description but was unable to remember the nurse's name.

There is a point to be made which wasn't exactly relevant at the time. I had called Dennis about something and he asked me why we hadn't found his nurse. He seemed more than a little annoyed by it and made the claim then, "I gave you her name."

While my search was diluted because of other avenues of investigation, V. G. Golubic, then living in Arizona, undertook the project. He again attempted to get the morning reports, and continued to communicate with those in St. Louis until he received the records from early 1946 through the end of 1947. He found an additional nineteen women assigned to the base as nurses, both military and civilian, and who were there are the right time. These were nurses who did not appear in the Yearbook or the base telephone directory and whose names surfaced in various documents recovered through Freedom of Information Act requests. Golubic expanded his investigation to include genealogical searches, and a list of nurses in the Army in the 1940s that contained 124,000 names. No one named Naomi Self, or any of the likely variations of the spelling of the name surfaced. No one named Naomi Self was an Army nurse at any time in the 1940s.

All this might not be overly significant if not for Dennis' reaction to this information. He told researchers that he had said originally that we all wanted a name and he'd give us a name but it wouldn't be the right name. He said that he had promised her that he wouldn't tell anyone who she was and he was standing by that agreement. But remember the conversation I'd had with Dennis. He was annoyed that we'd failed to find her after he had given us the name.[500] He gave no indication that he had invented the name. This was a fabrication by Dennis.

When we examine testimony provided by Dennis we find it filled with errors, changes, and less than candid statements. He said that the reason that she couldn't be found was because he had given us the wrong name. He then claimed that her real name was Naomi Sipes, which would be another change and would send us off searching for another woman who didn't exist. Remember Golubic had a rather comprehensive list of names of all Army nurses and Sipes wasn't on them either.

The important part of the Dennis story was his meeting with the nurse and her confession that she had been involved in some sort of preliminary testimony of these alien creatures. If that story was not true, and the evidence now suggests that it is not, then what he tells us about his involvement is rendered less than useless. It is just one more tale that can be categorized as a "friend of a friend" and when added to all the other stories he told that couldn't be verified, we must reject this testimony.

Before we leave this however, it must be remembered that Dennis was identified for us by Walter Haut. And as will be said time and again, if Haut was who he eventually claimed he was, he would have known that the Dennis tale was false. If he didn't know, then he obviously was not on the inside as he would eventually claim and his testimony collapses.

Walter Haut, Eyewitness

Walter Haut's story remained consistent for decades and while it is true that he might have been honoring an oath taken in 1947, by the time he began to change the tale, it was obvious that he had no clear memory of many of the events in 1947. Anyone listening to the taped interview conducted by Wendy

---

[500] Randle, Kevin D. *The Randle Report: UFO's in the "90s*. New York; M. Evans and Company, Inc. 1997 pp. 186 – 192, see also, Carey and Schmitt. *Witness to Roswell*, pp. 143 – 145; Klass, Philip J. *The Real Roswell Crashed - Saucer Coverup*. Amherst, NY: Prometheus Books, 1997, pp. 68 – 71, 131 - 133

Connors and Dennis Balthaser in 2000 realizes that Haut is lost. He contradicts himself inside of single paragraphs and sometimes single sentences. These new interviews lead to a new affidavit written from him by Don Schmitt. While it is a common practice for someone else to construct an affidavit for another, it is also clear here that this affidavit is constructed out of little of real value. And while it is true that Haut might have said those things and signed the affidavit; it is not clear if he actually understood what he was saying or that he had a firm grasp of those long ago events.

Haut had insisted for decades that he had done nothing other than write and distribute the press release and that he didn't know any more.[501] Although there had been newspaper reports that the officers at Roswell in general and Haut in particular had been "severely rebuked" by those in Washington, D.C., Haut denied this. He said, "If there was anything, I'll be honest with you, that would not have made any impression on me as a call from Washington would."[502]

He then said, "I think I told you the whole story. Everything I knew about it."[503]

Some researchers have believed for years that Haut was more heavily involved than he claimed and that he was honoring his promise to Blanchard and his obligation to remain silent. Then, in 2000, Haut's story began to change. Wendy Connors and Dennis Balthaser interviewed Haut after a French film crew had been to New Mexico to interview him about the UFO crash.[504] He

---

[501] Walter Haut affidavit signed May 14, 1993. Multiple interviews conducted by Kevin Randle, both in person and over the telephone. Interviews conducted by others including Karl Pflock, Stan Friedman and Bill Moore.

[502] Walter Haut, telephone interview with Kevin Randle, April 20, 1989.

[503] Ibid.

[504] Wendy Connor and Dennis Balthaser, videotaped interview with Walter Haut, November 15, 2000. See also http://www.truthseekeratroswell.com/haut-questions.html; Randle, Kevin D. *Alien Mysteries, Conspiracies and Cover-Ups*. Detroit: Visible ink, 2013: pp. 103 – 105.

said that he had seen the bodies... or body, depending on which statement by Haut is accepted. He later expanded these statements in his interviews with Connors and Balthaser. He even consented to having a video record made of that interview.

But the water is a still a bit more muddied. Gildas Bourdais, a French UFO researcher who has written his own book on the Roswell crash, said he had talked to the French crew director and that Haut said nothing about bodies on camera. Bourdais wrote:

> He [Vincent Gielly, the director] told me that, when he did his filmed interview of Walter Haut, with Wendy Connors, Haut looked like someone who wished to say more, but could not. This lasted a long time, and he finally decided, a little disappointed, to end the interview. But then, he found Wendy, alone in another room, extremely disappointed because, she told him, she felt Haut was just about to talk when he ended the interview. That's what Gielly told me. He did not tell me that Haut had talked about seeing the craft and bodies. If he did, he may have promised not to repeat it, I don't know.

Connors and Balthaser said that Haut said something to the French which inspired the two of them to seek another audience with Haut to explain his earlier years in the military, and, according to them, ask a few questions to clarify the situation Haut found himself in back in 1947.[505]

The Connors/Balthaser interview contained some very disturbing statements by Haut. He was either badly confused, he was deeply conflicted about revealing secret he had kept for more than sixty years, or he just couldn't keep his new story straight, probably because of his advanced age. He left a somewhat rambling mishmash of contradictory information in various statements he made after 2000.

---

505 Ibid.

In one of his confusing statements, Haut said:

> That's a rough one [about the appearance of the craft] I haven't even thought about it low these many years and I honestly can't even visualize it, whether still in its shape, but a lot of dings in it.... I do not remember... I would venture a guess that probably a diameter of, uh, somewhere around 25ft... To the best of my remembrance there was one body... it was relatively a small body comparable to uh, oh maybe a 11-year-old, 10 or 11-year-old child. It was pretty well beat up. I cannot come and give you, to be honest, anything other than that. I remember something about the arms and I am trying to visualize that and all of a sudden it starts going through my little head that that they show some of those long arms in the cartoons... I thought there was several bodies... for some reason I feel there were several bodies... the more I think about it the more I start to get an idea it was single body.[506]

And then to thoroughly confuse the issue even more if possible, Haut retreated to the line he had been using from the very beginning, in the late 1970s and early 1980s and into the 1990s as he talked with various UFO investigators about his role. He said, "I didn't even see one. I just wrote a press release."

Connors then said, "I am talking about when you saw a body in the hangar partly covered by the tarp. You only saw the one."

Walter said, "Yes."

But to really complicate the issue, Walter Haut also said, in that same interview:

> "I don't really know. I ...it hurts me to try and give an answer because I am not certain of the whole thing. I feel there has

---

[506] Wendy Connor and Dennis Balthaser, videotaped interview with Walter Haut, November 15, 2000. See also http://www.truthseekeratroswell.com/haut-questions.html

> been information released that uh maybe shouldn't have been released, maybe the information that we got in the operation of releases maybe something you can put out to anybody. I just... I don't know, I don't want to talk about a lot of the detail number one because I don't have a lot of knowledge about the detail, everybody thinks that I saw them, I didn't, I put out a press release that Colonel Blanchard told me what he wanted in the press release and I ran it into town and gave it to the news media and went home and ate lunch."[507]

Haut eventually agreed to create an affidavit, to be released only after his death.[508] According to Carey and Schmitt, this was akin to a "deathbed confession," which, they say, "carries more weight than a signed statement."[509]

Walter Haut's Final Affidavit

The statement was dated December 26, 2002 and Haut lived another three years. After 2000, Haut began to hint that he knew more and it was from these conversations that the affidavit compiled by Don Schmitt and Tom Carey and was witnessed by Julie Shuster, Haut's daughter, was drawn. Unlike the rambling statements he made on audio and video tape after 2000, the affidavit is a very clear and concise statement about all these events. Haut's statement said:

> (1) My name is Walter G. Haut.

---

[507] Ibid.

[508] For a comprehensive history of the affidavit, see Carey and Schmitt. *Witness to Roswell.* 245 – 254.

[509] According to the law, this, technically, was not a deathbed statement, but one created years before his death.

(2) I was born on June 2, 1922.

(3) My address is 1405 W. 7th Street, Roswell, NM 88203.

(4) I am retired.

(5) In July, 1947, I was stationed at the Roswell Army Air Base in Roswell, New Mexico, serving as the base Public Information Officer. I had spent the 4th of July weekend (Saturday, the 5th, and Sunday the 6th) at my private residence about 10 miles north of the base, which was located south of town.

(6) I was aware that someone had reported the remains of a downed vehicle by midmorning after my return to duty at the base on Monday, July 7. I was aware that Major Jesse A. Marcel, head of intelligence, was sent by the base commander, Col. William Blanchard, to investigate.

(7) By late in the afternoon that same day, I would learn that additional civilian reports came in regarding a second site just north of Roswell. I would spend the better part of the day attending to my regular duties hearing little if anything more.

(8) On Tuesday morning, July 8, I would attend the regularly scheduled staff meeting at 7:30 a.m. Besides Blanchard, Marcel; CIC Capt. Sheridan Cavitt; Col. James I Hopkins, the operations officer; Major Patrick Saunders, the base adjutant; Major Isadore Brown, the personnel officer; Lt. Col. Ulysses S. Nero, the supply officer; and from Carswell AAF [Fort Worth Army Air Field] in Fort Worth, Texas, Blanchard's boss, Brig. Gen. Roger Ramey and his chief of staff, Col. Thomas J. DuBose were also in attendance. The main topic of discussion was reported by Marcel and Cavitt regarding an extensive debris

field in Lincoln County approx. 75 miles NW of Roswell. A preliminary briefing was provided by Blanchard about the second site approx. 40 miles north of town. Samples of the wreckage were passed around the table. It was unlike any material I had or have ever seen in my life. Pieces, which resembled metal foil, paper thin yet extremely strong, pieces with unusual markings long their length were handled from man to man, each voicing their opinion. No one was able to identify the crash debris.

(9) One of the main concerns discussed at the meeting was whether we should go public or not with the discovery. Gen. Ramey proposed a plan, which I believe originated with his bosses at the Pentagon. Attention needed to be diverted from the more important site north of town by acknowledging the other location. Too many civilians were already involved and the press already was informed. I was not completely informed how this would be accomplished.

(10) At approximately 9:30 a.m. Col. Blanchard phoned my office and dictated a press release of having in our possession a flying disk, coming from a ranch northwest of Roswell, and Marcel flying the material to higher headquarters. I was to deliver the news release to radio stations KGFL and KSWS, and the newspapers the *Daily Record* and the *Morning Dispatch*.

(11) By the time the news had hit the wire services, my office was inundated with phone calls from around the world. Messages stacked up on my desk, and rather than deal with the media concern, Col. Blanchard suggested that I go home and "hide out."

(12) Before leaving the base, Col. Blanchard took me personally to Building 84, a B-29 hangar located on the east side of the tarmac. Upon first approaching the building, I observed that it was under heavy guard both outside and inside. Once inside, I was permitted

from a safe distance to first observe the object just recovered north of town. It was approx. 12 to 15 feet in length, not quite as wide, about 6 feet high, and more of an egg shape. Lighting was poor, but its surface did appear metallic. No windows, portholes, wings, tail section, or landing gear were visible.

(13) Also from a distance, I was able to see a couple of bodies under a canvas tarpaulin. Only the heads extended beyond the covering, and I was not able to make out any features. The heads did appear larger than normal and the contour of the canvas suggested the size of a 10-year-old child. At a later date in Blanchard's office, he would extend his arm about 4 feet above the floor to indicate the height.

(14) I was informed of a temporary morgue set up to accommodate the recovered bodies.

(15) I was informed the wreckage was not "hot" (radioactive).
(16) Upon his return from Fort Worth, Major Marcel described to me taking pieces of the wreckage to Gen. Ramey's office and after returning from the map room, finding the remains of a weather balloon and radar kite substituted while he was out of the room. Marcel was very upset over this situation. We would not discuss it again.

(17) I would be allowed to make at least one visit to one of the recovery sites during the military clean up. I would return to the base with some of the wreckage which I would display in my office.

(18) I was aware two separate teams would return to each site months later for periodic searches for any remaining evidence.

(19) I am convinced that what I personally observed was some type of craft and its crew from outer space.

(20) I have not been paid nor given anything of value to make this statement, and it is the truth to the best of my recollection.

THIS STATEMENT IS TO REMAIN SEALED AND SECURED UNTIL THE TIME OF MY DEATH, AT WHICH TIME MY SURVIVING FAMILY WILL DETERMINE ITS DISPOSITION.

Signed: Walter G. Haut

Signature Witnessed by: Chris Xxxxx (blacked out signature of witness.

Dated: December 26, 2002

Over the years Haut made many recorded statements. For decades, all he had done, according to those statements, was write the press release and that was all that he knew about the case.[510] Now there is a brand new statement in which he is in the middle of this with all the inside knowledge that anyone could hope for.

The problem is not that his earlier statements contradict his later statements, but that his later statements were highly confused, and highly contradictory even inside one interview, and inside one statement in that interview. Haut is on the record in too many places saying that all he did was write the press release. In fact, in one of those earlier interviews, as noted above, Haut said, "Colonel Blanchard told [me] to put out [a] news release concerning [the] flying disc but that [I] couldn't see it…"[511]

---

[510] For a reproduction of Haut's original affidavit, see Pflock, *Inconvenient Facts*, 261. Haut in various interviews with Kevin Randle and Don Schmitt including in person on April 1, 1990; with Kevin Randle, April 20, 1989. See also Carey and Schmitt, *Witness to Roswell*. pp. 251 – 254.

[511] *UFOs Are Real*, shot script, Tape Seven, Page One.

He is seen in the Connors/Balthaser interview giving that same story but wrapping it around tales of bodies and craft. He seems to be very confused and it seems that he is drawing on all the information that has been published over the years, not to mention all the documentaries that have been dedicated to the subject.

Some of the new statement makes no real sense based on what we know today. People were brought into it who had no need to know. While it could be argued that the primary staff had to be aware of the situation, some of the others could do their jobs without being told everything. The real problem seems to be the idea that they would announce that they had a flying saucer and mention only the site found by Mack Brazel. The other site, closer to town, would not be described. They were confirming that something had been found which would have excited more people than those few who might have said something.

The note that the press already knowing seems to be a reference to the Johnny McBoyle tale that had been described in *The Roswell Incident*. McBoyle had been out to the site and had seen what he described as a "big, crumpled dishpan" and burned spots on the ground.[512]

Lydia Sleppy, who worked at the radio station, said that "He just called me and said he had something for me to put on the line… He said that he had gone into the coffee shop in Roswell and this rancher walked in… He [the rancher] had been out on the ranch… when he came on this thing that was all smashed up."[513]

Sleppy also said that as she was attempting to put the story on the news wire, she was interrupted. A bell, signaling an incoming message sounded and

---

[512] Berlitz, Charles and William L. Moore. *The Roswell Incident*. New York: Grosset & Dunlap, 1980: pp. 14 – 16.

[513] Lydia Sleppy, telephone interview with Kevin Randle, February 5, 1993. See also, Randle, Kevin D. *Roswell Revisited*. Lakeville, MN: Galde Press, 2007, pp. 71 – 75.

she flipped over to receive on the teletype machine. She said the message was, "Do not continue this transmission."[514]

McBoyle was said very little about this in later life. Most of the information about this came from Sleppy. He did corroborate that he had been reporting on a crash when he changed it from a flying saucer to an aircraft accident. He did not talk about it with her or his boss at the time Merle Tucker.[515]

Had Haut's later statements been consistent inside the context of the interview and had they been consistent throughout that interview and other, later ones, then we could say that he was providing us with information that he'd had all these years. But that's not what we have here. We have contradictory statements that are all over the place.

Confirmation of Haut's Tale?

There is some confirmation of this new tale told by Walter Haut from another source that does provide some interesting and relevant information. Richard C. Harris, who was an assistant finance officer of the 509th, lived in Albuquerque.[516] By the mid-1990s he was a frail man who needed live-in help. He was, naturally, quite interested in the Roswell UFO case, having served at the base in 1947. He is in the Yearbook, so there is confirmation that he was there at the right time.

---

514 Slate, B. Ann and Stanton T. Friedman. "UFO Battles the Air Force Couldn't Cover Up." *Saga's UFO Report*, 2,2 (Winter 1974): p. 60; Sleppy telephone interview with Randle, February 5, 1993. In later interviews Sleppy would insist that it was the FBI who had issued the warning but there is nothing to suggest which agency had attempted to stop the transmission.

515 Johnny McBoyle, unrecorded telephone interview with Kevin Randle, December 17, 1990.

516 Harris' picture appears in the Yearbook produced in 1947 providing evidence that he was assigned to the base in the proper time fame.

But, in his living room was a small bookcase that held a stack of books dealing with UFOs, the Roswell case and MJ-12. Harris was a firm believer in MJ-12 and alien visitation.[517]

This simply means that he was familiar with the case as it had been written about in the various books over the years. He had seen many of the documentaries on the case so he could have been badly contaminated as a source. Having seen the documentaries, read the books and magazine articles doesn't mean that what he said was based on what he had read and seen and not wholly on his memories, but it must be considered that his memories could be colored by those other sources.

Anyone who has served in a command position or a position of responsibility in the military knows that everything must not only be paid for but it must all be properly documented. There are all sorts of funds that are designated for all sorts of purposes and it is considered illegal to take funds appropriated for one purpose and use them for another. This means that funds meant to pay for a unit's flight training, for example, can't be used to transport alien bodies and craft from one location to another. Funds must be designated for that purpose. A cross country navigation problem and therefore training could also carry wreckage from the crash to Wright Field but would be accounted for by the training flight because it was listed as navigational training. The money had been successfully juggled and hidden under authorized training and while not an illegal transfer of funds, is certainly somewhat questionable.

Harris said that they worked hard to find the money from legal sources, that they worked hard to cover the real purpose because there would be audits and there would be examinations that had nothing to do with the crash but everything to do with looking for fraud. So the money spent to house those brought in, for the aircraft flights to and from various locations, for the special equipment and to pay the soldiers were all juggled around so that it was

---

[517] Personal observations by Kevin Randle at Harris' Albuquerque home during the videotaped interview.

properly annotated and properly spent. Harris was proud of the job they had done covering the paper.

The key point of Harris' story was this little anecdote. He said that he had been out near one of the hangars in early July 1947 and ran into Walter Haut. Haut told him what was on the other side of the door, meaning one of the dead aliens was there and told Harris he could take a quick look if he wanted. Harris said that he put his hand on the door knob, but didn't turn it. For some reason his curiosity failed him at that point. He didn't take the look that Haut had told him to.[518]

This, of course, suggests that Haut had deeper knowledge than he had claimed for decades and Harris mentioned this more than a decade before the Haut affidavit was published so he couldn't have known that Haut would so radically alter his story. It is some corroboration for Haut's new story provided by someone who was not in communication with Haut but it is very flimsy corroboration.

Walter Haut – The Real Father of Roswell

The problem is that all these stories surfaced after the publication of *The Roswell Incident*. Although it could be argued that the book was a national publication, there is no evidence that it was big news in Roswell or anywhere else except for the UFO community. It could be argued that many of those who told of seeing the bodies had not heard about the book when they began to share their stories with UFO investigators. It could also be argued that many of those stories were inspired by that book and without it as a guide, they would have never said anything.

But since the investigation of Roswell expanded to the point that it has, there has been one source that could be counted on to provide all sorts of information on many aspects of the case and who could provide many new witnesses. Walter Haut met with everyone with an interest in the case who came to Roswell and Walter Haut provided the names of some of the best witnesses

---

[518] Richard Harris, personal interview with Randle in Albuquerque, August 1996.

with the best stories to tell. It might be said that nearly all roads in Roswell led back to Walter Haut.

True, the beginning in 1947 was Mack Brazel, but he was never interviewed by any UFO researchers and never said all that much about what he had seen. His son, Bill Brazel, another of the important witnesses, had only seen some bits and pieces strange metallic debris found in a pasture some 70 miles or so from Roswell. He surrendered that to the Air Force in the late 1940s. He could talk about what his father said, and he could talk about what he had picked up, but according to him that was very little and Mack Brazel never mentioned bodies to his son.

And Jesse Marcel said that he had seen metallic debris spread out in that pasture that, to him, was unusual but he saw no craft or bodies. Stories surfaced from Marcel relatives that he had talked of bodies to them, but those tales are dubious at best. They were told after Marcel had died and couldn't contradict what they said. There is nothing in the record to suggest that Marcel ever saw bodies or that he even knew about them even though he was the intelligence officer at the Roswell Army Air Field.

That sort of information, about the recovery of the bodies, came indirectly from Walter Haut. He gave us the names of some of the worst of the witnesses but were those who could talk about bodies. If he was who he claimed to be, which was an unofficial aide to Blanchard and that he was on the inside of all the information about the Roswell crash, having attended the various staff meetings and briefings, he would have known what each of these people really knew about the crash. Interestingly he had been friends with both Glenn Dennis[519] and Frank Kaufmann before the Roswell case exploded in the late 1970s. And if he was who he claimed to be, he would have known that Kaufmann had no role in this and that Dennis, while an embalmer at Ballard's in 1947, might have received the telephone call about the small caskets, and, by the late date of these interviews, Haut should have known that no nurse named Naomi Self

[519] To quote from Philip Klass, Haut said, "I happen to personally believe Glenn. He's a friend of many, many years." See Klass, *Real Coverup*, p. 68.

existed.[520] Walter Haut should have known that many of these people knew nothing of importance about the crash or the recovery of the object but he still pointed us toward them.

Haut, along with Dennis, and another longtime friend Max Littell, banded together to create the International UFO Museum and Research Center in Roswell. Suddenly the tale that brought researchers into town was now also a tourist destination and eventually a yearly festival. Littell was fond of telling of bigger plans that included some sort of an amusement park located on the west side of Roswell. He even had some sort of a scale model of what the site would look like when it was completed. Those plans seem to have died with Littell.

But each time it seemed that interest in the case flagged, Walter had a suggestion of someone else to interview or another bit of information to generate new interest. These new witnesses had grandiose tales of alien craft, alien bodies, clandestine units and special flights from Washington, D.C. And while there were men and women ready to share that allegedly highly classified information, there was never any documentation to prove it. Sure, there were pictures of them in the Yearbook or orders that proved them to have been in Roswell in July 1947, but they just never had any other documentation. It was Walter Haut who facilitated much of this, providing clues to witnesses and filtering reports to those who wanted to hear about alien creatures. Walter Haut gave birth to the Roswell story as we all know it today.

---

[520] As an aside, during my search for the nurse, I found four women named Naomi Self, one each in Alabama, California, North Carolina and Oklahoma. None of them had served in the military and none of them had ever been trained as a nurse.

Chapter 11:

The Final Analysis

The whole purpose of this exercise, this reexamination of all the information available on the Roswell UFO crash case, and these new investigative avenues that were opened and explored, was to try to provide a concise history of this report and draw a line to the truth about it. In reviewing that mountain of data, there were some new discoveries that do shed light on what happened. The end result is something that will please virtually no one, but I believe it is the best that can be said about the Roswell case today. I have found some little known information, interview notes that have been overlooked in the past, and some evidence that tells us a great deal about what actually happened in Roswell in July 1947.

I have broken all this down into three categories: What do we think, what do we know and what can we prove. As can be imagined, the "What can we prove" category is the smallest because we can prove very little of what we think we know about the Roswell case and most of that is in the tangential areas. The what do we think is the largest because we can rely on bits of data and testimony that stand alone. In other words, there is very little in the way of hard evidence available even after decades of work, other than testimonies of those who claim participation, and in the past, a large number of those people have been found to be lying about it.

What Do We Think?

Because the standard of evidence here is lower here than the other two categories, we can discuss the discovery of the "escape pod," the retrieval of alien creatures, and draw other conclusions based solely on the testimony of those who were there though most of it was gathered decades after the fact. According to some of the witnesses, there were bodies of alien creatures found to the southeast of the debris field. Although very old when he first mentioned he had seen the bodies, retired Air Force NCO Eleazar Benavides said that he was assigned to the guard detail inside a hangar on the base. Once inside the hangar, he had seen creatures he described as having grayish features, faces that looked swollen and a hairless head with large black eyes. His description fit, generally, with that provided by witnesses to alien creatures in other UFO cases and to those who claimed to having been abducted by aliens.

Anna Willmon told of seeing only two bodies near their wrecked UFO north of Roswell but her tale is virtually stand alone. The descriptions match, generally, what others said, though she talked about a skin of dark brown, and uniforms that seemed to match the Boy Scouts rather than those of alien aviators. The remains of the craft, the mention of something like a washtub are similar to that offered by Johnny McBoyle,[521] but there is no real corroboration for her tale.

While most of those who claimed to have seen bodies can now be discounted, there are a few first-hand testimonies to them. Edwin Easley said nothing specifically about it, but when Roswell was mentioned to him, said, "Oh the creatures." When asked about it, he said that following the extraterrestrial path was not the wrong path to follow.[522]

Others had talked with those who had seen bodies. Brigadier General Arthur Exon told of hearing that the bodies had been flown to Wright Field and that one had eventually been sent to Lowry Air Force Base. He had no descriptions and his information was second hand at best.[523]

---

[521] McBoyle, in a telephone call to Lydia Sleppy at the radio station, McBoyle said that the crash looked like a crushed dishpan, meaning it was circular in shape.

[522] In a telephone interview with Kevin Randle on February 2, 1990, Easley was asked if he, Randle, and Don Schmitt were following the right path. Easley asked what that meant and Randle responded that "We think it was extraterrestrial." He said, "Well, let me put it this way. It's not the wrong path." The interview was not recorded.

[523] Arthur Exon, audiotaped telephone interview by Kevin Randle, May 19, 1990, audiotaped telephone interview by Don Schmitt, June 18, 1990, personal interview by Kevin Randle at Wright-Patterson AFB, unrecorded.

Frank Joyce didn't see any bodies himself but mentioned that Mack Brazel told him of bodies, saying that they had smelled terrible and that they weren't green. He said that Brazel made the comments to him after he had completed another radio interview in which he, Brazel, suggested that what he had found was little more than the remnants of a weather balloon.[524]

Frankie Rowe told of her father's (Dan Dwyer) experience when he drove out to see the wreck for himself. He told her that there had been bodies but one of the creatures had survived the crash. He provided a rough description of the bodies, suggesting that face looked something like an insect and the color matched that of the Jerusalem Cricket. Dwyer mentioned this to his married daughter Helen Cahill as well and to some of his fellow firefighters including J. C. Smith.[525]

Several of the children of those assigned to Roswell in 1947 have suggested that fathers had told them of alien bodies. These include Beverly Bean, daughter of Melvin Brown, who said her father had told her about bodies on several occasions. Carlene Green said that her father, Homer Rowlette, just before he died, told her about them. He didn't say much, only that the ship was rounded and he had seen little people.

In keeping with this examination of tales of the body, the theory is that there were three separate sites, one that Mack Brazel found that was filled with the metallic debris and a second where wreckage and bodies were found that is referred to by some as the Dee Proctor site and a third thought of as the bodies site. It was the first site that Marcel described. It might have been this third site that Anna Willmon saw.

---

[524] Pflock, Karl. *Roswell: Inconvenient Facts and the Will to Believe*. Amherst, NY: Prometheus Books, 2001, pp. 121 – 123; Carey Tom and Donald Schmitt. *Witness to Roswell*. Pompton Plains, NJ: New Page Books, 2009, pp. 55 – 56.

[525] Frankie Rowe, personal interview with Kevin Randle, January 2, 1993, affidavit signed November 22, 1993; Helen Cahill, affidavit signed November 22, 1993. Affidavit published in Pflock, *Inconvenient Facts*, p. 267.

There are some other things that we think about all this. Although Irving Newton said that Marcel was in Ramey's office pointing to parts of the balloon and asking questions about that, it seems that he might have been mistaken. Major Charles Cashon, the 8th Air Force PIO might have been present and he is the one asking the questions. Marcel had been ordered not to speak and it is unlikely that he would have disobeyed orders from Ramey especially with the general sitting right there. Newton did say that he knew neither Marcel nor Cashon.[526]

Marcel said, at one time, if he was in the pictures, then it was the real stuff he had found and if it was anyone else, it was the substituted debris. All the pictures taken in Ramey's office are of a weather balloon and degraded rawin target. Marcel, when shown the pictures that appear in *The Roswell Incident*, told reporter Johnny Mann that it wasn't the stuff that he had found.

Mann isn't the only source on that. During the filming for *UFOs Are Real*, Marcel had said that if he was in the picture it was the real stuff, but then, shown the pictures from Ramey's office, said, "There's a picture in the same room. It's not the material I brought here."[527]

There are, in fact, lots of these stories now in circulation and it is difficult to keep them all straight. In the aggregate it seems to be an impressive amount of testimony and there seems to be no reason for anyone to invent a tale and assign it to a deceased relative, but in the end that just what has happened. Much of it is just second-hand testimony that does little to advance our knowledge. It contributes to what we think but does nothing for adding proof to the Roswell case.

This does show however, what we think based on all that testimony. There were three sites, there were bodies recovered, and the military, with elements of the government thrown in, orchestrated a cover up that persists to today. It is unfortunate that none of this can be proved because there simply isn't any

---

[526] Irving Newton, telephone interview with Kevin Randle March, 24, 1990.

[527] *UFOs Are Real*, shot script, Tape One, page two.

authentic documentation for these beliefs. If we remove the second-hand testimony, then we are left with very little.

What Do We Know?

What we know is somewhat more than what we can prove and less than what we think in volume of testimony and is based on testimonies gathered decades after the events and, of course, based on the newspaper articles published at the time. We know, for example, that Mack Brazel went into Roswell to report his discovery, was interviewed by newspaper reporters, and was more than annoyed at his treatment by the military and said so to many of his neighbors and his son, Bill. Bill Brazel collected bits of the strange debris in the months after the event and described the unusual properties of it to UFO researchers. These properties include the great strength of the very thin metal, debris that sounds like fiber optics and the "memory" metal which would return to its original shape when bent or folded.[528] He did show some of this to friends such as Sallye Tadolini, who, after a fashion, provided a description that matched some of what Bill Brazel said. Part of Tadolini's affidavit would be quoted in an Air Force report, but they edited the transcript to give a false impression of what she said.[529]

There were many others who saw the metal or who handled the metal and mentioned the strange properties. Jesse Marcel, Sr., said that there was a parchment like debris that he couldn't burn and a thin metal that was so tough that a sledgehammer couldn't dent it. Robert Smith said that another NCO had a small piece of metal that he showed him. He said that when it was wadded up, it would unfold itself. Frankie Rowe said that she saw a piece of debris that a

---

[528] Bill Brazel, audiotaped personal interview with Kevin Randle and Donald Schmitt, February 19, 1989 in Carrizozo, New Mexico.

[529] For the affidavit see Pflock, *Inconvenient Facts*, pp. 285 – 286. See also Weaver and McAndrew, *The Roswell Report*: Executive Summary, p. 23.

state trooper had. Her description was that it unfolded itself in a fluid motion like quicksilver.[530]

The members of Colonel Blanchard's staff, who were interviewed, reported with a single exception, that what was found was alien in nature. These included Major Patrick Saunders, Major Jesse Marcel, Major Edwin Easley and Lieutenant Colonel Joe Briley. Lieutenant Colonel Robert Barrowclough wrote to Kent Jeffrey that nothing had happened and he hoped that people would stop calling him about it. The weight of the testimony suggests that Barrowclough was less than candid about this.

There were several flights out of Roswell that had to do with moving debris and bodies to other installations. Obviously there was a flight to Fort Worth, again based on testimony and newspaper articles. Robert Slusher suggested that he had been on a flight with a destination of Fort Worth but they carried large crates that held more than the couple of items that Marcel took to Texas. O.O. "Pappy" Henderson told his wife that he had flown some of the material and possibly the bodies to Wright Field in Ohio. Robert Porter talked of a flight out of Roswell carrying some of the remains.

All of this suggested that something very unusual happened, but as noted, it is based only on the testimony of those who were there. Taken at face value, this is important testimony, but that is the problem. How much of this testimony has been contaminated, how much is confabulated, and how much is invented is the question that can't be answered?

What Can We Prove?

We're now down to what we can prove and there isn't much of it. We can prove that something fell outside of Roswell and no one disputes that. It was found by a rancher named William W. "Mack" Brazel and he took fragments to the Chaves County Sheriff, George Wilcox in Roswell. In turn, the sheriff

---

[530] In an interview conducted for a television documentary in 2014 she said that her father had the piece of debris that she saw. There is no reason for her to have changed what she had said to so many others for so long but change it she did.

alerted the officers at the Roswell Army Air Field and Major Jesse Marcel, Sr., the air intelligence officer, responded to the call.

We can prove that the story was given to the local media by the base public relations officer, Walter Haut, and that the story went out over both the UP and the AP wire service lines based on the statements by Haut, George Walsh and Frank Joyce. This is established by the article printed in the *Daily Illini* on July 9, 1947. We can prove that they all, military and media, referred to it as a flying saucer or a flying disk. We can prove that in 1947 flying saucer and flying disk were not synonymous with alien spacecraft. In 1947 that was only one of the interpretations and not always the preferred one.

We can prove that Mack Brazel had returned to Roswell midweek and gave an interview to the local newspaper as well as the AP reporters sent to Roswell from Albuquerque. Brazel's picture was taken while in Roswell and Major Edwin Easley confirmed that he had been held at the base in the guest house during that time.

We can prove that Marcel flew to Fort Worth and that he met with Brigadier General Roger Ramey. In Ramey's office, he was photographed with the remains of a degraded weather balloon and rawin radar reflector, as were Ramey and his Chief of Staff Colonel Thomas DuBose. Ramey, his intelligence officer Major Edwin Kirton and one of his weather officers, Warrant Officer Irving Newton, identified the debris as a weather balloon based on what they had seen and this was reported in newspapers the next day.

We can prove that Ramey, as well as other members of his staff, said that the debris was the remains of a weather balloon and a rawin radar target. The news media, that is radio and newspapers, reported the identity of the material in Ramey's office as those items saying that this was what had been found in New Mexico and that interest in the story waned at that point.

We can prove that there was a document held by General Ramey during the meeting with J. Bond Johnson of the *Fort Worth Star-Telegram* and that parts of the document can be read. We know, then, who took the photograph, when it was taken and where it was taken. The documentation for all this is rock solid and comes from the testimony of Johnson, DuBose, Marcel, and documents created at the time and recovered from the primary sources decades later. There are no disputes about any of this. The interpretation of the memo held by Ramey is not universally accepted and is still quite problematic.

We can prove that Blanchard went on leave on Tuesday afternoon. That story was given to the newspapers and was printed in many of them. This was also confirmed by the 509th Bomb Group unit history for July 1947.[531]

We can prove that on July 9, 1947, both the Army and the Navy began efforts to end the speculation about the flying saucers and that they attempted to suppress stories about them. This was reported in newspapers from around the country on that day.

And, we can prove that this explanation for the find near Roswell was reinforced on July 10, 1947, when an array of balloons was launched from Alamogordo. This was, of course, part of the New York University constant level balloon project. Based on the photographs published in the newspapers and the testimony of Charles Moore, one of the engineers on the project who identified the ladder in one of the photographs as the one he had purchased with petty cash to facilitate the balloon launches, there is no dispute about this.

We can prove, based on a statistical analysis of the number of newspaper articles that the number of flying saucer stories dropped off significantly after the Army and Navy worked to end the reporting. This strategy was significant in ending discussions in public about the Ghost Rockets seen during the summer of 1946 over Scandinavia.

But none of this, no matter how solid the evidence and documentation for it may be proves that what fell outside of Roswell was an alien spacecraft. Based solely on this information and documentation, it would appear that the recovery was of a balloon and rawin radar target. The photographs of the debris in Ramey's office establish that. It is only when testimony, gathered decades after the fact come into play that we move in another direction.

---

[531] This actually makes little sense. Why would Blanchard suddenly go on leave on a Tuesday afternoon? If he waited until midnight to sign out, he wouldn't lose a day of leave as he did by signing out on Tuesday afternoon. In fact, why begin a leave in the middle of the week unless it was an emergency leave but the records do not show that.

The Documentation

There is some documentation that relates to all this, but none of it supports the idea that what fell was an alien spacecraft. There had been some high-level discussion of the flying saucer reports but most of it had originally been classified. It was only later that the information made it into the public arena. One of those was a letter originally classified as secret written General Nathan F. Twining, wrote that was entitled "AMC [Air Materiel Command] Opinion Concerning "Flying Discs." Interestingly, it did make a comment about UFO crashes. The letter said:

> 1. As requested by AC/AS-2 there is presented below the considered opinion of this Command concerning the so-called "Flying Discs". This opinion is based on interrogation report data furnished by AC/AS-2 and preliminary studies by personnel of T-2 and Aircraft Laboratory, Engineering Division T-3. This opinion was arrived at in a conference between personnel from the Air Institute of Technology, Intelligence T-2, Office, Chief of Engineering Division, and the Aircraft, Power Plant and Propeller Laboratories of Engineering Division T-3.
>
> 2. It is the opinion that:
>
> a. The phenomenon reported is something real and not visionary or fictitious.
>
> b. There are objects probably approximating the shape of a disc, of such appreciable size as to appear to be as large as man-made aircraft.
>
> c. There is a possibility that some of the incidents may be caused by natural phenomena, such as meteors.

d. The reported operating characteristics such as extreme rates of climb, maneuverability (particularly in roll), and action which must be considered «)9»evasive«):» when sighted or contacted by friendly aircraft and radar, lend belief to the possibility that some of the objects are controlled either manually, automatically or remotely.

e. The apparent common description of the objects is as follows:

(1) Metallic or light reflecting surface.

Page two began with the heading, "Basic Ltr fr CG, AMC WF to CO, AAF, Wash. D.C. subj "AMC Opinion Concerning "Flying Discs:".

(2) Absence of trail, except in a few instances when the object apparently was operating under high performance conditions.

(3) Circular or elliptical in shape, flat on bottom and domed on top.

(4) Several reports of well kept formation flights varying from three to nine objects.

(5) Normally no associated sound, except in three instances a substantial rumbling roar was noted.

(6) Level flight speeds normally about 300 knots are estimated.

f. It is possible within the present U.S. knowledge --provided extensive detailed development is undertaken -- to construct a piloted aircraft which has the general description of the object in sub-paragraph (e) above which would be capable of an approximate range of 7000 miles at subsonic speeds.

g. Any developments in this country along the lines indicated would be extremely expensive, time consuming and at the considerable expense of current projects and therefore, if directed, should be set up independently of existing projects.

h. Due considerations must be give the following:

(1) The possibility that these objects are of domestic origin - the product of some high security project not known to AC/AS-2 or this command.

(2) The lack of physical evidence in the shape of crash recovered exhibits which would undeniably prove the existence of these objects.

(3) The possibility that some foreign nation has a form of propulsion possibly nuclear, which is outside of our domestic knowledge.

3. It is recommended that:

a. Headquarters, Army Air Forces issue a directive assigning a priority, security classification and Code Name for a detailed study of this matter to include the preparation of complete sets of all available and pertinent data which will them be made available to the Army, Navy, Atomic Energy Commission, JRDB, the Air Force Scientific Advisory Board Group, NACA, and the RAND and NEPA projects for comments and recommendations, with a preliminary report to be forwarded within 15 days of receipt of the data and a detailed report thereafter every 30 days as the investigation develops. a complete interchange of data should be effected.

4. Awaiting a specific directive AMC will continue the investigation within its current resources in order to more closely define the nature of the phenomenon. Detailed Essential Elements of Information will be formulated immediately for transmittal thru channels.

N. F. Twining
Lieutenant General, U.S.A.
Commanding

The problem for researchers today is that one paragraph about the "lack of physical evidence in the shape of crash recovered exhibits." Clearly Twining and the highest-ranking members of his staff would have been notified if a flying disc had crashed. The laboratories and facilities to examine and exploit such a find were at the Wright Field, Patterson Field complex in Dayton, Ohio. Reverse engineering of captured foreign rockets, aircraft and weapons was accomplished at the Foreign Technology Division there.

The letter was in response to a request by Brigadier General George Schulgen that had outlined nearly two dozen UFO reports, many of them made by those who would be considered credible witnesses.[532] The argument can be made that no mention of the crash debris from Roswell was made because it had not been included in the reports forwarded to Twining. Since it wasn't included among those sightings, Twining's staff saw no reason to officially inform Schulgen and others of its existence, especially since they could accomplish their task without mentioning it to them. While those in Washington were expecting AMC to tell them that no further investigation was necessary

[532] For a complete analysis of Brigadier General Schulgen's see Randle, Kevin D. *The Government UFO Files*. Canton, MI: Visible Ink Press, 2014, pp. 64 – 69. See also, Swords, Michael and Robert Powell, UFOs and Government. San Antonio, TX: Anomalist Books, 2012, pp. 36 – 44.

because the answers were held by those at the top, Twining was telling them that such was not the case.

The counter argument would be that Twining would not wish to create an investigation with a high priority that had the mission of exposing the Roswell crash and recovery. Rather than encourage an investigation, Twining could have easily ended the interest, at least in the military, by suggesting that the information reported was insufficient to demonstrate that something real was flying through American airspace. That would have prevented an inadvertent compromise of the secret.

While he was suggesting that the flying disks represented a real phenomenon, he was also suggesting that there had been no crash recovered debris. This might not be the definitive answer for either side, but the indication that the Air Force had no crash recovered debris is an interesting comment and certainly not the last word on it.

Twining's letter was not the only document created in that time frame that reflects on the reality of the Roswell case, at least tangentially. Another study of the unusual sightings, even more highly classified than the Twining letter was prepared about a year later. *Air Intelligence Report No. 100-203-79*, was officially entitled "Analysis of Flying Object Incidents in the U.S.," and dated 10 December 1948. It was originally classified as Top Secret, and apparently was a joint effort between the Directorate of Intelligence of the Air Force Headquarters and the Office of Naval Intelligence.[533]

The purpose of the study, according to the document itself was "To Examine patterns of tactics of 'Flying Saucers' (hereinafter referred to as flying objects) and to develop conclusions as to the possibility of existence."

Under facts and discussions, the report said, "THE POSSIBILITY that reported observations of flying objects over the U.S. were influenced by previous sightings of unidentified phenomena in Europe... and that the observers

---

[533] See *MUFON UFO Journal* 207, July 1985, pp. 3 -19: To read the report, see http//www.project 1947.com/fig/1948air.htm. See also Klass, Philip. *Sun #26* (March 1994).

reporting such incidents may have been interested in obtaining personal publicity.... However, these possibilities seem to be improbable when certain selected reports such as the one from U.S. Weather Bureau at Richmond are examined. During the observations of weather balloons at the Richmond Bureau, one well trained observer has sighted strange metallic disks on three occasions and another observer has sighted a similar object on one occasion.... On all four occasions the weather balloon and the unidentified objects were in view through a theodolite...."[534]

The report included an interesting paragraph about the origins of the objects. It said, "THE ORIGIN of the devices is not ascertainable. There are two reasonable possibilities: (1) The objects are domestic devices, and if so, their identification or origin can be established by a survey of the launchings of airborne devices... (2) Objects are foreign, and if so, it would seem most logical to consider that they are from a Soviet source..."

The conclusions drawn by the authors, at the bottom of page two, and marked top secret, were, "SINCE the Air Force is responsible for control of the air in the defense of the U.S., it is imperative that all other agencies cooperate in confirming or denying the possibility that these objects are of domestic origin. Otherwise, if it is firmly indicated that there is no domestic explanation, the objects are a threat and warrant more active efforts of identification and interception."

And finally, the report said, "IT MUST be accepted that some type of flying objects have been observed, although their identification and origin are not

---

[534] Under a section entitled, "Descriptions of significant incidents," these sightings were described as, "During April 1947, two employees of the Weather Bureau Station at Richmond, Virginia reported seeing a strange metallic disk on three occasions through the theodolite while making PIBAL observations. One observation was at 16,000 feet when a disk was followed for 15 seconds. The disk appeared metallic, shaped something like an ellipse with a flat bottom and a round top. It appeared below the balloon and was much larger in size. The disk appeared to be moving rather rapidly, although it was impossible to estimate its speed. The other observations were made at 27,000 feet in like manner." Although unidentified in the Air Intelligence report, one of the observers was Walter Mineczewski. See also, Swords, Michael and Robert Powell. UFOs and the Government. San Antonio, TX: Anomalist Books, 2012, pp. 32 – 33.

discernible. In the interest of national defense it would be unwise to overlook the possibility that some of these objects are of foreign origin."

This was a report created to brief high-ranking officers in the military on the unidentified flying object situation as it stood in the fall of 1947. It seems that the officers creating the document would have access to all the classified information needed to accurately assess the situation. They would be in a position tell their superiors everything they knew, or could discover, about UFOs, regardless of how highly classified that information might be. And, according to the thinking of many, if Roswell represented the crash of an alien spacecraft, it would be mentioned in this report. Because there was no mention of a crash many skeptics have concluded that this alone proves there was no flying saucer crash at Roswell.

But there is one fact that is important when reviewing this document. There are events left out because, according to the officers, they did not have access to all areas of military secrecy. The authors admitted, subtly, that they did not have all the privileged information they needed. The report, by itself, does not prove that Roswell UFO crash didn't happen, or that these officers were lying to their superiors if it did.

There is another document that affects all of this. It is a report about the Scientific Advisory Board Conference that was held on March 17 – 18, 1948 in Room 3E-869 in the Pentagon. Colonel Howard McCoy was discussing the creation of Project Sign to investigate the flying saucer reports. He said, "We have a new project – Project SIGN – which may surprise you as a development from the so-called mass hysteria of the past Summer when we had all the unidentified flying objects or discs. This can't be laughed off. We have over 300 reports which haven't been published in the papers from very competent personnel, in many instances – men as capable as Dr. K. D. Wood, and practically all Air Force and Airline people with broad experience. We are running down every report. *I can't tell you how much we would give to have one of those crash in an area so that we could recover whatever they are*. [Emphasis added.]"[535]

---

[535] Pflock, *Roswell: Inconvenient Facts,* pp. 77 – 78.

McCoy made a similar statement in letter he sent to General Cabell on November 3, 1948, that was classified as "secret." McCoy wrote, "The possibility that the reported objects are vehicles from another planet has not been ignored. However, tangible evidence to support conclusions about such a possibility are lacking."[536]

McCoy, because of who he was at the time, and where he worked, would have been in the loop. Had there been a crash and recovery of debris, McCoy would have been one of the few officers brought in on this. He seemed to be saying, in arenas where those involved had the clearances and the need to know that there was no crash recovered debris. His statements are quite strong on the subject.

These documents, especially those classified as "top secret" were probably considered to be beyond the reach of the public. The officers and scientists creating them thought that no one outside a very small group would ever see them and because of that spoke the truth. For those who believe that there was a crash of an alien craft outside of Roswell, they are truly devastating. They suggest that the answer, whatever it might be, had nothing to do with alien visitation.

The Final Analysis

Many years ago, Dennis Stacy who was the editor of the *MUFON Journal* at the time, thought that Roswell case should become clearer with all the investigative activity centered around it. Instead it was becoming muddier. But now we've been able to clarify the situation. What we see here is that the evidence, once thought to be so robust is based on little more than testimony. True, there is some very interesting first-hand testimony, but all of it was gathered decades after the fact. We know that some contamination has influenced those witnesses because we have seen the testimonies evolve over time. It wasn't a conscious attempt to fool anyone, it was just the natural tendency of witnesses to

---

[536] Swords and Powell. *UFOs and the Government*, pp. 494 – 495.

incorporate elements of other witness testimony into their own. A good example is the tale told by Lydia Sleppy. When first reported in 1976, she said that her attempt to put the story on the radio station's teletype was interrupted by a warning to cease the transmission. Eventually she would say that the warning included the line, "This is the FBI." That detail was not mentioned in the original report she gave in 1974 but added sometime after 1990.

There had been some very interesting and exciting information offered in Jim Ragsdale's tale of seeing the object fall from the sky to the point where he had seen the bodies in the distance. Later he would claim there were sixteen of them wearing helmets made of solid gold which is a tip off. Gold is a soft, heavy metal that is unsuited for helmet. The Air Force would use his descriptions of the bodies as a way to prove their anthropomorphic dummies theory, never realizing that if Ragsdale was making up his tale, then their explanation failed at that point.

Glenn Dennis told an exciting tale of a missing nurse, but research destroyed the credibility of Dennis' story. There was no nurse who was killed in an aircraft accident as he had claimed and no nurse by the name he had given us. It became clear that his tale was untrue especially when he attempted to revitalize it by claiming he had warned us the name he provided was not her real name.

We are left with the stories told by members of Colonel Blanchard's primary staff. Those leaving records were Jesse Marcel, Edwin Easley, Joe Briley, Patrick Saunders, and more than a few of the other officers assigned to the base including Pappy Henderson, Walter Haut, and Chester Barton. There were many enlisted men who contributed to the story including Robert Smith, Lewis Rickett, Thaddeus Love, Robert Porter, E. L. Pyles, Eleazar Benavides and Robert Sluster to name but a few.

Attempting to look at the whole of the Roswell case dispassionately, all we have left, after all the work, after all the investigation, and after all the research in various libraries and archives, are the witness testimonies, many of them recorded on audio tape and some of them also recorded on video tape. What we have then is a record of what they said decades after the fact and that has been preserved in various locations including CUFOS and FUFOR. Some of those testimonies have been uploaded to YouTube (often in violation of copyright). We can see and hear what these witnesses had to say about what they did and saw in 1947. But in the end, at the close of business, only testimony is left without any sort of documentation hinting at the crash, without physical evidence whether it is pieces of the craft, biological samples of the bodies or pictures of either or both to support these claims and that isn't quite enough for skeptics who routinely say, "Extraordinary claims require extraordinary evidence."

And while some are arguing that Roswell was a case of overwhelming proof of alien visitation that was covered up by the government others say that it is made up of only testimony and not very good testimony that proves nothing about alien visitation. It is contaminated at best and complete fabrication at worst.

And while it is logical and appropriate to require extraordinary testimony, I respond with what Sherlock Holmes said, which is 'When you have eliminated the possible whatever is left, no matter how improbable is most likely the answer."

With the Roswell case we have eliminated all that was possible in 1947 but are left with no real answer. Such is the state of the Roswell investigation today. Recorded testimony of men and women who are no longer with us and not a shred of physical evidence or documentation of any sort. While the testimony of so many might be persuasive in a court, in science more evidence is required and although we have searched diligently for it for decades, we have been unable to produce it.

The Roswell case comes down to those testimonies and for those who look at it dispassionately, that is all there is. The number of witnesses suggest something quite unusual but the lack of physical evidence and documentation argues for some sort of error. Without a breakthrough, without an admission by government officials backed up with documentation and photographs, that is where the case will remain. We have the crash of something that might have origins on another planet or we might have the remains of something constructed on here on Earth that for some reason can't been located today. Right now there isn't sufficient evidence to prove either and that, unfortunately, drops us back into the mundane. Without that breakthrough we are going to be left with questions that can't be answered and it begins to look as if they might never be.

While some might find that testimony significant, others will not. As for me, I find myself drifting toward those who reject the extraterrestrial. At one time I was sure but that was when we had all that robust testimony, much of which is now thoroughly discredited. I have hope that we'll find an answer, and it might be extraterrestrial, but in today's world we just can't prove it.

We are back to Sherlock Holmes and his belief that "when you have eliminated the impossible whatever remains, however improbable, must be the truth." Here we seem to have eliminated everything, except that improbable, which suggests something alien. Unfortunately, we need something a little more tangible than a clever saying to make that leap of logic. Maybe someday we'll have it, but today is not that day and I fear tomorrow won't be that day either.

Appendices

Appendix A: The Myth of MJ-12

Although the story of the Majestic-12 (MJ - Twelve, MJ-12) is often reported to have started in December 1984,[537] it seems that the first comprehensive description of it came years earlier and was in the context of an unpublished novel rather than a government report.[538] The first public mention of MJ-12 with the names of the original members attached was in an issue of *Just Cause* in December 1985.[539] That information came from Lee Graham, a California technician[540] at Aerojet ElectroSystems Division in the aerospace industry and

---

537 Stan Friedman, *Top Secret/Majic*, 1996 p.20, Jerome Clark, *The UFO Encyclopedia, Second Edition*, 1998, p. 310, Philip Klass, "The MJ12 Crashed-Saucer Documents," *Skeptical Inquirer*, Winter 1987-88, p. 38; Stan Friedman, "MJ-12: The Evidence So Far," *International UFO Report*, Sept/Oct 1987, p.14; Vallee, Jacques. *Revelations*, New York: Ballantine Books, 1991, p. 38

538 Brad Sparks and Barry Greenwood, "The Secret Pratt Tapes and the Origin of MJ-12" in *2007 MUFON Symposium Proceedings*, p. 95 – 100; Kevin Randle, *Case MJ-12*, 2002, p. 207, Clark, *UFO Encyclopedia*, 1998, p. 308, Linda Moulton Howe, *An Alien Harvest*, 1989.

539 Larry Fawcett (publisher) and Barry Greenwood (editor), "MJ-12: Myth or Reality, *Just Cause*, December, 1985, p. 1-2.

540 He has been described by various sources as an employee, technician and an engineer working in the aerospace industry including Philip Klass' *SUN Newsletter* available at: www.csicop.org/specialarticles/show/klass_files_volume_44.

a UFO researcher. He said that his information had come from a military source.[541]

Tracing the history of the first mention has become difficult with most of those still researching MJ-12 forgetting about or ignoring the Aquarius Telex.[542] This was a single page document that was allegedly classified as secret and was a "request for photo imagery interpretation your msg [message] 292030Z Oct 80 [meaning 29 Oct 80 at 8:30 in the evening, Greenwich Mean Time]."[543] There is a single line in this official looking document that said, "Results of Project Aquarius is still classified Top Secret with no dissemination outside official Intelligence channels and with restricted access to MJ-Twelve."[544]

This provided, according to the various researchers, the first hint of some sort of secret project, handled at the highest levels of both the civilian government and military commands. It was designed to exploit the find of the alien craft outside of Roswell and has operated in secrecy until documents about it began to leak into the public arena. The original documents are now thought of as the Eisenhower Briefing Document, the Truman Memo and the Cutler-Twining Memo. These documents are wrapped in controversy and a complete examination of the history of them, how they entered into the public arena, and

---

541 Stan Friedman in *Top Secret/Majic*, p. 57, reported that, "As far as I know, the source was actually Bill Moore, who was not in the military and does not work for the government, but has a fondness for playing games." See also, Dolan, *Cover-Up*, p. 407

542 There is no mention of either the Aquarius Telex or Project Aquarius in Friedman's *Top Secret/Majic*, Ryan Woods *Majic:Eyes Only* or Friedman's 1991 report on Mj12, *Final Report on Operation Majestic-12* published by the Fund for UFO Research in April 1990. In a February 2001 letter, Friedman wrote, "I am not sure what you mean by the Aquarius Telex and don't even mention it in *Top Secret/Majic* or *Final Report*..."

543 For copies of the document, see William Steinman and Wendelle Stevens, *UFO Crash at Aztec*, 1986, p.625: William Moore and Jaime Shandera, *The MJ-12 Documents: An Analytical Report*, 1990, Appendix C.

544 Ibid.

what steps have been taken to authentic them will provide the answers about the reality of them.

The History of the Aquarius Telex – An Early Form of MJ-12

According to William Moore, he had been handed the Aquarius Telex in February, 1981 with the intention that he would give it to a New Mexico businessman and physicist, Paul Bennewitz. Bennewitz was convinced that he had detected electromagnetic signals which extraterrestrials were using to control and monitor the humans they abducted.[545] He was also seeing, and filming, what he believed to be alien spacecraft over the Manzano Nuclear Weapons Storage Facility and the Coyote Canyon test area near Kirtland Air Force Base, Albuquerque, New Mexico.[546]

Bennewitz contacted the Air Force at Kirtland AFB, and was taken seriously. On October 24, 1980, he contacted Richard Doty of the Air Force Office of Special Investigation, suggesting that he had evidence that something might be threatening the Manzano Storage Facility. His information was taken down, and eventually resulted in a report signed by Major Thomas A. Cseh, who commanded the base investigative detachment.[547] His report, obtained under the Freedom of Information Act (FOIA) said:

> On 26 Oct 80, SA Doty, with the assistance of JERRY MILLER, GS-15, Chief, Scientific Advisor for Air Force Test and Evaluation Center, KAFB, interviewed Dr.

---

[545] Jerome Clark, *The UFO Encyclopedia*, Second Edition, 1998, p. 303; Vallee, *Revelations*, pp. 78 – 82; Richard Dolan, *UFOs and the National Security State: The Cover-Up Exposed, 1973 – 1991*, 2009 pp. 225 - 231.

[546] Timothy Good, *Above Top Secret*, 1988, p. 406

[547] Clark, *UFO Encyclopedia*, 1998, p. 303

BENNEWITZ[548] at his home in the Four Hills section of Albuquerque, which is adjacent to the northern boundary of Manzano Base. Mr. MILLER is one of the most knowledgeable and impartial investigators of Aerial Objects in the southwest. Dr. BENNEWITZ has been conducting independent research into Aerial Phenomena for the last 15 months. Dr. BENNEWITZ also produced several electronic recording tapes, allegedly showing high periods of electrical magnetism being emitted from Manzano/Coyote Canyon area. Dr. BENNEWITZ also produced several photographs of flying objects taken over the general Albuquerque area. He has several pieces of electronic surveillance equipment pointed at Manzano and is attempting to record high frequency electrical beam pulses. Dr. BENNEWITZ claims these Aerial Objects produce these pulses… After analyzing the data collected by Dr. BENNEWITZ, Mr. MILLER related the evidence clearly shows some type of unidentified aerial objects were caught on film; however, no conclusions could be made whether these objects posed a threat to Manzano/Coyote Canyon areas. Mr. MILLER felt the electronical [sic] recording tapes were inconclusive and could have been gathered from several conventional sources. No sightings, other than these, have been reported in the area.[549]

[548] Although addressed throughout as "doctor" and Bennewitz studied physics, he never completed the work for his doctoral degree according to Jim Campbell who interviewed Bennewitz. The interview is published online at: http://ufoexperiences.blogspot.com/2005/11/final-of-interview-with-paul-bennewitz.html.

[549] Clark, *UFO Encyclopedia*, p. 303

All this is important because it put Bennewitz into touch with Doty. As it would play out later, Doty seemed to be the conduit from the military to Bennewitz, acting through Moore. The Aquarius Telex with its early mention of MJ-12 according to Moore was given to him so that he would pass it along to Bennewitz with the hope that Bennewitz would take it to the media. Once he had done that, then the Air Force could denounce it as a hoax, ending any credibility that Bennewitz had.[550]

The Aquarius Telex then, and according to Moore, was a disinformation ploy to damage Bennewitz. Bennewitz never used the document and didn't reveal it to anyone and didn't take it to the media. If it was truly a disinformation ploy by the Air Force, it failed.

Moore did circulate the memo inside the UFO community.[551] Or it was through his carelessness that it entered into the mainstream media. Some of the information that Moore had was passed to Tracy Torme, who in turn passed it along to members of the Canadian UFO Research Network (CUFORN) who published it in a report they called, "Information For Those With A 'Need To Know' Clearance Only".[552]

According to the CUFORN report, "Bill Moore told us that when he left his briefcase unlocked for a minute or so in San Francisco International Airport in early 1983, a copy of the stolen teletype mentioning Aquarius was taken from the briefcase. Peter Gersten, the attorney representing Citizen's [sic] Against UFO Secrecy [CAUS], must have taken it from the case, because only a few days later, Gersten was on ABC-TV's 'Good Morning, America' show and showed the teletype on camera in a brief interview…"[553]

---

[550] Moore and Shandera, *MJ-12: An Analytical Report*, p. 8

[551] W. Todd Zechel, "The MJ-12/Aquarius Hoax," privately published, 1989, p. 4. It is in this document that Zechel claims that Bennewitz came up with the term, "Extraterrestrial Biological Entity" (EBE), which would figure prominently later in the Eisenhower Briefing Document.

[552] Zechel, *MJ-12/Aquarius*, p. 8

[553] Ibid. p. 9.

Moore himself had a different version of this, according to what he wrote. He said, "In September of 1982, insofar as I was aware, there were only three copies of this document in existence. One of these I had passed on to Bennewitz, a second was in safekeeping, and a third was in my briefcase during a trip I had made to San Francisco. While there, I had a morning meeting with a man who turned out later to be an associate of UFOlogist Peter Gersten of New York. That same afternoon, my car was broken into and my briefcase was stolen. Four months later, a copy of that same document, complete with notations I had penciled on it, turned up in the hands of none other than Gersten himself. To this day I have never received a satisfactory explanation of how he received that document."[554]

The controversy didn't stop there. Moore had suggested, in his later explanations about the Aquarius Telex, that he had seen another AFOSI telex about the same event. An Air Force officer, only identified at the time as "Falcon" showed Moore the original. Moore wrote, "At a meeting with 'Falcon' on that date [March 2, 1981][555] , Moore was shown the original 'AD' [Aquarius Document], which appeared to be a typical government telex on thin computer paper with perforated edges. After examining it, Moore asked if he could keep it. 'Falcon' said no, that Moore was only being given the opportunity to read it; at which point Moore proceeded to re-read it while making a conscious effort to imprint as much of it as possible upon his memory in the process. Immediately following that meeting, he wrote out some notes on a legal pad. (For the record, Moore had already been told of the alleged existence of Project Aquarius during an earlier meeting with 'Falcon' in December, 1980 and had filed a FOIA request on it with HQ/USAF dated 29 December, 1980…"[556]

---

[554] Moore and Shandera, *MJ-12 Analytical Report*, p. 9

[555] Although it was widely reported that this meeting took place in February 1981, Moore claimed that he found the correct date in his notes. In his *MJ-12 Analytical Report*, page 10, he wrote it was March 2, 1981.

[556] Moore and Shandera, *MJ-12 Analytical Report*, p. 10

That wasn't the last time that Moore would see a version of the document. According to him, at a meeting with both Falcon and Doty in Albuquerque weeks later, Moore was given a different and what he called, "a retyped" version. He noticed that it had some differences in it and pointed those out. He was told this was to "sanitize" it and then was told that he could keep that copy and that perhaps Bennewitz might like to see it.[557]

Moore claimed that he knew that the version he now held had been retyped and he didn't give it to Bennewitz for a number of months. He marked up a copy so that he could identify it later and then, with a caveat to Bennewitz that the document might not be authentic, handed it over. Bennewitz never did publicize the document.[558]

The story takes another turn here. Since Bennewitz didn't use it, no one had heard about it until Gersten began to circulate a copy inside the UFO community. Moore said that his briefcase was stolen on September 13, 1982 and that the crime was reported to the San Francisco police.[559] Now Moore suggests that Gersten became a target because of the various FOIA requests he had made and that those on the inside might have targeted him in the same way that they had targeted Bennewitz. Gersten, however, didn't do anything to make the document public, so his credibility remained intact.[560] This, of course, suggests that Zechel's claim that Gersten showed the document on national television is not true.

There is a corollary to this story, however, and that revolves around who retyped the original AFOSI document and inserted the line, "Results of Project Aquarius is still classified Top Secret with no dissemination outside official

---

[557] Ibid. p. 10

[558] Ibid. pp 10 – 11.

[559] Ibid., p. 11

[560] Ibid. p. 11

Intelligence channels and with restricted access to MJ-12." Dick Hall, in a letter dated March 20, 1989, wrote, "Unfortunately, I think Bill [Moore] has engaged in some minor 'manipulations' of his own which may be coming back to haunt him (e.g., censoring the uncensored MJ-12 document without saying so; retyping the AFOSI document without saying so)…"[561]

Hall was describing a meeting held by the Fund for UFO Research in which Moore had said that the original copy of the telex was very poor and difficult to read. He had retyped it and pasted on the headings to that it had the appearance of the original message. Hall wrote, "My recollection is that the original simply was of poor quality for reproduction and so he re-typed it, the 'cut and paste' job referring to various rubber stamps and symbols from the original document."[562]

Hall would modify this later. He would write that he didn't remember if Moore had actually said that he had re-typed it, or if the version he had had been retyped by someone else. It was clear, however, that the document that was being circulated inside the UFO community was an actual AFOSI telex that had been modified to include the statement about Project Aquarius and the Majestic Twelve.

The important fact in this is that Moore said that he received the altered memo on March 2, 1981, some three years before the MJ-12 documents that included the Eisenhower Briefing and the Truman memo arrived at Shandera's home. It resets the clock on these matters and suggests that Moore, Shandera and Doty all had seen a reference to MJ-12 in 1981. And it is something that has now disappeared from the discussions of the MJ-12 documents.

---

[561] Letter in the files of Kevin Randle.

[562] Ibid. See also, Robert Hastings, "The MJ-12 Affair: facts, Questions, Comments," *The MUFON Journal* No. 254, June 1989, p. 8.

Project Aquarius (MAJIK 12) – The Novel

Bob Pratt had been on the *National Enquirer* UFO beat for a number of years. He was a careful researcher and though he had once worked for a grocery store tabloid, he was careful about what he published, verifying the information before writing.[563] His expertise about UFOs ranged far beyond the six years he worked for the *National Enquirer*, and saw him author books with J. Allen Hynek, and with Brazilian UFO researchers. Pratt was respected throughout the UFO field.

According to Pratt in a letter to UFO researcher Robert Todd, "The original idea behind the book was Project Aquarius. In January 1982, I happened to be in Houston and flew out to Phoenix to visit Bill at his request. He wanted to talk to me about something he couldn't discuss on the phone… I sat in a chair and took notes as he told me about Project Aquarius, MJ-12 and a number of other things."[564]

Pratt wrote, "My working title was MAJIK 12, but I wanted to call it I. A. C., for identified alien craft, supposedly the name government insiders use for UFOs. However, when I got the finished manuscript back from Bill, he had put The Aquarius Project on the title page."

Ironically, Pratt wrote, "When this MJ-12 business broke in 1987 or whenever, I wrote to Bill saying that if there was anything to it, we ought to dust off our manuscript and try to sell it again. He never answered."[565]

---

[563] Richard Donlan, *UFOs and the National Security State: The Cover Up Exposed*, 2009, p. 146; Pratt had also served as the editor of the *MUFON UFO Journal*. See also: http://www.mufon.com/bob_pratt/obit.html.

[564] Letter from Pratt to Robert Todd, February 20, 1989, copy in Randle's files. See also, Sparks, "The Secret Pratt Tapes and the Origins of MJ-12," *MUFON UFO Journal*, September 2007, p 6.

[565] Ibid.

Pratt described, in some detail for Todd, the plot of the book, which was going to be a novel because they couldn't prove Moore's claims about the reality of the information. Pratt wrote:

> Our "hero" was an AFOSI agent who scoffed at UFO reports, etc., but comes to believe because he has to investigate what you [Todd] call the "infamous Ellsworth AFB incident"[566] and later has his own UFO encounter. Because he realizes UFOs are real – knowledge that most AFOSI agents don't have – he is assigned to keep tabs on people and things in the UFO world from his new posting at Kirtland. To give him a thorough understandings of the phenomenon, the Air Force sends him to Bolling Field in Washington, DC, where he is given access to a number of secret UFO files, and it is there he first learns about Project Aquarius.
>
> Hero goes back to Kirtland and in carrying out his job he does a number on a number of people, making sure some people, including ex-military types, don't talk about their experiences, masterminding disinformation plots, and so on. Somewhere in doing all of his dirty deeds, the hero's conscience takes over and he finds himself rebelling against official policies ("the people have a right to know…"). He winds up a dead hero, his body shipped off to planets unknown on a UFO operated by aliens in cahoots with the government.[567]

---

[566] This is a reference to a carbon copy of an alleged official Air Force incident report that was sent to the *National Enquirer* and eventually to Bob Pratt about an event at Ellsworth AFB, South Dakota. Pratt noted in his letter to Todd, "As for the Ellsworth incident, it never happened (although I understand Doty claims it really did – which I take to be more disinformation on somebody's part). See also, Bob Pratt, "Truth about the Ellsworth Affair," *MUFON UFO Journal*, January 1984, pp. 6 – 9

[567] Pratt Letter, February 20, 1989

All of this sounds suspiciously like that tales that Moore would be telling over the next few years as he chased the insider information he was promised for his cooperation with the AFOSI. But Project Aquarius was about to move from fiction into nonfiction. In early 1983, Moore would have an opportunity to photograph documents that were consistent with Aquarius. In April 1983, Linda Moulton Howe, a documentary film maker and UFO researcher would see something similar to that shown to Moore, which she would attempt to authenticate.

Project Aquarius – The Presidential Briefing

Bill Moore would say that in early March, 1983, he received a telephone call telling him that he would have an opportunity to see some important information, but that he would have to travel to get it. He was told that he would receive instructions and that he had to follow them carefully or there would be no revelation.[568]

According to Moore, he made his way across the United States until he ended up in a motel in up-state New York. At 5:00 p.m., a person, Moore didn't reveal the gender of the "courier," came to his door and handed him a manila envelope. He was given nineteen minutes to examine the contents. He could do anything he wanted with them, but at the end of the time he was required to return the material to the person who had brought it to him.[569]

There were eleven pages in the envelope and the document was labeled "Executive Briefing, Subject: Project Aquarius." It was dated July 14, 1977. Moore took photographs of the pages, using a quarter at the bottom of each page to provide perspective. When he finished that, he began to read the document into his tape recorder, saying "line" at the end of each line, and verbally

[568] Moore and Shandera, *MJ-12 Analytical Report*, p. 31; Randle, *Case MJ-12*, pp 205 – 207.

[569] Ibid.

noting each punctuation symbol.[570] When the time was up, the courier collected all the pages, counted them, stuffed them back into the envelope, and left.[571]

Moore reported, "The documents themselves seem to be transcription of notes either intended for use in preparing a briefing, or taken down during one and typed later. Much of the information therein is highly controversial in nature and is of such esoteric quality there seems to be no real way to verify much of it."[572]

Portions of the documents have been released and have appeared in a number of books and articles.[573] These have been identified as the Carter Briefing and suggests that Aquarius Project began in 1953 "by order of President Eisenhower, and was under the control of MJ-12. … The purpose of Project Aquarius was to collect all scientific, technological, medical and intelligence information from UFO/IAC sightings and contact with alien life forms. This orderly file of collected information has been used to advance the United States Space Program."[574]

Within weeks, Linda Moulton Howe, who had a contract with HBO to produce a documentary on UFOs, would have a chance to review the handwritten notes of what was clearly from the same the Carter Briefing as had

---

[570] Moore and Shandera, *MJ-12 Analytical Report*, p. 31

[571] This is a ridiculous scenario. It is more appropriate in a movie than in real life. Moore was left with nothing that could be verified as authentic and he had no provenance for the documents because he didn't know who the courier was or where the documents originated. In other cases of leaked classified documents, there is enough information available for an independent review of the authenticity. With MJ-12, the trail stops before any of the insiders are identified and the documents authenticated.

[572] Moore and Shandera, *MJ-12 Analytical Report*, pp. 31 – 32.

[573] See, for example, Steinman and Stevens, *UFO Crash at Aztec*, pp. 612 – 618.

[574] Quote from both Moore and Shandera's released documents, and from example of documents circulated in the UFO field, including a representation in Steinman and Stevens, *UFO Crash at Aztec*, pp. 616 – 618. Also see *Just Cause*, June 1987, p. 3

been shown to Moore.[575] While in New York to sign the HBO contract, Howe met with Gersten and Patrick Huyghe, a science writer and publisher of *The Anomalist*, for dinner. Gersten mentioned he knew of an AFOSI special agent named Doty who claimed some very interesting things about UFOs and said that Doty might say some of them on camera if Howe wanted to interview him. Gersten said that he would attempt to set up a meeting between Howe and Doty.[576]

Howe and Doty met on April 9, 1983 in Albuquerque and they drove out to Kirtland AFB, to a small building that Doty described as his boss' office. Inside, he asked her to sit in a specific chair because "eyes can see through windows."[577]

He then said to her, "My superiors have asked me to show you this." He took an envelope from a drawer, opened it, and handed Howe several sheets of paper. He told her that she couldn't copy them, she couldn't take notes, but she could read them.

The document was called, "A Briefing for the President of the United States on the Subject of Unidentified Flying Vehicles." There was no date on the report, and the president wasn't named. Howe didn't know for which president it had been prepared, though later information from others, would suggest that this was part of the Carter Briefing.[578]

The documents read by Howe confirmed the Roswell UFO crash in 1947[579] and it suggested that one of the extraterrestrial crew had survived to

---

[575] Clark, *UFO Encyclopedia*, pp. 307 – 310; Randle *Case MJ-12*, pp. 208 – 212, Linda Moulton Howe, letters dated May 10, 1983, November 15, 1987, March 16, 1988; Robert Hastings, "The MJ-12 Affair: Facts, Questions, Comments," March 1, 1989

[576] Clark, *UFO Encyclopedia*, pp. 307 – 310.

[577] Randle, *Case MJ-12*, p. 209; Timothy Good, *Alien Contact, Top-Secret UFO Files Revealed*, 1993, p. 114 – 115; Vallee, *Revelations*, pp. 41 - 43

[578] Clark, *UFO Encyclopedia*, pp. 307 -310; Dolan, *Cover-Up Exposed*, 299 - 309.

[579] This doesn't mean there is documentation for the Roswell crash because it is clear that the document is fraudulent. Roswell was mentioned with other crashes known to

be taken to a "safe house" at Los Alamos National Laboratory. According to this document, the alien bodies were those of small, gray humanoids about three and a half to four feet tall, with oversized heads, large eyes and no nose. The survivor was befriended by an Air Force officer and the creature died on June 18, 1952.[580]

Howe, in her descriptions of what she saw, mentioned other projects such as Snowbird,[581] Sigma,[582] and Aquarius, which were the same projects that Moore had mentioned in his partial release of information later, and, of course, were part of the documents that have since been printed.[583]

There was also a mention of Project Garnet which had been created to investigate alien influences on human evolution. Unnamed government sources suggested that the "grays," aliens from a nearby solar system had manipulated the DNA of primates at various intervals including 25,000; 15,000; 5,000 and 2500 years ago. It was also claimed that the aliens had created an

---

be hoaxes including the December 6, 1950 El Indio – Guerrero crash, the Aztec crash of March, 1948 and the Kingman, Arizona crash of May, 1953.

[580] Clark, *UFO Encyclopedia.* p. 308; Randle, *Case MJ-12*, pp. 207 – 209; Vallee, *Revelations*, pp. 41 -43; See also, *UFO Universe*, July 1998, p. 21.

[581] Snowbird was an actual project and has been described by Barry Greenwood as a "Joint Army/Air Force peacetime military exercise in the sub-arctic region in 1955, according to Gale Research's *Code Name Dictionary*, 1963." See *MUFON UFO Journal* Number 236, December 1987, p. 12

[582] Project Sigma, according to Greenwood in *MUFON UFO Journal* Number 236, December 1987, p. 12, had nothing to do with UFOs, but was a top secret project, by the Air Force in conjunction with Rockwell International.

[583] Moore and Shandera, *MJ-12 Analytical Report*, pp 32 – 34, Steinman and Stevens, *UFO Crash at Aztec*, pp. 612 – 618.

individual some two thousand years ago to teach humanity about love and non-violence. In other words, according to these documents, Jesus Christ had been created by the aliens.[584]

According to the documents seen by Howe, Project Blue Book had been a public relations operation that was supposed to divert attention from the real investigative projects. Doty, in his conversations with Howe at that time, mentioned MJ-12 but suggested that "MJ" stood for "Majority" rather than "Majestic." Whatever the real name, it was a committee of twelve high-ranking government officials, military leaders, and top scientists who set policy for the exploitation, cover-up, and dissemination of UFO information and the government interest in UFOs. They were also responsible for the disinformation about UFOs that was circulating through the media and the UFO community.[585]

All of this could be seen as corroboration for Moore's tale of his clandestine meeting with the stranger in a hotel room in New York. If Moore had been inventing a tale, complete with a briefing in an envelope, surely Howe wouldn't invent a similar tale. It is clear from the timing of events that Howe didn't know about Moore's meeting or what Moore had learned. Moore's meeting had taken place only a couple of weeks earlier, at least, according to what Moore had reported.

But there is a link. Moore, of course, knew Doty. They had been working together for some time. Although Moore claimed that Falcon was an Air Force colonel, he said that Doty, known as sparrow, was his go-between to the colonel. Doty, as Ronald L. Davis, was a silent partner in the book that Moore was writing with Bob Pratt about the AFOSI agent and UFOs. Doty was mentioned by Pratt as a silent third partner, so that even if Doty wasn't Davis, there is a connection.[586] Much of what Doty was giving to Howe was part of the novel

---

[584] Howe, Linda Moulton, *An Alien Harvest: Further Evidence Linking Animal Mutilations and Human Abductions to Alien Life Forms*, 1989. Also mentioned in November 15, 1987 letter from Howe.

[585] Vallee, *Revelations*, 63 – 75; Randle, *Case MJ-12*, p. 211

[586] Letter from Bob Pratt to Robert Todd, February 20, 1989, p. 1

about Project Aquarius. Since both Moore and Howe knew Doty, that is the connection that suggests the information wasn't corroborative, but simply the same information from the same source given to two different people.

All of this suggests that the Carter Briefing as shown to Moore, the presidential briefing shown to Howe, and Project Aquarius, is nothing more than a hoax. Some have argued that it was all part of a government disinformation scheme to lead UFO investigators away from the truth.[587]

That was where everything stood for a long time. Howe, promised many things for her documentary, was unable to obtain any of them. Inside the UFO community there were was little talk of MJ-12, and attempts to prove this with documentation, such as the Aquarius Telex, failed to gain any sort of public release.

But there were many indications that someone knew about it long before the Eisenhower Briefing Document arrived. As early as 1981, there were mentions of both Aquarius and MJ-12. It wouldn't be until the end of 1984 that the critical document allegedly arrived, and it would be some three years before the documents themselves would be released, though rumors would circulate as early as 1983.

The Eisenhower Briefing Document

On December 11, 1984, Shandera, whose association with UFOs was through Bill Moore, received a taped, manila envelope that had no return address, but was post marked in Albuquerque on December 8.[588] Inside that envelope was

---

[587] For a comprehensive analysis of the disinformation theory, see Sparks, "The Secret Pratt Tapes and the Origins of MJ-12, *MUFON UFO Journal*, September 2007, pp. 3 – 10, and *MUFON UFO Journal*, October 2007, pp. 3 – 11.

[588] Moore and Shandera, *MJ-12 Analytical Report*, p. 43; Timothy Good, *Above Top Secret*, 1988, p. 257; Good, *Alien Contact Top-Secret Files Revealed*, 1993, p. 124; John Alexander, *UFOs Myths, Conspiracies, and Realities*, 2011, pp. 122 – 123, Clark, *UFO Encyclopedia*, p. 310; Vallee, *Revelations*, pp. 38 - 39.

a second that was heavily taped, and when that was opened there was a third envelope with the Marriott Hotel logo on it. Inside that was a canister of undeveloped Kodak Tri-X, black and white, 35 mm film.[589] When developed, the film showed eight pages of a document that would become known as the Eisenhower Briefing Document and a one page memo on White House stationery, now known as the Truman memo. All the documents were duplicated a second time and the rest of the roll was blank.[590] They took the negatives to a photo lab to be printed and both Moore and Shandera were there, in the darkroom when the prints were made.[591]

There is a second version of this story. As Howard Blum told it, Shandera was at home shortly after noon, when he heard the mail arrive. He found the package, as described, and opened the three envelopes until he came to the plastic container that held the film cartridge. Since he was supposed to meet Moore for lunch, he took the canister with him. When he showed it to Moore, they left the restaurant without eating.[592]

"Bill Moore used a kitchen glass to measure the developer," according to Blum. "He [Moore] didn't have much photographic flair, but when the two friends left the restaurant, it was agreed that it would be quicker – and more secure – if he developed the film himself."[593]

---

[589] This is based on their displaying of a mailing envelope with a December 1984 date on it. There is no way of proving that envelope actually contained the film. There is nothing to tie it to the film and the EBD.

[590] Stan Friedman, "Final Report on Operation Majestic 12", Fund for UFO Research, April 1990, p. 2, Friedman, "MJ-12: The Evidence so far, *International UFO Reporter*, September/October 1987, pp. 13 – 20.

[591] Moore and Shandera, *MJ-12 Analytical Report*, p. 44. It should be noted that this makes no logical sense. The danger is in developing the negatives. If an error is made, there is no way to recover. If a print is ruined, another can be made from the negative. It would be more logical to have the film developed professionally and then make the prints at home.

[592] Howard Blum, *Out There*, 1990, pp. 239 -241.

[593] Ibid., p. 241

Moore, with Shandera's help, then made contact sheets for the negatives.[594] Apparently they didn't know how long to expose the photographic paper or how long to leave it in the fixer so that they had to make guesses. Once they had washed them, they took them into another room and using clothes pins, hung them up to dry.[595]

After they had examined what they had, they called Stan Friedman to tell him. They all agreed that they couldn't release the documents immediately because they didn't know if they were legitimate. They would have to search for confirmation, and that was something that could take years.[596]

Moore, it seems might have had some advance knowledge of the Eisenhower Briefing Document. Moore wrote that:

> "More recently (1984) a confidential informant who is still in government employ, has stated that there were four bodies recovered at Roswell, but that all were badly decomposed and had been attacked by predators before their discovery. According to this source, who claims to have seen the bodies, the occupants of the craft apparently ejected just before it exploded, and were killed when they struck the ground. (The assumption being that the ejection apparatus failed to operate properly.) There were discovered by aerial reconnaissance several miles southeast of the Brazel ranch crash site, and

---

594 Friedman, in a letter to Joe Nickell on October 9, 1987, wrote, "To the best of my knowledge there is no set of contact prints anywhere. What purpose?"

595 Ibid. p.241

596 Friedman, *Top Secret/Majic*, p. 56

> were recovered in a somewhat separate operation. This individual has also stated that the propulsion and control systems of the craft were almost totally destroyed by the explosion."[597]

The MJ-12 document, which would allegedly arrive at the end of 1984, said, "During the course of this operation, aerial reconnaissance discovered that four small human-like beings had apparently ejected from the craft at some point before it exploded. These had fallen to earth about two miles east of the wreckage site. All four were dead and badly decomposed due to action by predators and exposure to the elements during the approximately one week time period which had elapsed before their discovery."[598]

Although Moore refused to name the source for the revelation about the four alien creatures killed, it is clear that he had the information prior to the receipt of the Eisenhower Briefing document. The two paragraphs, while not an exact match, contain the same information.

There is additional information about this. At the UFO Expo West in Los Angeles on May 11, 1991, Jaime Shandera was lecturing about the situation concerning crashed saucers in general and about the Plains of San Agustin specifically which suggested something important about the sequence of events leading up to the receipt of the EBD. He had this to say:

> The people that supposedly found stuff in Socorro did not find stuff in Socorro. The party of archaeological people and the Barney Barnett part of the story; they were at the Corona site, not in Socorro [Plains of San Agustin]. I know [this is] the way you understand it because it's the way it's always been

---

[597] Bill Moore, "Crashed Saucers: Evidence in the Search for Proof," in Crashed UFOs: Evidence in the Search for Proof, William Moore Publications; June 1985, June 1986, June 1987, pp. 50

[598] From the Eisenhower Briefing document, p. 1

> written and even the way it was written in *The Roswell Incident.* That's wrong. There is new evidence that it was all in the Corona site. The way it happened was this – there were not two sites that were more than one hundred miles or so apart ... and the so-called Roswell site was just outside of Corona. The archaeologists and Barney Barnett part of it, that was over in Corona. There was no person that found anything in San Agustin.[599]

On that same day in Los Angeles, that is May 11, 1991, Antonio Huneeus and Javier Sierra interviewed Bill Moore about some of the things that Shandera had said earlier. Moore mentioned the Gerald Anderson tale and why he did not accept it as authentic. He confirmed that he was on board with Shandera about the Plains, saying, "There is no reason to believe anything occurred on the Plains of San Agustin on that particular date..."

Moore then said something that is quite revelatory. He said, "The original hypothesis was that the object had come down in two places, the first being the Brazel site, the second being the Plains of San Agustin, and that in 1985 I abandoned [it] simply because the only witness who put the thing in the Plains of San Agustin at all was Barnett's boss, Danley, [who] it turned out, was not sure of the place, and it turned out that Barnett could have been up at the Brazel site..."[600]

The timing is such that, based on what Moore claimed, he had the information months before the EBD arrived. This was information that most others

---

[599] Shandera, Jaime, UFO Expo West, notes from Antonio Huneeus and Javier Sierra. Copy in Author's files.

[600] Moore's statement that he had rejected the idea of a Plains of San Agustin crash in 1985 is interesting. He said he has rejected it because Danley couldn't actually provide a location or date for Barnett's story. When interviewed in October, 1990, Danley was unable to verify the time, date or location.

inside the UFO community didn't have. It suggests once again that the information in the EBD was the state of UFO research in the mid-1980s, and not what it was in 1952 when the document was allegedly created.

Moore began sharing information that came through the EBD inside a small circle of UFO researchers. According to Barry Greenwood, just a year after Shandera had received the undeveloped film, Moore was told by Lee Graham he had heard about some of it, and Graham began to make inquiries into MJ-12, which had been mentioned on the Aquarius Telex.[601]

Greenwood noted that as the final stages were completed on *Clear Intent*, the book he wrote with Larry Fawcett, they had collected documents relating to UFO sightings around Kirtland AFB. Greenwood wrote:

> What wasn't inserted in to the book was a document, allegedly a government document, which had made the rounds in the so-called UFO grapevine. It told of an analysis performed by the Air Force of the "Dr. Bennewitz" photographic data, which had been part of the subject matter in the confirmed AFOSI file release on Kirtland. It concluded that some of the photos were "Legitimate negatives of unidentified aerial objects". The document went on to discuss, very briefly, a top-secret "Project Aquarius" and something called "MJ12". These terms meant little to us at the time so inquiries were launched to determine the origin of the terms and the document.[602]

[601] Greenwood, "Just Cause," December 1985, p. 2

[602] At last, at least according to the various accounts of the Aquarius Telex and the mention of MJ-12, someone thought enough of it to make some inquiries. Prior to Greenwood pointing it out in 1985, it seemed that everything revolved around the authenticity of the document and no one seemed to see the other avenue for investigation, that is, to find out if there was a Project Aquarius and an MJ-12.

> No government agency had any knowledge of the matter and the alleged originator, AFOSI (Air Force Office of Special Investigations), said that the document was a "fabrication". Since we had no evidence to prove otherwise and since the document came from a source other than the FOIA I chose not to use it in CI [*Clear Intent*]. My copy of the "Aquarius" document contained several misspellings and irregularities in format, further adding to my suspicions but I did not rule out the possibility of this perhaps being a retyped version of the original, legitimate document by a "mole" in the military. The piece was filed and remained in limbo for a time.[603]

Greenwood also mentioned that during the summer of 1985, Graham was also doing some research into MJ-12. Greenwood, "... California UFO researcher Lee Graham was conducting inquiries into MJ12, apparently based on specific information from an unknown individual in the government."[604]

Friedman later identified Graham's source as Moore. He wrote, "Graham... claimed he had been given *Just Cause* the MJ-12 list, which had been shown by a source in the military. He had not been allowed to copy the list but he had taken notes. (As far as I know, the source was actually Bill Moore, who was not in the military and does not work for the government, but has a fondness for playing games). As a joke Bill once pulled out a MUFON identification card, flashed it at Lee, and indicated that he was working for the government. Lee bought it."[605]

It wasn't until 1987 that the Eisenhower Briefing Document began to circulate throughout the UFO community. In the spring of 1987, Moore and

---

603 Greenwood, "Just Cause," December 1985, p. 2

604 Ibid.

605 Friedman, *Top Secret/Majic*, pp. 56 – 57

Shandera learned that the document was about to be published in Europe.[606] According to Friedman, one of the insiders who warned Moore was Doty. On May 31, 1987, Martin Bailey, wrote an article, "Close Encounters of an Alien Kind – And Now if You've Read Enough About the Election, Here's News from Another World."[607] It showed some of the Eisenhower Briefing Document.

When Moore learned that the documents had been released in Europe, with the story having been picked up by Reuters and then published in North American newspapers, he reacted. Using *Focus*, his publication of the Fair Witness Project, he published the documents, but these were heavily censored, including black markings through the security classifications on each page. He would later suggest that the markings were made based on instructions from his inside contacts.[608]

Apparently it was British author and UFO researcher Timothy Good who was responsible for the leak. He showed a copy of the Eisenhower Briefing Document to the press which resulted in the *Observer* article.[609] He wrote, "My copy of the document was sent to me by an intelligence source in the United States in March 1987, specifically for inclusion in *Above Top Secret*. I published the document in the conviction that it was probably authentic."[610]

Good clarified this in his book, *The UFO Report – 1990*. He said that the source was CIA.[611] That would mean that Good had received the documents

---

606 Ibid.

607 Martin Bailey, "Close Encounters of an Alien Kind – And Now if You've Read Enough About the Election, Here's News from Another World," *London Observer*, May 3, 1987, p. 1

608 Friedman, *Top Secret/Majic*, pp. 58 – 59

609 Clark, *UFO Encyclopedia*, p. 310.

610 Good, *Alien Contact*, p. 123

611 Barry Greenwood, "On Timothy Good's Source of the MJ-12 Briefing Paper," dated February 18, 1990; see also, Timothy Good, *The UFO Report – 1990*, 1990.

from someone other than Moore, and that there were two sources leaking material into the civilian world.

Barry Greenwood, however, believed that the source of Good's copy was Bill Moore. Greenwood noticed what he called a "telltale fingerprint" on the copies of the Eisenhower Briefing Document. These are markings left by a dirty copy machine when it duplicates documents. The machine Moore used left a characteristic smudge on the page and since the smudge "floated" it can be deduced that it was an artifact of the machine and not something on the original copy. Greenwood identified the smudge on the copies used by Good and that meant the Good's copies came from Moore… or someone who used the same machine as Moore.[612]

The problem then, was that there was no provenance for the Eisenhower Briefing Document. It had arrived, anonymously in the mail with only a postmark to indicate the source. Because it arrived on film, rather than as direct copies, or even original documents, there wasn't much in the way of forensic evidence available. Kodak film was ubiquitous, and had nothing distinguishing about it. The wrappings, meaning the sealing of all the flaps on the envelopes was typical of the way classified information was sent through the mail, but provided no hint as to who had sent it. Without a provenance, any attempt to authentic it was greatly reduced. Without a live source to be questioned, without original documents, with nothing but the photographs of the documents, there was nothing to be learned.

Friedman thought that the lack of provenance was not a big issue. On February 13, 2001, he wrote, "Lack of provenance is bothersome, but understandable. Whoever filmed the EBD and/or planted the CT [Cutler-Twining memo] was violating security by the filming and the release. Having a classified document is not against the law. Being an authorized recipient who leaks it to

[612] Letter from Barry Greenwood to Kevin Randle, 1993, including copies of Eisenhower Briefing Document. Greenwood explained that the "large format copies are from Moore's Apr 87 *Focus* and the small format from Good's book."

uncleared personnel is very much a violation. One might suggest that the lack of provenance is an indication of genuineness."[613]

John Alexander, a retired Army colonel who worked with classified material and on classified projects disagreed with this assessment. He wrote:

> Begin with the source. As far as is publically available today, the source of the documents remains unknown to everyone, including the people who released them. Rather than meeting under a bridge late at night, the perpetrator sent an anonymous envelope, not even indicating what the film related to, even the general topic. There have been no face-to-face meetings between the reporters [Moore, Shandera and Friedman] and their confidential source, nor was there any way to vet him or her.[614]

This lack of provenance might have led directly to the next document in the MJ-12 mystery. This was the Cutler-Twining memo that had been found in a box of declassified documents held by the National Archives. That would be an attempt to end one avenue of discussion.

A final problem is that these appear to be classified documents and as such are subject to restrictions on who should be able to access them and who is authorized to see them. In September 1988, the FBI received, from the Air Force Office of Special Investigation, an alert that classified material might have been compromised. In the text of the FBI message, it was noted that over the six weeks prior, there had been publicity in the Dallas regarding "Operation

---

613 Randle, *Case MJ-12*, pp. 170 – 171; Letter from Friedman to Randle, February 13, 2001.

614 Alexander, *UFOs: Myths*, pp. 125 – 126

Majestic – 12." The Special Agent wrote, "It is unknown if this is all part of a publicity campaign."[615]

The concern seemed to be that they had a document that still was classified. They noted that it was over thirty-five years old, but they didn't know if it had been properly declassified. The Bureau had been requested to "discern if the document is still classified."

In an FBI message dated December 2, 1988, labeled as "Possible disclosure of classified information regarding Operation Majestic – 12," they were advised by the AFOSI the document was fabricated. They wrote, "Document is completely bogus." At that point, the FBI investigation ended.

The Cutler-Twining Memo

The one document that seems to have a provenance is an onion skin copy of a short letter that has become known as the Cutler-Twining memo. What appeared to be the corroboration for the MJ-12 committee was found by Shandera and Moore at the National Archives as they reviewed Record Group 341, which had just been declassified.[616] Located, between two file folders, in a dusty box of recently reviewed material was a memo that mentioned a rescheduling of an MJ-12 briefing. It was a memo for General Nathan F. Twining, then Air Force Chief of Staff, from Robert Cutler, the Special Assistant to the President. Dated July 14, 1954, it established the existence of the MJ-12 committee and that the committee was still active in the mid-1950s.

---

[615] Copies of the FBI documents were accessed on January 1, 2015 and are available at for viewing at: http://vault.fbi.gov/Majestic%2012/Majestic%2012%20Part%201%20of%201/view

[616] Moore and Shandera, *MJ-12 Analytical Report*, pp. 92 -94; Friedman and Berliner, *Crash at Corona*, pp. 68 – 69; Alexander, *UFOs: Myths*, pp. 127 – 128; Friedman, *Top Secret/Majic*, pp. 86 – 102.

According to the story as told today, in July 1985, Moore and Shandera flew to Washington to review records that had recently been declassified. They were searching in what was labeled as Records Group 341 which are Headquarters Air Force Intelligence files. They were working in Entry 267 and had reviewed some 120 boxes of papers. While putting one file folder back into the box and pulling out the next, Shandera found a piece of paper between the files. This was the Cutler/Twining memo.[617]

It was less than a page long and was labeled as a "Memorandum for General Twining. The subject was the NSC.MJ-12 Special Studies Project. The text said:

> The President has decided that the MJ-12 SSP briefing should take place during the already scheduled White House meeting of July 16, rather than following it as previously intended. More precise arrangements will be explained to you upon arrival. Please alter your plans accordingly.
>
> Your concurrence in the above change of arrangements is assumed.[618]

It was written for Robert Cutler who was a Special Assistant to the President to cover circumstances while he was out of the country. It was not signed which is somewhat problematic. Without a signature, it does little to help authenticate MJ-12.

According to Friedman, the onionskin paper had the proper watermark, the carbon was blue ink and the structure of the memo, or rather the last line,

---

[617] Friedman, *Top Secret/Majic*, pp. 86 – 102

[618] Copies of the memo have been printed in several books, including Moore and Shandera, *MJ-12 Analytical Report*, Appendix I; Friedman, *Top Secret Majic*, p. 87 – 88

was consistent with others prepared by Cutler or his staff. In other words, there was nothing to suggest that it was fraudulent, at least on the surface.[619]

But that isn't the whole story of the Cutler/Twining memo. Given this is MJ-12, there had to be some sort of foreign intrigue involved. Friedman made it clear that in March 1985, or about four months before Moore and Shandera made their search, he had been at the Modern Military Branch of the National Archives and talked with Ed Reese. His discussions with Reese, and the possibility that some UFO related material might be in those documents now being reviewed for declassification and release into the public area, lead to Jo Ann Williamson, who thought the review Entry 267 from Air Force Records Group 341 would be completed in about a month.[620]

While that was going on, Moore and Shandera received a number of strange postcards. One with a postmark of New Zealand had a return address of Box 189, Addis Ababa, Ethiopia.[621] Another mentioned Reese's Pieces and Suitland, which was a name of an annex to the National Archives.[622] These "clues" meant nothing to Friedman or Moore until Shandera found the memo in Box 189 of Record Group 341. These postcards seemed to be pointing the way to the memo and would have made the search easier had anyone been able to decode them.

This would seem to have ended any discussion about the lack of provenance because the document was found at the National Archives in material that was in their collection. It is the only document ever located in an independent area that made a reference to MJ-12. Every other document seemed

---

[619] Friedman, *Top Secret/Majic*, p. 91; Friedman, Final Report, pp. 24 – 27.

[620] Friedman, *Top Secret/Majic*, pp. 88 – 89.

[621] This is more of the "movie" intrigue that seems to fascinate Moore. To point him in the right direction, those insiders attempting to leak information arrange for postcards to be sent from exotic locations with obscure references that mean nothing until later. According to Friedman, he has never seen the postcards but mentions that two others have. He fails to name them in published sources.

[622] Moore and Shandera, MJ-12 Analytical Report, pp. 92 - 93

to have arrived in the mail of various UFO researchers without any provenance for it. The problem, however, is that even though the Cutler/Twining memo was found in a box of recently declassified material, and it was held by the National Archives, it didn't belong where it was found. There is no record of that memo being in the box prior to Moore and Shandera finding it and though the listings on the boxes are not comprehensive, they are complete.

Reese, who had been talking to Friedman, said, in a letter to the late UFO researcher Robert Todd:

> In none of these reviews was the Cutler-Twining memorandum identified as present and requiring any special attention. But the declassification guidelines used by both the Air Force and the National Archives would not have permitted them to declassify National Security Council documents. If discovered in the files during any of these reviews such documents [sic] would have been withdrawn and provided to a National Security Council declassification specialist for final determination. It was never so identified.[623]

The purpose of such a review was to ensure that no document that was properly classified was released to researchers, historians, or journalists and in this particular case, UFO investigators. The Air Force only had a general idea what was in each box and that meant that an officer had to review every sheet of paper in that box. In such a review, the memo would have been found and removed for the special handling required.

Questions about the Cutler/Twining memo were regularly sent to the Archives. On May 9, 1988, Jo Ann Williamson, the Chief of the Military Reference Branch produced a letter with the subject of "Reference Report on MJ-12 (Revised). She wrote:

---

[623] Robert Todd, "MJ-12 Rebuttal," *MUFON UFO Journal*, No. 261, January 1990, p. 17

1. The document was located in Record Group 341, entry 267. The series is filed by a Top Secret register number. This document does not bear such a number.
2. The document is filed in the folder T4-1846. There are no other documents in the folder regarding "NSC/MJ-12."
3. The Military Reference Branch (Edward Reese) has conducted a search in the records of the Secretary of Defense, the Joint Chiefs of Staff, Headquarters U.S. Air Force, and other related files. No further information has been found on this subject.
4. Inquiries to the U.S. Air Force, the Joint Chiefs of Staff, and the National Security Council failed to produce further information.
5. The Acting Director of the Freedom of Information Office of the National Security Council informed us that "Top Secret Restricted Information" is a marking which did not come into use at the National Security Council until the Nixon Administration. The Eisenhower Presidential Library also confirms that this particular marking was not used during the Eisenhower Administration.
6. The document in question does not bear an official government letterhead or watermark. The NARA conservation specialist (Mary Ritzenthaler) examined the paper and determined it was a ribbon copy prepared on "diction onionskin." The Eisenhower Library has examined a <u>representative sample of the documents</u> in its collection of the Cutler papers. All documents <u>in the sample</u> created by Mr. Cutler while he served on the NSC staff have an eagle watermark in the <u>bond paper. The onionskin carbon copies have either an eagle or no watermark at all.</u> Most documents sent out by the NSC were prepared on White House letterhead paper. For the brief period when Mr.

Cutler left the NSC, his carbon copies were prepared on "prestige onionskin."

7. The Judicial, Fiscal and Social Branch searched the Official Meeting Minute Files of the National Security Council and found no record of a NSC meeting on July 16, 1954. A search of all NSC Meeting Minutes for July 1954 found no mention of MJ-12 or Majestic.
8. The Judicial, Fiscal and Social Branch (Mary Ronan) searched the indices for the NSC records and found no listing for: MJ-12, Majestic, unidentified flying objects, UFO, flying saucers, or flying discs.
9. The Judicial, Fiscal and Social Branch (Mary Ronan) found a memo in a folder titled "Special Meeting July 16, 1956" which indicated that NSC members would be called to a civil defense exercise on July 16, 1956.
10. The Eisenhower Library states, in a letter to the Military References Branch, dated July 16, 1987:

"President Eisenhower's Appointment Books contain no entry for a special meeting on July 16, 1954 which might have included a briefing on MJ-12. Even when the President had 'off the record' meetings, the Appointment Books contained entries indicating the time of the meeting and the participants…

The Declassification Office of the National Security Council has informed us that it has no record of any declassification action having been taken on this memorandum or any other documents on this alleged project…

Robert Cutler, at the direction of President Eisenhower, was visiting Overseas military installations on the day he supposedly issued this memorandum --- July 14, 1954.

> The Administration Series in Eisenhower's Papers as President contains Cutler's memorandum and report to the President upon his return from the trip. The memorandum is dated July 20, 1954 and refers to Cutler's visits to installations in Europe and North Africa between July 3 and 15. Also, within the NSC Staff Papers is a memorandum dated July 3, 1954, from Cutler to his two subordinates, James S. Lay and J. Patrick Coyne, explaining how they should handle NSC administrative matters during his absence; one would assume that if the memorandum to Twining were genuine, Lay or Coyne would have signed it."

When certifying a document under the seal of the National Archives we attest that the reproduction is a true copy of a document in our custody. We do not authenticate documents or the information contained in a document (underlining in original.)[624]

Given all this, and the searches made, it is clear that the Cutler/Twining memo had been planted in the Archives, probably to provide a provenance for it and the other MJ-12 documents.[625] There was nothing found in any of the

---

[624] Jo Ann Williamson, For the Record Memo dated May 9, 1988 about "Reference Report on MJ-12 (revised). Reproduced here as it was sent to researchers with the same underlining and emphasis as the original. See also "National Archives MJ-12 Response," *MUFON UFO Journal*, September 1987, pp. 17 -18.

[625] Even Friedman believes that the Cutler/Twining memo was planted in the National Archives. In *his Final Report on Operation Majestic Twelve*, he wrote on page 56, "Whoever planted it at the National Archives…" Moore and Shandera, in their Analytical report, wrote on page 103, "All things considered, we conclude a 95% probability that the document is genuine, and that it was deliberately planted in the National Archives by person or persons unknown who then systematically undertook to be sure that we would discover it there." See also, Dolan, *Cover-Up*, p. 405

searches that would suggest the document was legitimate. It had been smuggled into the Archives and it was claimed that it was found in the specific location mentioned.

Friedman suggested that such activity was impossible. There are rigid inspections of briefcases, files, notebooks and other items. In fact, he suggested that all such items had to be secured in lockers prior to entering the various research rooms. He wrote:

> Archives not only don't let you bring any original documents out, they won't allow you to bring anything into secure areas. All of your belongings – briefcases, notebooks, jackets, even scrap paper – are placed in a locker outside the secure areas. If you require notepaper while you're working, the clerks give you some – from inside.[626]

Except, of course, the security going in is not quite that strict and notepads and legal pads are allowed in. They might flip through them, but they do allow them in. Coming out, the rules are tighter, especially after it was discovered that some researchers in other historical areas were carrying out documents a couple of pages at a time.[627]

Friedman, Moore and Shandera, in their analysis, conclude that the document is authentic, but that it was planted where it didn't belong for Moore and Shandera to find. They suggest it might be disinformation, but that it is disinformation with a nefarious purpose behind it. In fact, Friedman in a letter to

---

[626] Friedman, *Top Secret/Majic*, p. 90

[627] Based on Randle's experience at the National Archives in the Washington, D.C., including research there as late as 2002. See also Burkett, *Stolen Valor*, pp. 435 – 443; Greenwood, "MJ-12 Magic Act," *MUFON UFO Journal*, 236, December 1987, pp. 14 – 15.

Randle wrote, "Of course I am convinced by the evidence that the CT memo is indeed genuine and was planted by an insider..."[628]

The Truman Memo

Accompanying the Eisenhower Briefing Document, as found on the 35 mm film developed by Moore, is a one page memo, written on White House stationery, which notes Truman's plan on how the recovery at Roswell was to be treated at the highest levels of the government.[629] It is, in essence, though not in fact, an executive order by the president to the Secretary of Defense, James Forrestal, telling him that he is authorized to "proceed with all due speed and caution upon your undertaking. Hereafter this matter shall be referred to only as Operation Majestic Twelve."[630]

It is signed by President Truman, and it was part of the Eisenhower Briefing Document, arriving on the same role of 35 mm film.[631] If this document is a fake, then it too would suggest that the entire EBD is a fake as well.

There are a number of problems with this memo, one of which Friedman frequently points out. It was typed on two typewriters and Friedman believes that this adds to the authenticity of the document. Others, who have dealt with fraudulent government documents in the past, suggest that this is just one more proof that the both documents are faked.

The major flaw is the Truman signature. While no one is suggesting that it is a forgery, they are suggesting that it had been lifted from another, legitimate document, pasted onto the memo and then copied to remove any cut lines

---

628 Friedman in a letter to Randle, February 2001; Randle, *Case MJ-12*, p. 189

629 See Timothy Good, *Above Top Secret*, p. 551; Randle, *Case MJ-12*, p. 303; Friedman, *Top Secret/Majic*, p. 229

630 Cutler – Twining memo, dated September 24, 1947, unsigned. Original document in the custody of the National Archives.

631 Moore and Shandera, MJ-12 Analytical Report, p. 76 – 77.

that might appear.[632] In the age of copy machines and white out, it is done all too often with all too many documents. With Photoshop and other computer programs today, the process is even easier.

A second major problem on the Truman memo is that the signature on it matches, exactly, another Truman signature, this one from a letter dated October 1, 1947. There is a spurious ink up stroke on one of the bars in Harry that is unique to what is now called the donor document or donor signature.[633]

The positioning of the signature on the memo also makes it suspect. Truman habitually placed his signature so that the stroke on the "T" touched the bottom of the text on his letters, notes, memos and other written and signed communications.[634] On the disputed Truman memo that is not the case.

---

[632] "Government UFO Papers – Fact or Fiction?" *Search Magazine*, Winter 1987 – 88, No. 173, p. 48; Philip J. Klass, "New Evidence of MJ-12 hoax, *Skeptical Inquirer*, Winter 1990, pp. 135 – 140; Moore and Shandera, *MJ-12 Analytical Report*, pp. 89 – 92; Friedman, *Top Secret/Majic*, 83 – 85.

[633] It might be said that this extra stroke on the "H" in Harry makes this signature unique and for that reason easily identifiable. It has not been observed on other, authenticated signatures, which suggests that this signature is a transfer from another, authentic document that has been identified.

[634] Joe Nickell and John F. Fischer, "The Crashed-saucer Forgeries," *International UFO Reporter*, March/ April 1990, p. 8; Ted R. Spickler, "The Truman MJ-12 Letter," *International UFO Reporter*, May/June 1991 pp. 12-13; Ronald Story, *The Encyclopedia of Extraterrestrial Encounters*, New American Library, 2001: 322 - 324.

Questioned documents expert Peter Tytell,[635] who originally received copies of the EBD and the Truman memo from Friedman,[636] said about this problem:

> [Philip] Klass is the one who came up with the prototype signature. And that's an absolute slam dunk. There's no question about it. When you look at the points where it intersects the typing on the original donor memo [that is, the October 1, 1947 letter] for the transplant, you can see that it was retouched on those points on the Majestic-12 memo. So, it's just a perfect fit. The thing was it wasn't photocopied and it wasn't photographed straight on... The guy who did one of the photographic prints had to tilt the base board to try and get the edges to come out square so whoever did the photography of the pieces of paper was not doing this on a properly set up copy stand. It was done, maybe on a tripod, or it was done hand held. However it was done, the documents were not photographed straight on... There's a slight distortion of the signature but it is not enough to make the difference here. Nowadays it you could probably get it to fit properly with computer work but it's not that the signature is an overlay but it's that at those discrete points, and their dumb document examiners [Moore and Shandera whose "experts" attempted to explain these problems] talked about the thinning of the stroke

---

[635] Tytell is the expert that CBS sought out to validate the documents that alleged that President Bush had not finished his Air Guard service. When Tytell called them back, they told him that his services were no longer needed. CBS had decided that the documents, now considered forgeries by nearly everyone were authentic and passed on the chance to have them validated by Tytell.

[636] Peter Tytell, personal interview by Randle, August 20, 1996; Randle *Case MJ-12*, pp. 190 – 191; Philip Klass, "New Evidence of the MJ-12 Hoax," *Skeptical Inquirer*, Winter 1990.

> at this point. At that particular point, at the exact spot where it touches a typewritten letter and it has to be retouched to get rid of the letter."[637]

Moore and Shandera, in their *The MJ-12 Documents: An Analytical Report* wrote, after measuring the Truman signature, that it isn't an exact match as others had claimed. They suggest it is close, and, according to various handwriting experts, this makes the signature more consistent with authenticity.

The controversy over that memo wasn't ended there. Joe Nickell and John Fischer, two skeptics with what is now known as the Committee for Skeptical Inquiry (CSI), who were interested in MJ-12, received a copy of *The MJ-12 Documents: An Analytical Report* in 1991. Nickell and Fischer believed, according to an unpublished paper *Further Deception: Moore and Shandera's MJ-12 Report*, "... [The Moore and Shandera report] provides lessons in how not to investigate a Ufologically related questioned document case... Not only is neither a trained investigator, let alone a document specialist, but both are crashed-saucer zealots and one (Moore) has actually been suspected of having forged the documents."[638]

Then, to make their case, Nickell and Fischer, examined the status of the investigation of the Truman memo. They wrote that the document is "an incompetent" hybrid, and say that "no genuine memo has yet been discovered with such an erroneous mixture of elements."[639]

The important point here is that there is a mixture of errors on the Truman memo. It could be suggested, if there was but a single mistake, this might have

---

[637] Peter Tytell, audiotaped personal interview by Kevin Randle, August 20, 1996. See also Peter Tytell, personal interview by Philip Klass October 12, 1989.

[638] Joe Nickell and John F. Fischer, "Further Deception: Moore and Shandera's MJ-12 Report," Unpublished paper, January, 1991

[639] Ibid.

been a onetime occurrence. Unfortunately, as noted, there are other errors on the document as well and that suggests to many researchers that it is fraudulent.

Like Tytell, Nickell and Fischer had pointed out that the Truman signature "placed well below the text."[640] Moore and Shandera countered by saying, "The problem with this assertion is that those who make it used only letters signed by Truman as the basis for their study."[641] Nickell and Fischer responded, writing, "We did no such thing... here Moore and Shandera are guilty of outright misrepresentation... we studied typed letters and memos, handwritten notes, engraved thank-you cards, inscriptions and photographs... In every instance where Truman had personally signed the text... our observation of close placement applied."[642]

The only exceptions to this that have been found are documents that contain Truman's signature but were not signed by him. That is to say, his signature had been printed on the document. As a case on point, many veterans of the Second World War have a certificate signed by Truman but the signature was printed on the document rather than signed individually by Truman.

There is one other point that should be made. Friedman has repeatedly said that most of those examining the Truman memo have ignored the fact that the date was added by a different typewriter.[643] He believes the use of two different typewriters adds a note of authenticity to the document[644]. However, those versed in examining forged documents suggest that the use of two typewriters on a single document is a red flag suggesting forgery.

---

640 Nickell and Fischer, "Forgeries," p. 7.

641 Moore and Shandera, *MJ-12 Analytical Report*, p. 90

642 Nickell and Fischer, "Further Deceptions," p. 3

643 Stan Friedman, "Debunking a Debunker," Privately Published, 1988, p.4

644 B. G. Burkett and Glenna Whitley, *Stolen Valor*, 1998.

Fatal Errors

It could be argued that the questions raised by various individuals have been, if not answered, at least addressed. It could be argued that none of them, alone, is sufficient to invalidate the MJ-12 documents. It might be said that in the aggregate, those questions do suggest that MJ-12 is a massive hoax. But there are some things taken alone do more than suggest a hoax. They seem to prove it.

In both Barry Greenwood and Brad Sparks' MUFON Symposium presentation in 2007, and later in an article adapted from the paper and published in the *MUFON UFO Journal* under Sparks' by-line, there is a discussion of what they see as a fatal error in the Eisenhower Briefing Document.[645]

To explain what they mean by a "fatal error," they quote Stan Friedman, who had said that one way of determining if "the document is a phony [is] on the basis of any mistaken information in it."[646]

Both Bill Moore and Jaime Shandera echoed his concerns at one time or another by suggesting the same thing. Erroneous information in a document does suggest that it has been forged. They all are suggesting, as did Sparks and Greenwood, that these sorts of fatal errors would prove that the EBD, at best

---

[645] Brad Sparks, "The Secret Pratt Tapes and the Origins of MJ-12," *MUFON UFO Journal*, September 2007, pp. 3–10; Sparks, "Secret Tapes, Part II," *MUFON UFO Journal,* October 2007, pp. 3 – 11; Brad Sparks and Barry Greenwood, "The Secret Pratt Tapes and the Origins of MJ-12," *MUFON Symposium Proceedings 2007*, pp. 95 -159; Stan Friedman, "Friedman Rebuttal to: Sparks-Greenwood Paper Regarding Bob Pratt, Roswell & MJ-12, *MUFON UFO Journal,* October 2007, pp. 12 – 13: Brad Sparks, "Brad Sparks Response to Stan Friedman's Rebuttal to Sparks-Greenwood Symposium Paper, *MUFON UFO Journal*, November 2007, pp. 11 -14

[646] Sparks, "Secret Pratt Tapes," September 2007, p. 7. See also, Friedman, Comments on CSICOP/Majestic-12, August 26, 1987, p. 3; Moore, Shandera, Friedman, "Debunkers Ignore the Evidence," September 11, 1987; p. 5; September 5, 1987*, Focus*, p. 5a.

was disinformation and at worse a hoax, that diverted attention from more important areas of research.[647]

The error that Sparks and Greenwood point out in the EBD is that the distance to the debris field near Corona, New Mexico, is sufficiently inaccurate that this constitutes a major mistake. Those creating the report for review by a president would not make that sort of mistake. Such an error, even over something as minor as the distances involved, should throw the entire document into question.

According to Sparks, "The EBD wrongly claimed that the Roswell crash site [the Mack Brazel debris field] was 'approximately 75 miles from the Roswell base, when in fact it was only 62 miles away.'"[648] This is an error that Sparks has been pointing out from 1987. He computed the actual distance at 62 air miles and the distance by road it is over 100 miles, again not 75 miles mentioned in the EBD.[649]

Sparks suggested that the origin of the 75 mile figure was the 1980 book, *The Roswell Incident*. It is, at best, an estimate that is not based on the facts that should have been available to an aviation unit. Their navigation needed to be precise and even a miniscule error made at the beginning of a flight could result in missing the destination by dozens of miles. Those in Roswell would have known the precise distance to the Brazel debris field and that should have been reflected in the EBD.

There is a secondary problem that neither Sparks nor Greenwood considered and that was a site where more of the craft and the bodies had been found. In the late 1970s and the early 1980s, researchers had identified only the Brazel

---

[647] Ibid.

[648] There were various figures given for the distance to the ranch. The *Roswell Daily Record* of July 9, 1947, reported, "He [Brazel] returned to his ranch, 85 miles northwest of Roswell…" The *Albuquerque Tribune*, on July 9, uses the same 85 mile figure. Others used various figures giving the distance from Corona, New Mexico, one suggesting Brazel lived 30 miles southeast and another reporting 25 miles.

[649] Sparks, "Secret Pratt Tapes," September 2007, pp. 7 – 8.

debris field and another site in the western part of New Mexico along the Plains of San Agustin. Oddly, the EBD does not mention that site, nor does it mention the other sites closer to Roswell.[650] It would seem that a document prepared for the president-elect, and prepared to brief him on the recovery operation would contain all the relevant information. But, in the early 1980s, no one was talking about a body site near Roswell.[651]

These are not the only errors in the short paragraph in the Eisenhower Briefing Document which said, "In spite of those efforts, little of substance was learned about the objects until a local rancher reported that one had crashed in a remote region of New Mexico located approximately seventy-five miles northwest of the Roswell Army Air Base (now Walker Field)."

In 1947 it had been the Roswell Army Air Field that had later become Walker Air Force Base. It seems unlikely that military men would get the designation of the base wrong in both places that it is mentioned.[652] These two mistakes are the sort that a civilian would make when talking about a military installation and who was not familiar with the way the military named its facilities. It is not the sort of mistake that a military man would make.

That meant there were three errors in that one short paragraph that shouldn't have been made. It could be argued that 75 miles was close enough for "government work" as they say, but it would seem that the name of the

---

650 Thomas J. Carey and Donald R. Schmitt, *Witness to Roswell: Revised and Expanded Edition*, 2009, pp. 107 – 108.

651 In the 1980s, the conventional wisdom was a debris field on the ranch managed by Brazel and a second site on the Plains of San Agustin on the other side of New Mexico. Curiously, that second site is not mentioned in the MJ-12 papers. It wasn't until Randle, Schmitt and Carey developed information of a body site closer to Roswell, with Carey and Schmitt finding additional witnesses who provided a more precise location.

652 Survey of various official documents including the Unit History of the 509th Bomb Group in Roswell, June 1947 –
November 1947.

base in Roswell would have been accurate, given who was allegedly preparing the document and who would be reading it.

Another Fatal Error

These are not the worse of the errors that appeared in the EBD, however. In a paragraph on page five it was reported, "On 06 December, 1950, (sic) a second object, probably of similar origin, impacted the earth at high speed in the El-Indio – Guerrero area of the Texas – Mexican boder [sic] after following a long trajectory through the atmosphere. By the time a search team arrived, what remained of the object had been almost totally [sic] incinerated. Such material as could be recovered was transported to the A.E.C. facility at Sandia, New Mexico, for study."[653]

The problem here is that this one paragraph reflects the UFO situation as it existed in the 1980s and not as it was in 1952 when the document was alleged to have been prepared and certainly not as it exists today. To understand this, it is necessary to understand UFO research as it existed in the late 1970s and the early 1980s.

In the 1970s, W. Todd Zechel, a researcher of limited talent for investigation, claimed to have discovered a reference in a 1968 newspaper suggesting that a military officer had been at the scene of a UFO crash in 1948.[654] Using that article as a base, he was able to locate Robert B. Willingham, reported by Zechel to be a retired Air Force colonel. In 1977, Willingham signed an affidavit about his UFO experience. This was the discussion of a UFO crash just south of the Texas border in Mexico which would become the basis of that paragraph in the EBD.

According to what Willingham said then:

---

[653] The Eisenhower Briefing Document, page 5, copies available in many books and on line.

[654] Noe Torres and Ruben Uriarte, *The Other Roswell*, 2008, pp. 2 -3.

Down in Dyess Air Force Base in Texas, we were testing what turned out to be the F-94. They reported on the [radar] scope that they had an unidentified flying object at a high seep to intercept our course. It came visible to us and we wanted to take off after it. Headquarters wouldn't let us go after it and it played around a little bit. We got to watching how it made 90 degree turns at this high speed and everything. We knew it wasn't a missile of any type. So then we confirmed it with the radar control station on the DEW Line (NORAD) and they kept following it and they claimed that it crashed somewhere off between Texas and the Mexican border. We got a light aircraft, me and my co-pilot, and we went down to the site. We landed out in the pasture right across the from where it hit. We got over there. They told us to leave and everything else and then the armed guards came out and they started to form a line around the area. So, on the way back, I saw a little piece of metal so I picked it up and brought it back with me. There were two sand mounds that came down and it looked to me like this thing crashed right in between them. But it went into the ground, according to the way people were acting around it. But you could see for, oh I'd say, three to five hundred yards where it had went across the sand. It looked to me, I guess from the metal that we found, chunks of metal, that it either had a little explosion or it began to disintegrate. Something caused this metal to come apart.

It looked like it was something that was made because it was honeycombed. You know how you would make a metal that would cool faster. In a way it looked like a magnesium steel but it had a lot of carbon in it. I tried to heat it with a cutting torch. It just wouldn't melt. A cutting torch burns anywhere

> from 3200 to 3800 degrees Fahrenheit and it would make the metal hot but it wouldn't even start to melt.[655]

In 1980, Moore published *The Roswell Incident*, which, of course laid out what he knew about the Roswell UFO crash at that time. In that book, Moore mentioned, briefly, the Willingham tale without identifying Willingham specifically or giving much in the way of detail.

Moore wrote, "Then a second group, Citizens Against UFO Secrecy (CAUS), was formed in 1978 under the directorship of W. T. Zechel, former director of GSW [Ground Saucer Watch] and one-time radio-telephone operator for the Army Security Agency [ASA]. CAUS's announced aim was nothing less than an 'attempt to establish that the USAF (or elements thereof) recovered a crashed extraterrestrial spacecraft' in the Texas-New Mexico-Mexico border sometime in the late 1940s."[656]

This information clearly came from Zechel and was, basically, attributed to him. It doesn't quite square with the affidavit that Willingham signed for him, but the affidavit has no date on it and there is no precise location, other than near the Texas-Mexican border.

The problem for Zechel and later for Willingham was what Willingham had originally said about the crash. The story was that Zechel had found, in the NICAP files, an article about UFOs in a newspaper clipping that had appeared in a small, weekly in Pennsylvania. A search for that article proved fruitless, but a one paragraph story about the crash did appear *in Skylook*, at one time the official publication of MUFON.

According to that story:

---

655 Affidavit on file with CUFOS. See also, Kevin Randle, *A History of UFO Crashes*, 1995, pp 192 -193 for the text of the Willingham affidavit.

656 Berlitz and Moore, *Roswell Incident*, p 131.

> Col. R. B. Willingham, CAP squadron commander, has had an avid interest in UFO's for years, dating back to 1948 when he was leading a squadron of F-94 jets near the Mexican border in Texas and was advised by radio that three UFO's "flying formation" were near. He picked them up on his plane radar and was informed one of the UFO's had crashed a few miles away from him in Mexico. He went to the scene of the crash but was prevented by the Mexican authorities from making an investigation or coming any closer than 60 feet. From that vantage point the wreckage seemed to consist of "numerous pieces of metal polished on the outside, very rough on the inner sides."[657]

This then, matches with the information published by Moore in his book. The paragraph in *The Roswell Incident*, puts Zechel in touch with Moore, and proves that Zechel had shared his UFO crash information with Moore. Zechel, then, according to what Willingham said later, was responsible for the exact location of the crash. Zechel believed the crash had taken place on December 6, 1950 and was documented by an alert that had greatly concerned the military on or about that date.[658] Researcher Dr. Bruce Maccabee had found the documentation for the alert in a batch of papers released by the FBI after a FOIA request.

Zechel also provided the location for the crash, putting it near the small Mexico towns of El Indio and Guererro, not all that far south of the Texas border. This was what Zechel believed in the late 1970s and was what he was telling Moore at that same time. In fact, Zechel complained that Moore had

---

[657] *Skylook*, March 1968, p. 3

[658] Len Stringfield, *The Crash/Retrieval Syndrome*, privately published, 1980, p.22.

put the material in the MJ-12 document to suggest the craft had burned to stop Zechel from publishing a book about it.[659]

When the Willingham story first surfaced, in the late 1970s, many inside the UFO community believed it. The source was identified as a retired Air Force colonel, and he was named. Nearly everyone believed that this credential, meaning he was a high-ranking Air Force officer, lent credibility to the story.[660] Even the Center for UFO Studies was caught up in it, recording an interview with Willingham and including it on an LP with other, high-caliber witnesses.

It wasn't until decades later that Willingham's military record was checked by UFO researchers. According to the documents available, Robert B. Willingham entered the Army in December 1945 and was discharged in January 1947. He rose to the rank of E4 and is technically a veteran of World War II.[661] The war was not officially declared over until 1946 and anyone serving in the military in that transitional period is considered a World War II veteran.

Investigating further it was learned that Willingham had been an officer in the Civil Air Patrol (CAP), an auxiliary of the Air Force. Members are civilians who serve without pay or benefits but do wear a modified Air Force uniform and are awarded military ranks. Willingham apparently rose to lieutenant colonel in the CAP.

---

[659] Len Stringfield, "Retrievals of the Third Kind," *MUFON Symposium Proceedings*," pp. 81 -82

[660] For another take on how this relates to the Eisenhower Briefing Document, see Greenwood, "MJ-12 Magic Act," *MUFON UFO Journal*, December 1987, p. 13. Greenwood links the Del Rio UFO crash, to that in Moore's book, *The Roswell Incident*, and to the El Indio – Guerrero crash. See also Jerry Clark, "Crashed Saucers – Another View," *UFO Report*, February, 1980, pp. 28 – 31, 50, 52, 54 – 56.

[661] Public information section of Willingham's service record came from the National Archives (NARA) in St. Louis that houses all military records except those still serving whether on active duty, in the reserve, or as part of the National Guard.

He did attempt to convince the Air Force that he deserved a military pension based on his service in the CAP. He enlisted the aid of his congressional representatives but searches of his records did not corroborate his claimed service on active duty in the Air Force, assignments in the Air Force Reserve, training as a military pilot or service in Korea.[662]

Noe Torres and Ruben Uriarte interviewed Willingham for a book about his experiences. Willingham told them that Zechel had made up the December 6, 1950 date and that he had picked a crash site closer to Del Rio, Texas.[663] Willingham knew the 1950 date was wrong because he had been serving in Korea on that date, or so he claimed.

In fact, Willingham changed the date again, telling Torres and Uriarte that it had happened in 1954, and later telling others it might have been 1955.[664] He also said the location was south of Lantry, Texas, just across the border in Mexico.[665] In other words, critical information about the crash had been changed a third or fourth time.

What this does, because there is no other witness to have talked about this particular event, is further invalidate the EBD. Willingham has tied the date to Zechel and Zechel provided Moore with that information in the hope of obtaining a publisher for his envisioned book.

This then, is a fatal flaw simply because no one had mentioned any crash that fit the facts in the EBD until the late 1970s. The paragraph that mentions

---

[662] Documentation about Willingham's service from the Air Reserve Personnel Center in Denver, the Military records stored in St. Louis, and documentation supplied by Willingham in his attempt to secure a military pension for his service with the CAP. It should be noted here that Willingham does not have the necessary twenty years of military service (active duty, reserve or National Guard) required for a military pension. Documents submitted as evidence show signs of having been altered.

[663] Willingham, personal interview by Kevin Randle, March 2009.

[664] Torres and Uriarte, *Other Roswell*, pp. 23 – 38; Willingham interview with Randle.

[665] Ibid.

the crash is based on faulty information that is the invention of a man who did not serve as an Air Force officer, did not fly fighters, and whose military career is with the CAP. No evidence has ever surfaced to corroborate the statements made by Willingham, to corroborate his claims of high military rank, or his suggestion he was an Air Force fighter pilot.

MJ-12 Operations Manual

On March 14, 1994, Don Berliner, an aviation writer and the co-author of *Crash at Corona*, received a small package in the mail, which contained an exposed but undeveloped roll of Kodak Tri-X 35 mm black and white film.[666] He wrote that he took the film to local photo processor, thought nothing more about it, and left for a writing assignment. Upon his return, he picked up the film but didn't bother to open up the envelope to examine the pictures until he was talking on the telephone to a fellow member of the Fund for UFO Research. It was only then that he saw the connection to MJ-12.[667]

Berliner then had a set of prints made, made several copies, and took a set to the General Accounting Office because they were starting their research into the Roswell UFO crash for Representative Steven Schiff.[668]

Berliner wanted the existence of the manual to be restricted to a few UFO researchers. According to him, "Others were carefully brought into the picture, but I was determined to keep the document secret from the general public until much more was known about it. I was particularly concerned to have it studied by persons who were not known as strongly pro-MJ-12 or anti-MJ-12."[669]

---

[666] Don Berliner, letter to the editor of the *International UFO Reporter*, dated March 12, 1996. The letter was printed in the Summer *IUR*, p. 29

[667] Ibid

[668] Schiff, a New Mexico representative, had asked the GAO to investigate the Roswell case after a number of inquiries from his constituents. That report, released later suggested they could find no records or documents relating to the crash.

[669] Berliner letter, March 12, 1996, p. 2

The manual was called, "Extraterrestrial Entities and Technology Recovery and Disposal." It was marked as "Restricted," but also as "Top Secret/Majic Eyes Only." Inside it was noted that "All information relating to MJ—12 has been classified MAJIC EYES ONLY and carries a security level 2 points about Top Secret." It is labeled as "SOM 1-01 Majestic—12 Group Special Operations Manual"[670] The manual is dated April 1954 and contains the seal of the War Department.

The Table of Contents lists six chapters and three appendices, the last being photographs. None of the photographic pages are available. The manual itself is quite short, with the photographic appendix beginning on page 31. Most of the circulated copies end on page 23.[671]

There are some statements and notations in the manual that seem to be anachronistic. On page 7, in chapter three, "Recovery Operations," it suggested, "It may become necessary to issue false statements to preserve the security of the site. Meteors, *downed satellites* [emphasis added], weather balloons and military aircraft are all acceptable alternatives."[672]

Later, there is an "Extraterrestrial Technology Classification Table" that indicates where recovered material should be sent. Most of the references are to "Area 51, S—4."

The problem with all these items is that they suggest the document is fraudulent. The War Department had ceased to exist in 1947 when it was replaced, along with the Navy Department, with the Department of Defense. A document created in 1954, if it contained a seal, would have had the DoD seal on it.

---

[670] This information is derived from the manual as it was released into the public arena. It has also been stamped as "Not an Official USAF Document, Not Classified, Suspected Forgery or Bogus Document," be a reviewing committee of the Air Force and other government agencies.

[671] SOM 1-01 Majestic—12 Group Special Operations Manual, cover and Table of Contents.

[672] Ibid. p. 7

The problem with suggesting a downed satellite to a reporter in 1954 was that there had been no launches of any artificial satellites. It wouldn't be until October 1957 that the first satellite would be placed in orbit. Had someone attempting to cover up a UFO crash suggested a downed satellite in 1954, he would have been inviting the reporters to dig deeper to find out just who had been launching these satellites, from where they had been launched, and how it had been done in such secrecy.

Finally, Area 51 didn't exist in 1954. According to Aerospace Archaeologist Peter Merline, the place was not called Area 51 in any documentation prior to 1960. Before then it was referred to as Groom Lake and the area didn't begin to be developed until April 1955, or one year after the manual was published.[673]

Herbert L. Pankratz, an archivist at the Dwight D. Eisenhower Library, in a letter wrote:

> The classification markings on the alleged MJ-12 document are not consistent with federal regulations for the marking of classified materials as of April 1954. The "Restricted" classification category was terminated by executive order in November 1953 and would not have been used on a document in April 1954. Federal regulations also require that the cover page of a document reflect the highest level of classification for any material in the document. Since "Top Secret" is a higher category than "Restricted," only "Top Secret" should have appeared on the cover of the document.
>
> In addition, we have seen no evidence in our files that a security classification referred to as "MAJIC EYES ONLY" ever existed. Executive Order 10501 was signed by President Eisenhower on November 5, 1953. It set up three classification

---

[673] Email from Tom Printy, downloaded on October 22, 1998.

> categories: "Top Secret," "Secret," and "Confidential." A fourth category, "Restricted Data," (not the same as "Restricted")' was established by the Atomic Energy Act of 1954; and it is used only with regard to nuclear weapons matters.[674]

Dr. Robert Wood and his son Ryan have taken to the defense of the MJ-12 documents in general and the Operations Manual in particular. Wood looked at the list of criticisms and responded to each. He wrote, for example:

> "The 'downed satellite is a possible anachronism." In the first place, the entire strategy is that of deception – it is even the title of the paragraph in question! Deceptive statements are not usually true. Furthermore, it was just one of five choices offered to keep nosy people away. The big argument, though comes from those who say, "Why would anyone be impressed by a known false statement?" Actually, most people were aware of our plans for satellites in April of 1954, as a result of enormous coverage of this new space thinking. There are prominent public references to satellites before this date, including a Time Magazine article just the previous month speculating on whether a satellite had already been covertly launched. So, satellites were on the public's mind and "downed satellites" were a very credible concept.[675]

Jan Aldrich, who had worked with Berliner as one of those selected to review the manual, wrote, on the Current-Encounters email list:

---

[674] Letter from Herbert L. Pankratz to Kevin Randle, dated February 9, 1995. See also, Kevin Randle, "The MJ-12 Operations Manual: Another Forgery?", *International UFO Reporter*, Spring 1996, pp. 9 – 10.

[675] Wood, *Majic Eyes Only*, pp. 264 – 266.

> In 1954 if you told the public you had a downed satellite, when none had been launched yet, that would have caused a sensation and attracted every reporter around. They would have chartered planes to fly over the site and tried every means to get inside and have a look. There would have been demands in Congress to know what was going on. Of course, every spy would [be] in North America [to] try to find out about a downed satellite... Instead of diverting attention from the site it would have attracted much unwanted attention which could lead to additional security violations and possible total loss of all security.[676]

An independent study was made of the manual by several men who had a background in working with military and classified documents or who were familiar with the military and UFOs. These included Don Berliner, Jan Aldrich who had been top secret control officer and a COMSEC custodian in the Army; Tom Deuley who was a retired lieutenant commander and who had Nuclear Power and Naval Submarine Weapons program, seven years as a Naval Cryptologic Officer, one year as a Communications Officer in charge of a 24-hour secure communications facility and six years as a Communications Security (COMSEC) Custodian; and both Dick Hall and Mark Rodeghier examined the manual. Altogether they wrote about the manual:

> We believe this to be a hoax document; a deliberate fake designed to mislead the public and to plant false information in the UFO research community by person or persons whose motives are unknown. Deliberately planted false information ("disinformation") such as this forces investigators to waste their time checking on its validity rather than on more produc-

---

[676] Jan Aldrich, Current Encounters Email, downloaded April 2, 1999.

tive efforts. Our general reasons for concluding that the manual is a hoax are outlined below (for more specific details, contact the individuals):

(1) Documents and materials with high classifications have special provisions attached to them to ensure the ability to trace them at all times and to verify their integrity, until they are destroyed or declassified. The security markings on the SOM 1-01 document do not conform to required security procedures established for all agencies by presidential executive orders. In some instances they are totally contrary to established security procedures. No internal evidence exists in the document to show that proper accountability was exercised by the document's custodians.

(2) The inclusion of some accurate information has been cited as proof of authenticity, whereas it could equally well be interpreted as a cut-and-paste job to lend an air of authenticity. Partially legitimate but altered UFO-related documents are already known to exist.

(3) The content of the manual is strikingly inappropriate for its stated purpose. A field manual for dealing with crashed craft and alien bodies would have no reason to include (a) information on UFO history, (b) a chart of UFO types, (c) information concerning radar detection of UFOs, (d) a list of natural and artificial aerial phenomena which can be mistaken for UFOs.

(4) Military manuals of this type establish standards and define tasks which must be performed to accomplish the mission. The manual fails to establish such standards and is completely silent on personnel qualifications and equipment

> requirements. Furthermore, the methods of recovery and site security described in the manual are inadequate and tactically unsound. Regulations, materials, and training publication references cited are grossly inadequate or completely missing.
>
> The undersigned parties take UFO reports seriously and advocate thorough scientific investigation. However, when it comes to analysis of 2nd and 3rd and nth generation copies of documents, forensic analysis is almost impossible. Content analysis already has shown serious problems with MJ-12 related documents.
>
> The only way SOM 1-01 and other alleged "documentary proof" of MJ-12 could conceivably be authenticated would be by locating a documentary paper trail of certifiably original documents in government archives, or in private papers of important people. Even then, allegedly authentic documents would need to be subjected to forensic examination to determine such things as the age of the paper. And document experts would need to examine them for internal accuracy and style. Given the track record of fake documents and shoddy scholarship, rigorous peer review is essential.[677]

Without the slightest evidence that the document is authentic, without a clear provenance, and without finding any paper trail that could lead back to the manual, it would seem that all questions about it have been answered. While Bob and Ryan Wood play down the lack of provenance as relatively minor, the truth is that like the other MJ-12 documents, this is a major stumbling block. Unlike other leaks of documents into the public sector, these came

---

[677] The full Joint Statement is available on the CUFOS.org website. Last accessed on August 13, 2012.

without a pedigree, and without that, it is difficult to draw any other conclusion. There is, however, one other important piece of evidence.

Tim Cooper and the New MJ-12 Documents

Since the first of the MJ-12 documents were released in 1987, there have been more than 100 different documents given to researchers that comprise more than 4000 pages of various kinds.[678] Many of these documents came from Tim Cooper, described as a UFO researcher with an interest in crashed saucers, Roswell and MJ-12.[679] Robert Wood speculated that he had received the documents because he had filed many FOIA requests for information on the Kennedy assassination and UFOs.[680]

According to Wood, the first of these new MJ-12 documents arrived in Cooper's post office mail box in June, 1996 and contained a note that suggested this was a "trial package" with 38 pages of other materials.[681] Cooper would later claim that the material came from Thomas "Cy" Cantwheel, at the time an "elderly man," but who would remain forever in the background. Cooper would meet him briefly, one time, but was unable to trace him any further.[682]

Robert Wood said that they had submitted some of the material in the first batch of Cooper documents, including an important letter allegedly written by one of four people, to a number of forensic examiners whose task it was to determine authenticity. This document is known as the Einstein/Oppenheimer

---

[678] Robert Wood, "Forensic Linguistics and the Majestic Documents," in *6th Annual UFO Crash Retrieval Conference*, 2008, p. 98.

[679] Friedman, *Top Secret/Majic*, p. 144

[680] Wood, *Linguistics*, pp. 98 – 99.

[681] Wood said that the first words were illegible and it was assumed that these were "trial package."

[682] Ryan Wood, *Majic Eyes Only*, p. 37

memo since it was supposedly signed by both men. According to Wood, the document might have been created by Einstein, Oppenheimer, a fellow named Salina who sent the material to Cooper, or Cooper himself.[683]

The results of this seem to suggest that neither Cooper nor Oppenheimer wrote the document. There was one word, which Wood labeled a "zinger," meaning a word so rare that it wasn't in the samples of the others but did appear in the documentation that came from the Einstein collection. That word, "supra-national," suggested that the document in question had been written by Einstein.[684] This, of course, doesn't mean that one of the others hadn't written it, or that none of them had written. It could only suggest that one of them might have written or another probably hadn't.

Another of the documents that Wood received from Cooper was what Wood now calls the "Burn Memo."[685] This, according to Wood, is the "most historic" and was saved from destruction by an officer who was cleaning out the safe of James Jesus Angleton, a "legendary CIA counterintelligence chief," and after Angleton died.[686]

Wood wrote that he had received the memo from Cooper. It had a McLean, Virginia post mark, and according to Wood, that post mark was traced to a meter at the CIA. It is suggested that the memo came from the Director of the CIA, sometimes said to be MJ-1, and it sent only to the first seven of the twelve. It also identifies a host of new projects including Eviro, Parasite and

---

683 Robert Wood, *Linguistics*, pp. 99 – 100. This is a rather confusing report on these new documents given that it is based on a slide presentation. It seems that the Einstein/Oppenheimer memo is a response to a question about how to handle the presence of aliens. There is little to suggest who the original author is, so Wood supplied the document to forensic linguistics to see if the syntax, sentence structure, vocabulary and style suggested a specific author.

684 Robert Wood, *Linguistics*, pp. 100 – 103.

685 Ibid. pp. 103 – 106.

686 Ryan Wood, *Majic* Eyes Only, pp. 246 – 247.

Parhelion. The purpose of the memo "is to establish policy with respect to UFOs and JFK."[687]

With this document, as with the others, there is no way to trace it back into the CIA independently. It came from Cooper, who apparently got it from Cantwheel, but Cantwheel is not available for discussion. But even with that, there is another problem. It is marked as, "Top Secret/MJ-12." If it was part of the MJ-12, the classification should have been "Top Secret/Majic," or Top Secret/Majic Eyes Only."

Both Ryan and Robert Wood responded to inquiries about this. Ryan Wood wrote, "The burned memo, is in Angleton's' counter intelligence group at the CIA at this time. Not sure what their procedures would be in Fall of '63. We have TS/MAJIC in Sept 47. We have just TS (Operation Majestic-12) on the EBD in '52, we have TS/MAJIC on SOM1-01 in 57 (latest change on change control page) and we have TS just MJ-12 in '61 on Kennedy memo."[688]

Robert Wood agreed with Ryan and wrote, "I would add that if the document was fake, the question is 'Is it more likely that the faker would have used the wrong classification, or that an overworked secretary would not have adhered to the classification guidance that existed[689] ...if it was ever written?' …My impression is that the higher the classification up the chain, the less is the adherence to procedures and more to intent… There are quite a few variations known. For example, MJ Twelve as opposed to MJ-12 appears."[690] This

---

687 Robert Wood, Linguistics, pp. 103 – 104.

688 Ryan Wood email to Kevin Randle, January 11, 2012.

689 It seems that as you move up in the secretarial pool, your level of competence increases and those at the top are the very best. To type a document that is classified as "Top Secret," the secretary must hold the proper security clearances. They also follow the rules and regulations in effect at the time, often ensuring their bosses follow the proper procedures as well. Errors such as those found in these MJ-12 related documents would not be tolerated at that level.

690 Robert Wood, email to Kevin Randle, January 11, 2012. It should be noted here that using questioned documents to attempt to validate another questioned document is improper. That Wood has seen a mixture of code words on the MJ-12 documents

assumption is in error as the higher up the chain, the more rigidly the regulations are obeyed.

Friedman was also a recipient of new MJ-12 documents from Cooper. According to him, Cooper received the first of the documents in late 1992.[691] Friedman launched into an investigation of the documents, including one from Rear Admiral Roscoe Hillenkoetter (which Friedman called the "Hillenkoetter Memo), that seemed to be authentic as to tone and historical context. In other words, it seemed to be authentic.[692]

Other documents followed and analysis seemed to suggest that they were written in the proper style of the times and included information that suggested an intimate knowledge of history that only years of specialized study could provide. One of these, a long letter from Carl Humelsine to General George Marshall on September 27, 1947, was historically accurate and did mention MJ-12. Dr. Larry Bland at the Marshall Foundation recognized it as a retyped and altered version of an actual letter from Humelsine sent to Marshall.[693] This meant that the document with its MJ-12 reference inserted was a fake.

At this point Friedman realized that many of the documents that Cooper had supplied including the Hillenkoetter Memo had been faked in the same fashion. Curiously, Friedman reported, "The first document [Hillenkoetter memo] was extremely poorly reproduced, and Tim was initially reluctant to pass it on for fear he would be accused of forging it."[694]

---

does not suggest various code words were used, but that someone who did not understand security classifications had attempted to create documents to validate MJ-12. Unfortunately, and contrary to what Dr. Wood has suggested, this is more in line with fraud than authenticity.

691 Friedman, *Top Secret/Majic*, p. 144. This probably refers to the first communication Friedman had with Cooper while that of Robert Wood came later.

692 Friedman, Top Secret/Majic, pp. 145 – 146.

693 Ibid., pp. 158 – 159

694 Ibid, p. 145

As research continued into these new documents, more of them were found to be faked in the same fashion. That is, a letter or document from the proper time frame was selected, retyped on a typewriter from the era, and then presented as if it was a new discovery. It gave the appearance of being genuine, but when the original document was found, the truth was known. As it stands now, few accept the documents that came from Tim Cooper and his secret and silent friend, Cy Cantwheel.

A Final Bit of Evidence

In the early 1980s, Bill Moore began to talk about creating a Roswell document in an attempt to gather more information about the crash. He told fellow researchers that he had taken the investigation as far as he could and felt he needed something to shake new information loose and to convince reluctant witnesses to talk. He mentioned this to various UFO researchers and colleagues, most of whom warned him against such a course of action.

Philip Klass reported, "On April 16, 1983 – less than two years before William L. Moore and Jaime Shandera claim they received the "Top Secret/Eyes Only" MJ-12 documents from an unknown source – Moore reportedly sought the reaction of his friend Brad Sparks, a respected UFO researcher, to the idea of creating such counterfeit government documents. Sparks strongly recommended against it. Later, when Sparks called Stanton T. Friedman, he was shocked to discover that Friedman defended Moore's idea."[695]

Klass continued, writing, "SUN [*Skeptics UFO Newsletter*] first learned of Sparks' involvement in mid-1991 but he was reluctant to speak out. Subsequent events have overcome his reluctance. These include limited disclosure of Moore's idea of creating bogus documents by Kevin Randle and Don Schmitt in their 1994 book, 'The Truth about the UFO Crash at Roswell.' In a brief chapter debunking the MJ-12 papers, the book reports, 'According to Friedman, among others, Moore had suggested as early as 1982 that he wanted

---

[695] Philip J. Klass, *Skeptics UFO Newsletters*, #44, March 1997, p. 1

to create Roswell documents thinking that it might open doors that were closed.'"[696]

Friedman eventually responded to these revelations. He wrote:

> The debunkers sit in their armchairs making false claims. I should add that Kevin [Randle], again, in the blog,[697] repeated the false claim that Bill Moore supposedly told me it was time to make up a false document and put it out to smoke out others who knew about MJ-12. Supposedly I agreed.
>
> This is totally false. Perhaps it is a twist on the fact that Bill in his FOCUS publication included a heavily censored version of part of the Eisenhower briefing document. I had not been asked about doing that and didn't give approval.[698]

It is clear from the available data that Moore's plan was to create a Roswell document in an effort to convince others to tell what they might know about the crash. It had nothing to do with the censored version of the Eisenhower Briefing Document that they released. It was about creating a document, mentioned in the long months prior to the alleged receipt of the film by Shandera.

---

696 Ibid.

697 See "A Different Perspective," April 2007 for a report on this incident. Randle, while in California with Friedman to interview other Roswell witnesses was told by Friedman about Moore's idea. Don Schmitt had heard similar claims earlier.

698 Stan Friedman, "More Attacks on MJ-12," *UFO* issue 147, p. 19

The Real Majestic

There are documents that are labeled with Majestic that have a proper provenance, which means the origin of the document can be traced by anyone who wishes to do so and there is no doubt it is authentic. The first page, which was classified as Top Secret is entitled, "Report by the Joint Logistic Plans Committee the Joint Chiefs of Staff on Joint Logistic Plan for 'Majestic.'"[699]

There are some interesting things on that page. It identifies the problem, saying, "1. Pursuant to the decision by the Joint Chiefs of Staff on J.C.S. 1844/126, to prepare the Joint Logistic Plan in support of MAJESTIC*."

The asterisk references the same document mentioned in the body of the text. It provides no more information about it, but it is interesting because it is a reference to another document which could be traced to provide additional authentication. It also suggests something about how these highly classified documents are created and how many of them are inter-related.

The rest of the document is merely other paragraphs that reveal very little about what Majestic is and everything that it does say could, in fact, be considered as evidence of MJ-12. This is a document that deals with logistics, which can be simply defined as the support needed for military operations. It could be said that this is a document that relates to the movement of an alien craft, the wreckage or debris, and the bodies of the alien flight crew from one location to another. This would be the plan to explain the mode of transportation, how many soldiers would be needed, how they would be fed and housed,

---

[699] The reference for Emergency War Plan, codenamed MAJESTIC, is highlighted in yellow in the military history book seen here:

http://books.google.com/books?id=SeeNA-QAAQBAJ&pg=PA164&lpg=PA164&dq=%22plan+majestic%22+1952&source=bl&ots=jB7mbVYG8S&sig=GU5KwjiTYMHUlgPGzenVyGL00uI&hl=en&sa=X&ei=uWhbVMehBYKgyATchYDADw&ved=0CDAQ6AEwAw#v=onepage&q=%22plan%20majestic%22%201952&f=false

the fuel supplies, weapons and ammunition, route information and bases where additional support could be found and anything else rated to all of this.

The second page is a list of those who will receive the information which is quite long. It is labeled, "Top Secret Security Information," and is stamped, "Special Handling Required, Not Releasable to Foreign Nationals," and is dated 25 September 1952. Please notice the dating format that is not 25 September, 1952.

It is at the bottom of page two where is noted, "Forward herewith is a copy of the Joint Outline Emergency War Plan for a War Beginning 1 July 1952 MAJESTIC. This plan supersedes Joint Outline Emergency War Plan MASTHEAD, which was forwarded by SM-1197-51, dated 14 May 1951, copies of it will be either returned or destroyed by burning."

This suggests that it has nothing to do with UFOs or the Majestic-12, but the argument could be made that this is "typical boilerplate," meaning that the paragraph is sort of standard without a specific meaning other than instructions for removing the obsolete plan and replacing it with the new one. In today's world it would be a "cut and paste" error. In 1952, such a thing is more difficult to explain.

The third page makes it clear what is being discussed and what Majestic really is and ends all our speculation. Stamped with a date of 2 OCT 1952 (as opposed to 02 OCT, 1952) and with "Top Secret Security Information, the letter, in paragraph one said, "Enclosure (1), with attached copies of Joint Outline Emergency War Plan "MAJESTIC', is forwarded." This is a war plan and has nothing to do with UFOs. The markings on it, made in 1952, show what they should have been as opposed to what they are on the MJ-12 documents and the EBD. Yes, there might be variations depending on military service branch and the level of classification, but here is something that shows what was being used at the time, how it was used and what the specific wording was and should have been. This does not bode well for MJ-12, not to mention the duplication of code words.

By duplication of code words, it means that all code words for classified projects come from a master list so that there is no accidental duplication[700] (The military sometimes uses civilian code words for projects, such as Project Saucer, but the real name, which was classified at the time was Project Sign). To use the same or similar code words would lead to compromise. Someone cleared to deal with the War Plan – Majestic - wouldn't be cleared for the MJ-12 material, but the duplication of code words wouldn't make that clear.

This is the same argument made for Majic. During WW II there was a highly classified project known as Magic. This similarity could lead to compromise, if you had two projects with such similar names.

The last page of the documents makes it clear that there is no reason to assume this has anything to do with the investigation of alien craft, alien bodies or the recovery of an alien spacecraft. Paragraph 4 says, "The estimate of the Soviet Union's capability to execute campaigns and her probable courses of action contained in the Enclosure does not take into consideration the effect of opposition by any forces now in position or operational, or of unfavorable weather or climate conditions."

This is also classified as "Top Secret Security Information," and is dated 12 September 1952 (again is relevant because it puts it into the time frame of the EBD and it shows the dating format as it should have been written), is signed by W. G. Lalor, Rear Admiral, U.S. Navy (Ret.), and is also noted as "Reproduced at the National Archives."

Conclusions

It is astonishing that in 1981 William Moore had a document, the Aquarius Telex, which mentioned the MJ-12. There was nothing in that telex to reveal

---

[700] It must be noted that the "real" code names are not just grabbed out of thin air. The code names are taken from a list to prevent duplication. See Gale Research's Code Name Dictionary, 1963, for more information. For additional information see also William M. Arkin's book, "Code Names: Deciphering U.S. Military Plans, Programs, and Operations in the 9/11 World."

the nature of MJ-12, other than to suggest it was some kind of a highly classified project that had to do with UFOs given the overall nature of the telex. It also mentioned Project Aquarius,[701] which was later allegedly identified as a project that was concerned with alien visitation. In the 21st century, it is odd that these documents are ignored by those who indorse MJ-12. It bears repeating that Friedman wrote, "I am not sure what you mean by the Aquarius Telex and don't even mention it in TOP SECRET/MAJIC or in 'Final Report on Operation MJ-12.'"[702]

The secondary problem with the telex is that Moore said that it was a retyped version of an actual telex. Dick Hall explained that Moore had said, in a meeting at FUFOR that the telex had been retyped and that the headings had been cut from the original and pasted on the retype. On March 20, 1989, Hall wrote, "...retyping the AFOSI document [Aquarius Telex] without saying so..." would create problems for Moore. Hall would later suggest that Moore might not have retyped it himself, because he said that he had seen the original but he didn't retype the version he have been given. Either way, the first mention of MJ-12 was in a document that was clearly faked and this is known to but ignored by the main proponents of MJ-12.

The next appearance of MJ-12 is in a novel that Bill Moore suggested to Bob Pratt. According to the information supplied by Bob Pratt, available in his UFO files that are now housed with MUFON, Moore had approached him with the idea of writing a book about MJ-12. Pratt feared there wasn't enough corroborative evidence to support a nonfiction book, but they could certainly write one that dealt with the topic in a fictional world. According to Pratt, "The original idea behind the book was Project Aquarius. ... I sat in a chair and took

---

[701] Several FOIA requests to the NSA eventually resulted in responses that confirmed there was a highly classified project named Aquarius, but no other information about it could be released. However, not much later, NSA retracted the statement, saying that its confirmation had been based on a false assumption. There is no confirmation of any project named Aquarius. Greenwood, "MJ-12 Magic, Act," *MUFON UFO Journal*, Sep 1987.

[702] Friedman letter to Randle, February 2001: Randle, *Case MJ-12*, p. 184

notes as he [Moore] told me about Project Aquarius, MJ-12 and a number of other things."[703]

Pratt wrote, "My working title was MAJIK 12, but I wanted to call it I. A. C., for identified alien craft, supposedly the name government insiders use for UFOs. However, when I got the finished manuscript back from Bill, he had put The Aquarius Project on the title page."

Ironically, Pratt wrote, "When this MJ-12 business broke in 1987 or whenever, I wrote to Bill saying that if there was anything to it, we ought to dust off our manuscript and try to sell it again. He never answered."[704]

There are now two separate cases in which MJ-12 is mentioned and they arise from a fraudulent telex which almost no one now acknowledges, and from a work of fiction created by Moore, Pratt, and a third writer who is quite likely to have been Richard Doty. Pratt even used, as his working title, "*Majik 12*."

Aquarius was still in play in 1983, first when Moore claimed he made a cross country trip so that he could spend nineteen minutes looking at something that seemed to have been a presidential briefing. He took photographs but they didn't turn out very good and are useless as evidence. He did read the material into a tape recorder so that he might be able to reproduce the documents if the pictures turned out too poor to use but his voice on the tape does little to corroborate the tale.

About a month later, Linda Moulton Howe was briefed by Richard Doty and given what is now thought of as the Carter Briefing. She was at Kirtland AFB, and was allowed to read the documents but not allowed to take notes or to have copies of the documents. These went so far as to claim that aliens,

---

[703] Letter from Pratt to Robert Todd, February 20, 1989, copy in Randle's files. See also, Sparks, "The Secret Pratt Tapes and the Origins of MJ-12," *MUFON UFO Journal*, September 2007, p 6.

[704] Ibid.

apparently the grays from Zeta Reticuli,[705] had been manipulating human DNA for tens of thousands of years, had created the image of Jesus to teach humans peace and brotherhood, and interfered with human destiny.

All that seemed to be forgotten, however, when, according to Moore and Shandera, Shandera received the Eisenhower Briefing Document anonymously in the mail. Here was the smoking gun that proved Roswell was alien, and that there had been visitation for decades, at least according to the document, which is to say the crash occurred in 1947, a second in 1950. The implication was that visitation continued.

Interestingly, the Eisenhower Briefing Document, which reflected the state of UFO research in the mid-1980s, did not mention a crash on the Plains of San Agustin,[706] which Friedman believed had happened, no mention of the Aztec UFO crash which has again gained some popularity,[707] but does mention

---

[705] This is a reference to the Barney and Betty Hill abduction and the star map that Betty Hill said she saw on the alien craft. Marjorie Fish, in a herculean effort found a match to the pattern of the stars which suggested that the alien creatures were from the Zeta Reticuli star system. Better astronomical data, including the types and distances to some of the stars suggest that this information is now out of date and may be inaccurate.

[706] This is the Barney Barnett tale that was loosely linked to the Roswell UFO case by Barney Barnett. He told friends and family about finding a crashed UFO. His boss, "Fleck" Danley told Moore that it had been in the summer of 1947, but a diary kept by Ruth Barnett did not support this conclusion. If Barnett had seen anything, it was not in 1947 and was not on the Plains of San Agustin.

[707] A new book by Scott Ramsey, *UFO Crash at Aztec*, suggests that this event, in March 1948, is real but his evidence is thin, nearly nonexistent. Although Friedman now endorses this story, it is not mentioned in the MJ-12 papers. Had it been real, and if the EBD is real, then Aztec should have been mentioned. That it was not suggests that either Aztec is a hoax or the EBD is a hoax. The only logical conclusion is that both are a hoax, but in 1984, when the Eisenhower Briefing Document surfaced, the best information was that Aztec was a hoax, a position endorsed by Moore. See Moore, "UFOs and the U.S. Government: Part II," *MUFON UFO Journal*, December 1989, p. 9; Moore, "Crashed UFOs," 1985 MUFON Symposium Proceedings."

one just over the border from Texas in Mexico in 1950 that is a demonstrated hoax.[708]

In the mid-1980s, Bill Moore, in papers for the MUFON Symposium, offered his opinions on several other UFO crashes.[709] He said that after his investigation, he had concluded that Aztec was a hoax. Aztec, though mentioned by Howe as part of the Carter Briefing she saw, does not appear in the Eisenhower Briefing Document.

The Aquarius Telex had a provenance but not a good one. It was either created by Moore or Doty and Moore had said it was a retype of a real AFOSI telex. That ended its usefulness. The Carter Briefing had a provenance, which was Doty, or so Linda Howe said. Doty denied it, but Howe signed an affidavit attesting to the reality of her claims. Given what Moore had said, what Peter Gersten had said, and given what is now known, it seems that Howe is telling the truth. The Carter Briefing was as useless as the Aquarius Telex, and these documents were dropped from the discussions of MJ-12.

But the Eisenhower Briefing Document had a feel that suggested authenticity and many in the UFO community thought that it was the smoking gun. Critics said that it had no provenance. It could be traced to Moore and Shandera and no further. There wasn't a government source behind it and searches for any sort of corroboration had failed.

Failed, that is, until the Cutler-Twining memo appeared at the National Archives, in a box of recently declassified material. This document had a provenance… until the National Archives got into the act. They said that those copies circulating were "true copies" of a document in their possession, but not that it was a legitimate document. There were things wrong with it, not the

---

[708] This is the Willingham story that has been thoroughly debunked. Willingham never served as an Air Force officer, and if he did not, then his whole story collapses. In mid-1980s, many inside the UFO community, including Bill Moore believed the story. See also, Greenwood, "The MJ-12 Magic Act," *MUFON UFO Journal*, December, 1987 p, 13.

[709] William Moore, "Crashed saucers: Evidence in Search of Proof," *MUFON Symposium Proceedings*, 1985.

least of which was that it had been found in a box of material that had been reviewed prior to declassification, that it had required special handling which it had not received, and that it was supposed to have been reviewed by the NSA which it had not. Everyone knew where it originated, but few believed it was a real document.

Stan Friedman, the man who has defended MJ-12 for decades, thinks it is real, but it is disinformation. He believes the Cutler-Twining memo was planted at the National Archives to be found by UFO researchers.[710] It seems more likely that it was planted there, not by insiders as Friedman believes, but by those attempting to create an MJ-12 document that had a provenance. And if there had been other documents found in similar fashion, this might be the case, but none have surfaced.[711] Several UFO researchers have sorted through boxes and boxes of records, many from those who are alleged to have been members of MJ-12, but not a single reference has been found in all this time.

It is safe to say that all attempts to validate MJ-12 have failed. Proponents overlook some startling evidence that suggests MJ-12 is a hoax. It is clear from Friedman's comments about Project Aquarius and the Aquarius Telex that he is ignoring the circumstances of its appearance. Here are the first mentions of MJ-12. That would seem to taint the entire body of MJ-12 documents. Instead, they are ignored as if they don't exist.

The novel that Bob Pratt talked about, and the name of that novel, created before the 35 mm film showed up at the home of Jaime Shandera, tells all that is needed to know. Moore was describing MJ-12 to Pratt prior to learning

---

[710] It bears repeating here. Friedman has said, repeatedly, that he believes the Cutler-Twining memo is disinformation planted at the National Archives for Moore and Shandera to find. Letter to Randle, February 2001; see also Randle, *Case MJ-12*, pp 183 – 189 for additional discussion.

[711] This includes the infamous MJ-5, CIA memo which was published in Moore's newsletter, *Focus*, but has not surfaced anywhere else. According to Greenwood, "The MJ-12 Magic Act," *MUFON UFO Journal*, December 1987, "Type style [of the memo], placement of security markings, use of CIA letter[head] stationery instead of internal forms is atypical of CIA standards." In other words, it is a fake and its disappearance from the MJ-12 discussion seems to be proof of that.

anything about it, or so it would seem given the timing of various events. Their novel was the blueprint for MJ-12 and supporters ignore that at their peril.

The timing here was suspect as well. Moore is telling Pratt about a secret committee to exploit the Roswell crash, gives it a name, but Pratt refers to it in the book title as *Majik 12*. No one seems overly concerned that Moore was talking about this prior to the appearance of the Eisenhower Briefing Document. Just where did he get this information, or was he just making it up as he went along.

There is also Moore's reporting on what he learned in 1984 from a source that he refused to identify. He did write, "The reliability of the source of this information is believed to be good; but since the man is unwilling to go on the record and since his testimony remains almost totally uncorroborated, his testimony, at best can only be regarded as very interesting hearsay. It is reported here for purposes of information only."[712]

Regardless of the reliability of the source, the point is that Moore was publishing information that had been contained in the EBD before he had received it. It could be suggested that the EBD provided a second source for the information, but since the EBD has no provenance, it can't be used in that fashion. In other words, Moore knew what would be in it before he received it, yet claimed to be unaware of that information until it arrived.

This confusing little episode does nothing to validate MJ-12 or the EBD, but does throw up another cloud of suspicion, pointing the finger at Moore as the ultimate source of MJ-12. He had the Aquarius documents with their early mentions of MJ-12, he was writing a novel with Bob Pratt and one other (probably Doty) about all these mysterious activities, which was going to be called *Majic 12*, and then we have Moore providing information about the bodies found near the Brazel debris field months, if not years, before the EBD arrived.

This doesn't even address the problems with the El Indio – Guererro UFO crash as described by Robert Willingham and Todd Zechel. It is now clear that

---

[712] Moore, "Crashed Saucers,"

Willingham was inventing his tale of witnessing a UFO crash,[713] but in the mid-1980s, many inside the UFO community believed that Willingham was reliable. As demonstrated, Zechel, who had interviewed Willingham, was in communication with Moore and because many accepted the 1950 crash as real, it had to be included in the MJ-12 document.

Even with all that, there are those who suspect that MJ-12, while a hoax, was disinformation, meaning it came from a government source. That has never been established, and given what it known today, it seems that the sources of MJ-12 are some of those who released it into the public arena in the first place. Now we have dozens of MJ-12 documents making up more than 4000 pages, but the problem here, as it always has been, there is no provenance for them. They appear, almost magically, in mail boxes. Many of the new ones are from Tim Cooper, who was originally concerned that Friedman might think that he, Cooper, had forged them.

These new documents, at first, seemed to suggest that the forger, if such was the case, was an incredible historian. Obscure references to events were found to be accurate. Names attached to some of the documents, nearly unknown outside of government circles, were real. The historical context of the documents was accurate. These were the proof that was needed for MJ-12.

It was Friedman's research that proved many of those documents were, in fact forgeries. He learned from Dr. Bland that one of the first of the Cooper documents was a retyped letter that had been slightly altered. Friedman wrote, "With this unambiguous fraud as background, I became convinced that several other items were retyped and slightly changed versions of old memos or letters."[714]

---

713 The whole sad tale of Robert Willingham is laid out in Randle, *Crash*, 138 – 145, including the documentation from both NARA in St. Louis, which houses the military records of those no longer on active duty for any reason, and the Air Reserve Personnel Center in Denver, which houses records for the Air Force. Zechel selected the December 1950 because it corresponded to some documentation from the FBI that had nothing to do with UFOs but seemed to corroborate Willingham's tale.
714 Friedman, *Top Secret/Majic*, p. 159

What this discovery did was explain how someone could replicate, with such accuracy, the style of the time, the style of the various authors of those memos and letters, and why it did not take years to learn the proper history. All that was needed was a copy of the earlier document and a typewriter that featured an old font.

In the world today, it is strange that MJ-12 still has any following at all. Those who understand the problems with it, but want to retain something of it, say that it is disinformation. There is a grain of truth in the EBD, the Truman memo and the Cutler-Twining memo here and there, they suggest. But reality suggests that the best course of action is to set MJ-12 aside, stop wasting time, effort and money on research to verify it and begin to follow paths that will provide some useful information. Reduce MJ-12 to a footnote, which is all that it deserves. After more than three decades without any independent corroboration of it, it is time to give it up. MJ-12 has led nowhere and will lead nowhere in the future. It was a hoax that begin in 1980 and that's all it is.

Appendix B: The Plains of San Agustin Controversy

There were fragments of the Roswell UFO crash story in widespread circulation almost from the time that the revived original case broke nationally in 1978 with a publication of Major Jesse Marcel's interview in the *National Enquirer*[715] and later in the 1980 book, *The Roswell Incident*. The information pointed to a location near the tiny town of Corona, New Mexico, but Stan Friedman had found other information that suggested a different site or an additional site, this one over on the Plains of San Agustin. Friedman said that on October 24, 1978, a couple, Vern and Jean Maltais, approached him after a lecture in Bemidji, Minnesota, to tell him of Barney Barnett and a UFO crash that had gone virtually unreported.[716]

According to Friedman, they told him of their friend, Barnett, who had found a flying saucer crashed on the desert with four bodies lying near it years earlier. The military arrived after him and told Barnett and members of an archaeological team who had stumbled onto the crash to leave the area immediately. Friedman wrote, "They [meaning Vern and Jean Maltais] had no date for the Barnett story."[717]

Later, however, the date was narrowed down when Vern and Jean Maltais said they visited the Barnetts in February, 1950.[718] It was during this visit that

---

[715] Stan Friedman learned the first part of the story on February 20, 1978 and talked to Marcel the next day. Friedman, Stan. *Top Secret/Maji*c. New York: Marlowe and Company, 1996: 17; Len Stringfield reported that on April 7, 1978, he arranged for NBC radio newsman Steve Tom to interview Marcel. Stringfield, Len. *The UFO Crash/Retrieval Syndrome, Status Report II: New Sources, New Data*. Seguin, TX: Mutual UFO Network, 1980: 16; Bob Pratt interviewed Marcel for the *National Enquirer* on December 8, 1979. Pflock, Karl. *Roswell: Inconvenient Facts and the Will to Believe*. Amherst, New York: Prometheus Books, 2001: 225.

[716] Friedman, *Top Secret/Majic*. 18 – 19.

[717] Ibid. 19.

[718] Berlitz, Charles and William L. Moore. *The Roswell incident*. New York: Grosset & Dunlap, 1980: 53 – 58.

Barnett said that he had seen the crashed saucer. Jean Maltais, when asked where the craft had crashed, said, "…I don't exactly, recall. It was somewhere out of Socorro. He may have said exactly, but I don't recall. I remember he said it was prairie – 'the Flats' is the way he put it… Barney traveled all over New Mexico, but did most of his work directly west of Socorro."[719]

According to what the Maltais told William Moore during his interview with them, Barnett was out on assignment near Magdalena, New Mexico,[720] which is west of Socorro. The story was that Barnett thought at first it was a plane crash but when he got closer, crossing a little more than mile of desert, he saw that it was a saucer-shaped craft, twenty-five or thirty feet in diameter. It was a flash from the metallic craft that caught his attention.[721]

Maltais said that Barnett saw some bodies. They were dead, according to what Barnett told Maltais who, in turn, relayed it to investigators. There were bodies both inside and out, and Barnett said that the ones outside had been tossed out by the impact. He described them as being like humans but they weren't human. The heads were pear shaped, their arms and legs were skinny, the eyes were small and they had no hair. They were smaller than humans and the heads larger than those on human bodies.[722]

There were other witnesses there, according to what Maltais heard. These were archaeologists who had been working nearby and apparently had seen

---

[719] Ibid. 57 – 58.

[720] Note here that the section begins with Vern Maltais saying that Barnett claimed he was out west of Socorro, but later Jean said he didn't actually provide a location. It might have been out in "the Flats" but she wasn't sure.

[721] Berlitz and Moore. *Roswell Incident*, 54 – 55.

[722] Randle, Kevin D. and Donald R. Schmitt. *The Truth about the UFO Crash at Roswell*. New York: M. Evans and Company, 1994: 149 – 150; Vern Maltais, personal interview by Randle, August 1989, July 1990; Randle, Kevin and Karl T. Pflock. "Barney Barnett's Crashed Saucer: Where Did It Come From?" *International UFO Reporter*, (Spring 2003). 15 – 18. The information was developed from interviews with Maltais by Randle and Pflock, as well as that by other researchers.

the object fall the night before. Barnett seemed to think they were from an eastern university, but Moore reported they were from the University of Pennsylvania.[723]

Not long after Barnett arrived, the military turned up, and took over. They condoned the area and escorted everyone off the site after warning them not to talk about it.[724]

Barnett apparently didn't keep the secret very long. In 1947, J. F. "Fleck" Danley was Barnett's boss. Bill Moore asked Danley about the story. Danley said that Barnett had come into the office one day and said that the flying saucers were real. Danley had been in a bad mood, said that they weren't and wasn't interested in discussing it further. But Danley, thinking about it, felt bad, so he asked him about it. Danley said Barnett mentioned something about the "Flats" but couldn't remember much else.[725]

Moore returned four months later and again talked to Danley. At that time Danley said that he remembered the date and was sure that it was sometime early in the summer of 1947.[726] Other interviews with Danley suggested that he didn't have a clear memory of when Barnett had told him about the crashed saucer, and in fact, didn't have a clear idea where Barnett had been on the day he told Danley about the saucer.[727]

Danley mentioned that Barnett was a soil conservation engineer who worked out of Socorro and a satellite office in Magdalena. He did mention that Barnett occasionally made it into Lincoln County, New Mexico but that was

---

[723] Berlitz and Moore. *Roswell Incident*, 54.

[724] Randle and Schmitt. *Truth*. 149.

[725] Berlitz and Moore. *Roswell Incident*, 61.

[726] Berlitz and Moore. *Roswell Incident*. 61.

[727] J. F. Danley, personal interviews by telephone conducted by Kevin Randle, Oct. 1990, June 1991.

rare. Interestingly, Danley remembered Barnett said something about Carrizozo. He said that Barnett told him about the crash but he didn't remember him saying anything about bodies or creatures.[728]

To make this even more complicated, Friedman said that he had "reinterviewed Danley and several others who knew Barney in 1990 [clearly Friedman means he interviewed the people in 1990 and not that they knew Barnett in 1990] and again was told 'in the Plains.'"[729]

These were not the conclusions of others. Jaime Shandera, at the UFO Expo West in Los Angeles on May 11, 1991, was lecturing about the Plains of San Agustin. He had this to say:

> The people that supposedly found stuff in Socorro did not find stuff in Socorro. The party of archaeological people and the Barney Barnett part of the story; they were at the Corona site, not in Socorro. I know [this is] the way you understand it because it's the way it's always been written and even the way it was written in *The Roswell Incident.* That's wrong. There is new evidence that it was all in the Corona site. The way it happened was this – there were not two sites that were more than one hundred miles or so apart … and the so-called Roswell site was just outside of Corona. The archaeologists and Barney Barnett part of it, that was over in Corona. There was no person that found anything in San Agustin.[730]

---

[728] Ibid. Interview conducted by Randle, May 14, 1991.

[729] Friedman, Stanton T. "Basis for Belief in Two UFO Crash Sites in New Mexico in July 1947." Unpublished article, March 17, 1991.

[730] Huneeus, Antonio and Javier Sierra. "Excerpts from Lecture by Bill Moore and Jaime Shandera." UFO Expo West. Unpublished notes. May 11, 1991.

There were others who talked to Barnett about the case. Stan Friedman interviewed a military reserve officer from New York, William Leed, who said that in "the early 1960s," a fellow officer had told him about Barnett. Leed arranged to go to New Mexico soon after to talk to Barnett about the crash. Leed did hear the story from Barnett, thought that Barnett was sincere and was impressed that Barnett wouldn't talk to him until Leed showed him a military ID. Leed made it clear that he was there on a personal quest and this had nothing to do with the military or official business.[731]

There is nowhere in the various interviews with Leed that provide a date or a location. It is assumed that the date is early July, often on the second, and the location is out on the Plains, not far south of Highway 60. Friedman wrote of this meeting between Leed and Barnett, "No reason to think it was other than 'on the Plains,'" but offers no quotes suggesting that Leed believed this.[732]

In the 1990s, Friedman placed an ad in the Socorro newspaper, asking for anyone who had information about the Barnett story to contact him.[733] One of those who did was Harold Baca who in the 1960s lived across the street from Barnett. He said that as he helped Ruth Barnett take care of an ailing Barnett, he heard about the crashed flying saucer from Barney who was convinced that his cancer was the result of breathing contaminated air near the wrecked saucer

---

[731] Bragalia, Tony. "The 'Other' Roswell Crash: The Secret of the Plains Revealed." The UFO Chronicles found at http://www.theufochronicles.com/2010/05/other-roswell-crash-secret-of-plains.html (accessed January 22, 2015); also Stan Friedman Internet Letter published by andrskondras@gmail.com (January 19, 2015).

[732] Friedman, Stanton T. "Basis for Belief in Two UFO Crash Sites in New Mexico in July 1947." Unpublished article, March 17, 1991.

[733] Friedman, Stanton and Don Berliner. *Crash at Corona*. New York: Paragon House, 1992: 88.

and the bodies.[734] Baca seemed to think that it had happened "out on the plains."[735]

Friedman found addition witnesses who suggested that Barnett had said that he was on the Plains. These included the late Marvin Ake, who said he had heard the story "many years earlier, but provided no date and an unidentified and retired postmistress from Datil who said the saucer had been trucked through Magdalena at night.[736] Others in the area also remembered discussion of a flying saucer crash on the Plains, but some of them couldn't remember if they had heard about it in the late 1940s, or sometime after the publication of *The Roswell Incident* in 1980.[737]

The problem with all this is that Barnett died in 1969 before anyone was talking about flying saucer crashes in New Mexico in 1947. The interviews are all with people attempting to remember what was said decades earlier and give it some sort of time frame. The contamination of the witnesses can be seen in the evolution of their stories. Vern and Jean Maltais, for example, and according to what Friedman has written, had no date for the story. It wasn't until later that they seemed to narrow it down and thought it could be July 1947.

A secondary problem was that all these witnesses were second hand. They had seen and heard nothing themselves, but had all gotten the story from the same source, Barney Barnett. Almost universally they believed Barnett to be a sincere man who wasn't given to the invention of tales, but that doesn't change the fact that it all derives from a single source.

---

[734] Harold Baca. Telephone interview conducted by Kevin Randle, June 1991.

[735] Friedman and Berliner. *Crash at Corona*. 88. See also Friedman, Stanton T. "Basis for Belief in Two UFO Crash Sites in New Mexico in July 1947." Unpublished article, March 17, 1991.

[736] Ibid. 88. See also Friedman, Stanton T. "Basis for Belief in Two UFO Crash Sites in New Mexico in July 1947." Unpublished article, March 17, 1991.

[737] Johnny Foard. Personal interview by Kevin Randle in Magdalena, February 9, 1992.

That all changed when the NBC Television Network aired their report on the Roswell UFO crash on *Unsolved Mysteries*. One of those who watched was Gerald Anderson, and after the show's second airing in January 1990, he wrote to both Friedman and me. I was the first to interview Anderson and we talked for fifty-four minutes on February 4, 1990.[738]

Anderson said that he had been six-years-old when he was on the crash site. He then said, "I also mentioned that a few were alive… I can visualize the crash site any time I want by closing my eyes…"[739]

He then went on to describe what he remembered about the events of early July, 1947. He said:

> When we got there, we came over, well, we parked, walking around and we came over a kind of small run. This was an arroyo… it's like the Plains created a basin. We were walking down into this thing. There was an object and it wasn't stuck in the side of the gully, it was kind of impacted on a small slope.
>
> …I'm a kid and this means nothing to me, except I remember my dad grabbing me and saying, "Oh my God, let's get the hell out of here." And my cousin said, "Wait a minute. Look at this damn thing."
>
> They went over toward it. My dad told me to stay right here. Don't move. He kept saying, "Well, it might be a bomb or something like that."

---

[738] Randle, Kevin D. and Donald R. Schmitt. "Missing Time." *International UFO Report*, 17,4 (July/August 1992) and based on interviews conducted by several including Randle, Friedman and newspaper and radio reporters.

[739] Gerald Anderson. Telephone audiotaped interview with Kevin Randle. February 4, 1990.

We, of course [Anderson and his cousins] were craning our necks… we wouldn't stay put. We went over there. They were… they [the alien creatures] were up underneath this thing. They weren't scattered around all over the area. This vessel was not torn open. The side was torn out of it and there was a lot of cables and junk like that… There were four occupants. Three were on the ground underneath the thing and one was sitting next to it, like was sunning himself…. This creature was alive and this craft wasn't torn open… You couldn't see inside this thing.
We hadn't been there more than a minute, two minutes at the most and a group of people came up from the southeast. There people were from the University of Pennsylvania. They were doing something at some ruins. There were a bunch of old cliff-type ruins up on the rim.

Anyway, they were up there doing some excavations to the southeast along this rim and it wasn't that far away. It was like… they saw this thing, this light, the night before and they thought it was something else. They didn't realize what it was. They came down the next morning to investigate it… they had walked so I'm assuming three or four hours walking from wherever…

…I was aware that this person or this creature or whatever you want to call it, humanoid or whatever, was terrified of its fate. Every move that anybody made it was like it expected to be hit…

It wasn't very long before the military types showed up. There was a captain… he had red hair. He was an asshole. He threatened everyone with the most incredible things you could possibly believe… First off, he told my father if he repeated this

> he would see to it that he would spend his entire life in a military prison and he would never see his children again... These people [the military] were heavily armed and they were very ill-tempered... To make a long story short, they ran everyone off. Now one thing they messed up on. Some of those people from those diggings up there, had pieces of the wreckage in their pockets.[740]

Anderson then tried to explain where, on the Plains of San Agustin, the crash had taken place. Using the Very Large Array, which straddles Highway 60 to the west of Socorro, he suggested that the crash was to the north of the highway, and looking over a ridgeline near Magdalena and away from the radio observatory antennae.

He also attempted to explain that the cliff dwellings were to the southeast of where the UFO had crashed. The cliff dwellings were not up on a high ridge they were up on a kind of a bluff.

He then suggested something that he couldn't have known. He said that the archaeologists who had taken some of the debris, some bits of the metal, had "conspired" to bury it. According to Anderson, they were afraid to say anything because the guy was talking about having people shot. He then began to talk about the aliens again. He said:

> That one individual was still alive when they took him away. He had a slightly perceptible nose. I never saw a mouth... I think the thing that struck me the most were the eyes. The size of the eyes and the gentleness of those eyes. His head was... a child with hydrocephalus... Have you ever seen that? ...The person was much paler than we are. This could be shock.... I didn't see hair on his head.

---

[740] Gerald Anderson audiotaped telephone Interview with Kevin Randle, February 4, 1990.

> He was wearing… all of them, all four of them, were wearing what appeared to be uniforms. They were small in stature. Their hands were the kind of hands you'd see on a violinist. They were almost effeminate… they were long and slender…

I asked him about blood in a very leading question to see if he would take the bait and he said that there seemed to be bruising on the one that was alive. The bruising was more of a scrape like he had been thrown against something. He didn't move much and when he did it seemed that he was in pain. He added to this by saying:

> He had one side of his face had been scraped… like if you took a nose dive off a motorcycle into the rocks… it was, it wasn't oozing blood.[741] It was bloody, like a scrape. And it seems to be it was red.
>
> But this guy was very pale. His eyes were almost… I don't know how to explain this… they were oval-shaped and very, very big and very, very gentle. They were bluish. Not blue like blue in human eyes but blue like… sort of a milky blue…
>
> Now I do remember one other thing… there was a very strong odor in the area. And it was the smell of acetone but it was like acetone, like acetone would be if you were around a lot of it. Almost solvent smell… I do remember my dad telling someone, "Put that damned cigarette back. Don't light that."

---

[741] In the discussion of Anderson's veracity, this became a point of contention. The transcript is confusing there, suggesting that the scrape was oozing blood but that it wasn't. Given the discussion, it seemed that the scrape was bloody, which suggested it was oozing blood, but that the bleeding had stopped.

He did describe the craft by saying that it was shaped like a discus. He didn't see the top part so wasn't sure if there was a dome on top. It was tilted away from them so that Anderson only saw the bottom. He didn't see any sort of hatch.

That was all that Anderson had to say during that first interview. The conversation had lasted for fifty-four minutes. Anderson suggested that if I had any additional questions to please call him back. I never spoke to him again.

Here, finally, was a first-hand witness to a crash on the Plains in the correct time frame. It was no longer hearsay from second-hand sources but the observations of someone who was there. No longer were researchers left with only the ambiguous comments made by Barnett that could have been misunderstood or misinterpreted. Now there was someone who had been there, seen the wreck, and who suggested that there was a family diary that would prove that he was telling the truth.

Because he was a first-hand witness, other researchers wanted to talk to him and hear for themselves what he had to say. Stan Friedman had also received the information about Anderson from the *Unsolved Mysteries* call center and interviewed Anderson not long after I did.[742] According to what Friedman learned, Anderson said that they had driven out into that area looking for moss agate. They had driven into the area, parked, and then began walking. Anderson told Friedman:

> What we saw was a silver object, a circular silver object stuck in the ground, kind of at an angle. It was jammed into a hillside. Seemed to me there was more than one tree that had been knocked down…
>
> Initially, I really didn't realize what everyone was hollering about. And then when they said, "Crash!," and then when I finally saw it… it crossed my mind that it was a dirigible, a

[742] Friedman interviewed Anderson for the first time on February 16, 1990.

> blimp that had crashed… my brother said, “That’s a goddamn spaceship! Them’s Martians!”
>
> …And there were three of these crewmembers laid out on the ground, under the edge of this thing, in a shaded area, and there was one sitting upright. The ones that were laying on the ground, two of ‘em weren’t moving at all, they were just laying there.
>
> They looked like they had some sort of bandages on ‘em. One of them had it over his arm… The one I touched had it around his midsection and partially over his shoulder. It appeared to be, as I think back on it now, the one that was still alive and moving had given first aid to the others. And the one next to him was breathing very erratically and its chest was heaving in an unnatural way. He was obviously in distress…[743]

Anderson continued to describe the situation. He though that one of the dead creatures had entered *rigor mortis* because he thought it had stiffened up. But he also said that he thought the creatures were dolls even though he had seen one of them walking around.

He described how cold it was near the craft. The metal was cold, or as Anderson said, “It was very cold!”[744]

Anderson said that he had walked around to see where his brother was and found him up on the edge, looking into the gash on the side. He said:

> …I could see, and was directly into it looking through the outer hull and into another bulkhead, and that bulkhead was

[743] Friedman and Berliner. *Crash at Corona.* 90 – 92.

[744] Ibid. 92.

> shaped exactly like the other hull… You could see through the rip and you could see what looked like components… They were all seemingly hooked together with these cables – clusters of threadlike material in the form of a cable – and one of these was hanging out of that rip. There were several hundred strands per one of these clusters… they blew in the wind like a horse's tail, except there were lights all over the ends of them… And down the center between these things – they were all very neat little rows – there was, like scribbling… [On] each one of these component boxes there would be scribbling, it was almost a pink color on sort of a brown, woodlike background. Like it was writing or symbols that explained what this thing was.[745]

The other part of the interview describes the arrival of others on the scene including the archaeologists, Barney Barnett and the US Army. According to Anderson, the leader of the archaeologists was Dr. Buskirk. Buskirk tried to speak to the living creature using a variety of Earth-based languages.[746]

He also said that another guy showed up. He told Friedman:

> And there was this guy with a pick-up, and it was one of those old model pick-ups and it had a whip antenna on it, like a police car has. He walked over there and he looked like Harry Truman: he had a real ruddy complexion, he wore glasses and he had sort of khaki work clothes on and he had a straw hat… and he got to talkin' to Dr. Buskirk and Ted and my brother and my Dad and he said that he had seen it from out on the Playa. He had apparently worked out there and he acknowledged that he made maps or something like that.

---

[745] Friedman and Berliner. *Crash at Corona*. 93 – 94.

[746] Ibid. 95.

Bob Oeschler interviewed Anderson on March 24, 1991 on the *Hieronimus and Company* radio program on the American Radio Network. Now Anderson expanded on what he had told others. He was saying that the event had taken place on July 5, 1947. The date was found in a diary that had been supplied to Anderson by another relative.

Oeschler asked about the size of the craft and Anderson said, "I've heard other members of the family that were there estimate that this thing was like 50 foot in diameter."

They then began to talk about the damage that had been done to the craft. There was a gash in the side and there was debris scattered around the crash site. Anderson said that then, in 1947, it wasn't his impression that the craft had collided with another but then said:

> Looking back on it as an adult… that would be my impression now… owing to the fact there was very little damage to this craft when it collided a boulder field on the side of this hill. It sets in my mind that it would take an object of equal strength to puncture the outer hull like that…
>
> What you could see through the gash and bear in mind that the craft was kind of tipped up to one side so that you could stand and look pretty much into the gash… ah, what you could see in there appeared to be a double bulkhead. And on this double bulkhead there seemed to be some kind of components set in arrays and there seemed to be symbols of some kind there… Now I describe them several times in the past as pink scribbling… ah, for the lack of a better term… I would think that it would look like it [the symbols] might have been painted on…
>
> [The gash] appeared to be puckered inward if that's a good description… just as if something had shoved or rather or a

> sharp edge had come against it, like the shape of the outside of that particular disk. And it just literally caved it in and ripped it open. There were a lot of sharp metal pointed inward from the impact… It was fairly sharp from my perspective and it… the metal wasn't flexible. It was very solid object.[747]

The discussion then turned toward the alien creatures that Anderson again described as looking like dolls but then they began to move. He said that they were laid out, under the rim of the craft. To his surprise, as he approached the craft, it began to get colder. He said that if you stood next to it, you "felt as if it was emitting cold air but there was no sensation of the wind blowing on you or anything like that. If you touched it, it felt as if you had just taken in out of the refrigerator. It was ice cold."[748]

He went on to describe the alien creatures. He said, "They were approximately three and a half to four feet high. They were about the same height I was. And their heads were much larger than ours. They had real slender features. They all seemed to be wearing the same kind of clothing. Sort of silver gray shiny type of material that seemed to be all one piece like the suits were continuous… You could tell that they came together in the front but there was no visible evidence of a zipper or anything like that."

The important part of the interview was his description of the eyes. He said, "Their eyes were enormous. That was the most startling thing about them. They were very black and very large. Almost oval shaped."[749]

The last part of the Oeschler interview was about the others who showed up at the crash site. He mentioned that it was about forty-five minutes to an hour when the archaeologists arrived. He said:

---

[747] Gerald Anderson interviewed by Bob Oeschler on the *Hieronimus and Company* radio program on the American Radio Network, March 24, 1991.

[748] Anderson interview by Bob Oeschler March 24, 1991.

[749] Ibid.

> ...there was an archaeological group... there was a professor and his students and we understood they came from the University of Pennsylvania... For a while there this professor, this Doctor Buskirk and my uncle and my father continued to try to communicate with the one creature that was apparently unharmed and apparently had been giving first aid... There was two that were obviously dead. One that was somewhere in between, obviously critically injured and one that had no apparent injuries at all.
>
> Well, there was a man... and we're not sure who this man was. He, I've described him repeatedly as having looked like Harry Truman... He came from the southwest, from out on the Plains... He worked out there... We believe now, these many years later, that that man was in fact Grady Barnett who was one of the people mentioned in the book, *The Roswell Incident*. I remember this man as being rather large.[750]

The last group to show up, according to Anderson, was the military and he said that it was invasion force strength. He said:

> These people got out. They were all armed. They were all wearing MP brassards on their arms. There were all wearing sort of dust brown uniforms. The man who was in charge was an Army captain who had this flaming red hair and this rudest attitude that you have ever seen in your life. The sergeant who was with him was kind of an anomaly, I guess at that time, looking back from my perspective as an adult. This was 1947. This guy was a sergeant and he was black. He was definitely

---

[750] Ibid.

> second in command. There was no question about that... all of them were very rude, very forceful, they pointed weapons at people, they threatened to shoot people. We were summarily rounded up like cattle…
>
> They threated my father, and they threatened my uncle with statements that if you ever want to see your children again and if you don't want to spend the rest of your life in a prison you will keep your mouth shut.[751]

The interview ended with them discussing Anderson's background. There had been other interviews that created some problems for Anderson and his tale. In an article published in the Springfield, Missouri *News – Leader*, Anderson said that the purpose of the trip was to search for moss agate. Anderson told reporter Mike O'Brien, "The upright creature 'turned and looked right at me, and it was like he was inside my head – as if he was doing my thinking, as is his thoughts were inside my head… I felt that thing's fear, felt its depression. I relived the crash. I know the terror it went through."[752]

Cracks in the Anderson tale appeared almost immediately. His account was at odds with that had been developed by other investigations. Anderson had said that he and his family had arrived at the crash site first.[753] But according to those who had heard the story from Barney Barnett, he had arrived first.[754] Of course it could easily be that the memories of a five-year-old were

---

[751] Anderson interview by Bob Oeschler March 24, 1991.

[752] O'Brien, Mike. "Fact or Fantasy?" *Springfield News – Leader*. December 9, 1990, 5 – 7.

[753] Friedman and Berliner. *Crash at Corona*. 90 – 91.; Anderson, telephone interview with Randle, February 4, 1990; O'Brien, Mike. "New San Agustin Crash Witness." *MUFON UFO Journal*, 275 (March 1991): 3 – 5.

[754] Randle, Kevin D. and Donald R. Schmitt. *UFO Crash at Roswell*. New York: Avon Books, 1991: 180 – 181.

flawed or that the testimony of those hearing the story from Barnett had assumed he was first.

There were other problems and these were an outgrowth of the tales told by Anderson. He had provided long interviews with various researchers and reporters so there was a record of what he had said. In the first interview conducted with him, he had said, "His [the alien's] eyes were almost… I don't know how to explain this… they were oval-shaped and very, very big and very, very gentle. They were bluish. Not blue like blue in human eyes but blue like… sort of a milky blue."[755]

However, when interviewed by John Carpenter some time later, Anderson said that the eyes were "almost black." According to Carpenter, "…we asked him again later (while he was making drawings) what he had meant. He stated (unfortunately not taped) that the black eyes had a bluish tinge, giving them a 'murky-blue' appearance. This might resemble the bluish shine of black satin or iridescence of a butterfly's wings."[756]

This is not a minor change in the description of the alien creatures because Anderson had made it clear, both in that first interview and in the later interviews. Anderson had been quite clear about the milky blue color of the alien's eyes and it wasn't an error on the part of the transcriptionist as Carpenter claimed.[757] Carpenter, as well as others, thought that the murky blue eye color explained this.

The question had been why Anderson would say, "milky blue?"[758] The answer was on the cover of *Transformation: The Breakthrough* by Whitley

---

[755] Anderson, telephone interview with Randle, February 4, 1990.

[756] Carpenter, John S. "Gerald Anderson: Truth vs. Fiction." *MUFON UFO Journal*, 281 (September 1991): 3 – 7, 12.

[757] Ibid. 7.

[758] It is clear from the tape of the February 4, 1990, taped interview that Anderson had said "milky blue" and not "murky blue." As it turned out, this was an attempt to explain a discrepancy between two very different descriptions.

Strieber.[759] On the cover of the hardback edition was a stylized illustration of a grey alien with large, light blue eyes. It could be said that these were "milky blue."

Had this been the lone problem with the Anderson testimonies, it could be overlooked. Others, however, began to develop. Anderson mentioned that there was a family diary in which the incident had been mentioned. This would be evidence of an event and provide an exact date. A document created in 1947, if it could be validated, would bring a new level of credibility to both the Anderson tales and the idea of a crash on the Plains of San Agustin.

Friedman reported, "The closest thing to a description of the post-crash activity comes from a diary kept by Jerry Anderson's uncle, Ted. While the copy of the diary acquired by Stanton Friedman through a complex series of negotiations is not an original (the ink did not exist in 1947), it may be one of several copies made by Ted or Victor [Anderson] in an effort to assure the information's survival."[760]

Carpenter supplied more information about the diary. He wrote, "Verification of Gerald's involvement comes in a letter sent directly from his cousin, a Roman Catholic nun, in Colorado to Stanton Friedman in Canada. She states; 'My family has been plagued by this incident for years and it is far beyond time that such should stop. Why Gerald would wish to reopen this is completely beyond me… My father (Uncle Ted) was obsessed with this unearthly horror and kept several journals to prevent others from getting to them… wreckage and debris from the crash… out there near the caves."[761]

---

[759] Strieber, Whitley. *Transformation: The Breakthrough*. New York: Beech Tree Books (William Morrow), 1988. Cover of the hardback edition.

[760] Friedman and Berliner. *Crash at Corona*. 113.

[761] Carpenter, John S. "Reliving July 5, 1947." *MUFON UFO Journal*, 275 (March 1991): 9. Letter to Stan Friedman from Sister Mary Tarrasas with the note, "Opened by STF October 2, 1990."

Friedman, to his credit, submitted the diary for forensic analysis. Richard L. Brunelle, a forensic chemist provided his results on October 18, 1991. He wrote:

> Samples of the ink from all six pages were examined visually and microscopically, by infrared reflectance and infrared luminescence, by thin-layer chromatography (TLC) and TLC densitometry. The following results were obtained:
>
> 1) The same black fountain pen ink formulation was used to write all six pages. This particular ink formulation matches a Sheaffer Skrip black fountain pen ink formulation that was first manufactured in 1974. The combination of dyes present in this ink was not used until approximately 1970…
>
> Conclusion: In the opinion of the undersigned the writing on the six pages could not have been written in 1947, because the ink used did not exist then. The writing was most probably made some time after 1970.[762]

The diary hadn't been written by Uncle Ted, or at the very least, that version of it hadn't because Ted had died prior to 1974. Excuses were offered, including the one that family members, or Ted, had made handwritten copies.[763] But seemed unlikely in an era of copy machines available in any number of places that anyone would hand copy them.

Another controversy erupted over the leader of the archaeological expedition. Anderson eventually said that the man's name was Buskirk and that he with his students had come from the University of Pennsylvania. Anderson

---

[762] Richard L. Brunelle, letter to Stanton Friedman, dated October 18, 1990.

[763] Friedman and Berliner. *Crash at Corona*. 113

when asked if he was sure they had come from Pennsylvania said, "Well, yeah, I think so, but it has been a long, long time. Anyway, yeah, I'm sure that's right. They worked with the university and I'm thinking Pennsylvania."[764]

Tom Carey decided he might be able to find Dr. Buskirk. Carey wrote:

> …I called… Stanton Friedman, to find out the current status of the search, if any, for the archaeologists. It was my belief because of my anthropological background and knowledge of the literature of the discipline, I might be able to trace the archaeologists via a paper trail, if one existed. All that I would need, I reasoned, would be a few scraps of information to point me in the right direction. Friedman provided me those scraps. He related that the archaeologists had not been located and that, with respect to Dr. Buskirk, "I have made hundreds of phone calls all over the place looking for him…" Friedman also gave me a photocopy of the "Indentikit" sketch of Dr. Buskirk that Anderson had apparently drawn by Gerald Anderson after one of his recall/regression sessions… Written above the sketch were the subject's vital statistics: his name was "Adrian Buskirk," and he was in his late-30s to mid-40s; he stood about six feet tall and weighed in excess of 200 pounds; though balding, he had brown hair… brown eyes, and a "clean, ruddy, weathered" complexion…[765]

Carey had been unable to locate anyone named Buskirk in the anthropology literature which suggested that Buskirk hadn't produced any articles for the various journals. Carey then began a search for other publications. He wrote:

---

[764] Anderson interview with Randle, February 4, 1990.

[765] Carey, Tom. "The Search for the archaeologists." *International UFO Reporter* 16,6 (November/December 1991): 4 – 9, 21.

> On the chance that Adrian Buskirk or Van Buskirk was a prolific writer, my first stop at the card catalog had been to see if there were any references listed under those names. There weren't any but there was a reference card for a book, *The Western Apache: Living with the Land Before 1950* (1986), written by a Winfred Buskirk...
>
> ...I had written to the publisher of his book for a dust jacket which, I hoped, would contain Buskirk's picture... About two weeks later a long brown cardboard tube from the University of Oklahoma appeared in my mailbox... when I peeled back the flap of the dust jacket, I saw, staring back at me, the gentle face of Anderson's Dr. Buskirk...[766]

With Buskirk identified and eventually located and with his denial that he had been anywhere near the Plains of San Agustin in July 1947 nor had he seen any crashed flying saucer, the question became, "How did Anderson identify this real man?" It turned out that both had lived in Albuquerque and the first thought was that they might have been neighbors, but that wasn't the case. I eventually learned that Buskirk had taught, among other things, anthropology at the Albuquerque High School. Anderson had attended the Albuquerque High School at the time that Buskirk taught there. This seemed to be the connection.

Buskirk was quite helpful there and he was curious about how Anderson had tapped him as the archaeologist on the Plains. He was able to contact friends in Albuquerque who accessed the microfilm records from the Albuquerque High School. He wrote:

---

[766] Ibid. 7 – 8.

> They went to the Public Schools microfiche records and came up with this information... Gerald Francis Anderson... in the fall of 1956 enrolled at Highland High School (south of East Central Ave.), but after less than a year moved to Indianapolis.
>
> Then on 2 (or 3) October 1957, he enrolled at Albuquerque High, stayed a little over one year, then checked out 3 (or 2) October 1958... Now at Albuquerque High he was enrolled for a semester of Anthropology. This was a course I taught in the fall, so he must have taken it in 1957 – 1958 and, I presume passed it with credit. (I failed no one if I could help it).[767]

Buskirk gave the names of those he had talked to so that the information could be verified. Two of those people had been contacted by Friedman as well, at least according to Buskirk, so he should have had the same information.[768]

Those supporting the Anderson tale suggested Buskirk had not seen the transcript himself and therefore that didn't directly contradict Anderson. However, Buskirk had provided the names of those who had helped him in case others wished to verify the information. He wrote, "You [Randle] will probably want to call [named three officials] for a verification."[769]

Given all this, and the Identikit sketch provided by Anderson, it was clear that Winfred Buskirk was the Adrian Buskirk of Anderson's flying saucer tale. Friedman even wrote, "There is no reason to reject the idea that Winfred

---

[767] Winfred Buskirk, Letter to Kevin Randle August 8, 1991.

[768] Ibid.

[769] Ibid. I did call and while I was on the telephone was told that he was looking at the transcript. He verified that Anderson had taken Buskirk's class.

Buskirk was indeed present and will go to his grave without saying anything about it."[770]

But Buskirk did say something about it. In a letter to Randle he wrote, "Considering the mentality of Friedman and all, there is no way I could prove I was in Arizona all of July 1947, and I am not inclined to try.... The ceremonial pictures in *The Western Apache* and fairgrounds picture were all taken around July 3rd, 6th or 7th. I was certainly too busy on the reservation to be engaged in any archaeological sideshows."[771]

A final bit of evidence that suggests that Buskirk wasn't on the Plains in 1947. The Identikit sketches of Buskirk showed the man as he appeared in 1957 and not 1947. Tom Carey assembled what might be considered a photo lineup of Buskirk, showing him as he appeared in 1947, 1957 and in 1986. The Buskirk of 1947 was clearly thinner than the man of 1957.[772]

Had the story ended at that point, the case against Anderson seemed to be quite strong but it wasn't over. Although the argument about Anderson in Buskirk's class seemed to lack the final bit of evidence even though Buskirk was quite clear on the point, circumstances provided another opportunity to verify something that Anderson said and the evidence he offered to prove it.

After I left a message on Anderson's answering machine, he called back. With his permission, I taped the call so that there was a physical evidence of the length of the call. John Carpenter, in his article, *Gerald Anderson: Truth vs. Fiction*, mentioned, four times that my call with Anderson was only 26 minutes long. He was so sure about this that on August 12, 1991, he wrote, "I

---

770 Friedman, Stanton T. and Don Berliner. "Yes, There Was a Saucer Crash in the Plains in 1947." *The Plains of San Agustin Controversy, July 1947: Gerald Anderson, Barney Barnett, and the Archaeologists*." Chicago: CUFOS and Washington, D.C.: FUFOR, June 1992: 11

771 Exhibits. "Yes, There Was a Saucer Crash in the Plains in 1947." *The Plains of San Agustin Controversy, July 1947: Gerald Anderson, Barney Barnett, and the Archaeologists*." Chicago: CUFOS and Washington, D.C.: FUFOR, June 1992: 74 – 75.

772 Ibid. 72 – 73.

would love to hear the tape, if you would be willing to share a copy of all 26 minutes... Here are some nice examples of 'good science' which may disturb you. As I stated in my paper, 'verified phone records', the call from Gerald to you on 2-4-90 was 26 minutes long. Since reading these words is not good enough for you, maybe seeing a Xerox copy of the phone record will help you to understand we have the hard evidence (Check your tape player – maybe it's running slowly)."[773]

Included in the letter was a copy of the telephone bill which did show the telephone call from Anderson had lasted 26 minutes. It listed my telephone number and the charge for the call. There was also another long distance call on the bill but that had nothing do to with the investigation.

This didn't answer the question of the discrepancy, it merely ignored it. Copies of the tape, first shared with Linda Howe for her use only, then copied without permission, were sent to Carpenter. In his paper for the CUFOS/FUFOR monograph, Friedman acknowledged the tape's 50 minute length but has never made an effort to explain the missing time.[774]

Anderson was served by Southwestern Bell, and a telephone call to the business office solved the dilemma. The response was a copy of the telephone bill from their records. This showed that the call was 54 minutes and the charge was more than that on the Anderson supplied bill.[775]

The only conclusion to be drawn was that Anderson had altered his telephone bill and then presented it as evidence that our telephone conversation had lasted only 26 minutes. Carpenter's claim of verification came from the altered bill he had seen and not a copy that had originated in the offices of Southwestern Bell.

---

[773] Letter from John Carpenter to Kevin Randle, August 12, 1991.

[774] Friedman and Berliner. "Yes, There Was a Saucer Crash in the Plains in 1947." *The Plains of San Agustin Controversy, July 1947: Gerald Anderson, Barney Barnett, and the Archaeologists*." Chicago: CUFOS and Washington, D.C.: FUFOR, June 1992.

[775] Randle, Kevin D. and Donald R. Schmitt. "Missing Time." *International UFO Reporter* 17,4 (July/August 1992): 21 – 23.

The publication of the article about the Anderson telephone bill forgery seemed to penetrate the wall built around Anderson. John Carpenter wrote, "The account of five-year-old Gerald Anderson and his family stumbling across a crashed silver disc and the four alien bodies has been slowly eroding over the past year. Attempts to verify various aspects of his life keep falling short; other problems fail to become resolved and only seem to breed others.... However, recent events have now cast grave doubts on Gerald's story and his own truthfulness with us."[776]

Carpenter then went on to describe some events that had taken place on September 19, 1992 at the Midwest Conference on UFO Research. Carpenter wrote:

> On that Saturday night Gerald Anderson asked to meet with a small group of researchers of my [Carpenter's] choosing that could witness several documents he wanted to present... Gerald presented his military papers for our inspection but had whited out his serial number [something I would have done as well]. Also listed were several secret operations in the South Pacific that he had been involved in as a member of the Navy Seals.[777] ...However, Gerald then apologized to Stan and myself for having a fake phone bill statement toward the goal of 'making Kevin Randle look bad.' Originally, Randle had indicated he and Gerald had a long friendly conversation on February 4, 1990. Gerald claimed it was much shorter and not all that friendly.

---

[776] Carpenter, John. Gerald Anderson: Disturbing Revelations. *MUFON UFO Journal* 299 (March 1993): 6 – 9.

[777] There is no independent evidence that Anderson ever served as a Navy SEAL. His name appeared on a Wall of Shame hosted by former SEALs and listed as a "wannabe."

> ...Within a couple of weeks he produced a xerox [sic] of a microfilm record demonstrating a 26 – minute phone call with Kevin Randle. It never seemed like any big deal and rather a minor side issue at best. The phone bill appeared authentic and nobody indicated any suspicion until Kevin Randle related that he had a tape recording of the 50+ minute phone call...
>
> I finally was able to learn that Gerald had indeed had a friendly 54 minute phone call just as Randle had claimed... at this meeting on September 19, he then produced a second "original" phone bill – this one indicating a 28 – minute phone call! I then announced I had the tape that runs 54 minutes. Everyone seemed puzzled.[778]
> On Monday Stan Friedman decided to ask Gerald to go with him to our local office of Southwestern Bell – but Gerald declined. Stan explained the situation to the phone company and had no trouble obtaining a copy of the original bill.[779] The call was clearly listed as 54 minutes![780]

Carpenter explained that Anderson responded with anger, suggesting that Friedman was somehow responsible for the mess, though Friedman has remained a supporter of the Anderson fable. Friedman, in fact, would attempt to blame me for Anderson faking the phone bill suggesting that my negative

---

[778] The whole sad saga of the telephone bill had been laid out in the *IUR* in the July/August issue several months earlier. It contained copies of the forged phone bill and of the real bill obtained by Randle from the Southwestern Bell office, including the supporting documentation to prove all that had been suggested.

[779] Ironically, Friedman and Carpenter are often given credit showing that the Anderson phone bill had been forged. When first mentioned, Carpenter had wondered if the tape was slow, not that there was something amiss.

[780] Carpenter, John. Gerald Anderson: Disturbing Revelations. *MUFON UFO Journal* 299 (March 1993): 6 – 9

statements about Anderson forced him into that position, but does it make any sense to attempt to support your statements by creating a faked telephone bill? If revealed as fake, it would underscore the statements I had been making as authentic and raise additional questions about him.

The revelations about Anderson and his tale were not ended. Carpenter had more to say. He wrote:

> We now know four new things about Gerald Anderson: (1) He was capable of constructing a very clever fake phone bill. (2) He had admitted lying to us about that first phone bill. (3) He had just been caught lying to all of the gathered researchers about this 28 – minute phone bill (which means he had just constructed another phony!) and (4) Gerald was now avoiding us – his main supports and acting quite guilty in my opinion. Having caught him in these lies and recognizing what clever forgeries he could have created immediately threw doubt on ever other document or claim he had made. And if faking a phone bill – hardly an essential part of the case – was easily accomplished, what else could this man be capable of faking?[781]

Carpenter detailed other parts of his investigation and then made another revelation. He wrote:

> And then another strange event occurred. Although Gerald had stated all along that he was acting independently of his family wishes, he conveniently "received" an unprecedented statement from the scattered, hard-to-reach or hard-to-locate relatives of Gerald's family, depicting an uncharacteristic

---

[781] Carpenter, John. Gerald Anderson: Disturbing Revelations. *MUFON UFO Journal* 299 (March 1993): 6 – 9.

> "unified position" recommended dropping all contact with UFO investigators – that only his lawyers would now speak for him.
>
> This was simply unbelievable. I suspected a faked document and found what would seem to be a telltale flaw. All the family signatures are dated on the same day, December 24, 1991, which simply did not allow for any round-robin passage through the mail system.[782]

All of this seemed to be beating the dead horse. There were so many failures of corroboration and so much evidence of creating fraudulent documents and even the claim of being a Navy SEAL that there was little credibility left in the Anderson tale. Carpenter wrote, "One thing I know for certain: I can no longer trust anything my old friend Gerald Anderson wishes to tell me."[783]

In January, 1993, Berliner and Friedman placed a statement in the MUFON Journal. It said:

> Continued investigation into the reported 1947 crashes of two alien craft in New Mexico has caused the authors of the newest book on the subject to re-evaluate their position on one major information source.
>
> Don Berliner and Stanton Friedman, authors of Crash at Corona (Paragon House), New York, 1992) no longer have confidence in the testimony of Gerald Anderson, who claims to have stumbled upon a crash site with members of his family.

---

[782] Ibid. 8.

[783] Ibid. 9.

> Anderson has admitted falsifying a document, and so his testimony about finding wreckage of a crashed flying saucer near the Plains of San Augustin [sic] in western New Mexico and then being escorted out by the U.S. military, can no longer be seen as sufficiently reliable.
>
> The authors regret the need to take this step, but feel it is absolutely necessary if they are to stand behind their book and the subsequent research into what continues to be the most important story of the millennium. This does not mean they feel there was no crash at the Plains of San Augustin [sic]; there is considerable impressive testimony to such an event. Nor does it mean that everything reported by Gerald Anderson is without value.[784]

Dennis Stacy, the editor of the *MUFON UFO Journal*, added a comment. He wrote, "Although it strongly suggests it!"[785]

But that wasn't quite the end of Gerald Anderson and the trouble around his tale of a flying saucer crash. In 1998, Stanton Friedman wrote:

> *However, despite the negative comment by Don Berliner about some problems with Gerald, at the beginning of the 2nd Edition of Crash at Corona, I am still a Gerald Anderson booster... though not so much of a Don Berliner Booster. I saw Gerald in Roswell in 1997, I think. I was in his home with an Argentinean reporter arranged by me.*
>
> *I was disappointed about the phone bill business as Gerald knows. But I have still defended him to various and sundry*

---

[784] Berliner, Don and Stanton Friedman. "Anderson Axed." *MUFON UFO Journal* 297 (January 1993) 20.

[785] Stacy, Dennis. Editorial comment. *MUFON UFO Journal* 297 (January

including fiction writer and anti-abduction propagandist Kevin Randle.

But the trouble extended beyond just the telephone bill, both versions that he offered. Uncle Ted's diary was problematic because it was clear that Ted couldn't have hand written the copies. He had died years before the ink formula used had been developed. There was the trouble with Anderson and the high school course that he took from Buskirk. Anderson had produced a transcript that showed he hadn't taken the class, but it was a Xerox copy that showed signs of having been altered. Asked to have the high school principal send one directly to researchers, Anderson refused.[786]

And not addressed directly were Anderson's shifting stories. In the original interview, he'd placed the crash site north of the highway running east and west through that part of New Mexico. This is nowhere near the final site of Horse Springs that Anderson identified for Friedman. Even Friedman pointed out that a Horse Springs location is in direct conflict with Uncle Ted's diary.[787]

Even the diary has the date wrong. When it was determined that Barnett had been in Socorro on July 5, Anderson said that it had been on July 3. Later still, Berliner and Friedman said that the date was actually July 2 but according to the diary, the family had yet to arrive in New Mexico.[788]

---

[786] Mark Rodeghier noted that in his "transcript evaluation work, a photocopy of an American high school (or college) transcript was never accepted for official purposes. An official copy of such a transcript is always readily available, and until one is supplied, our policy was (as is that of any reputable agency or college) that no transcript had been submitted whatsoever." See Rodeghier, Mark. "Commentary of the Reports of Randle/Schmitt and Berliner/Friedman." *The Plains of San Agustin Controversy, July 1947: Gerald Anderson, Barney Barnett, and the Archaeologists*." Chicago: CUFOS and Washington, D.C.: FUFOR, June 1992.

[787] Randle, Kevin D., Donald R. Schmitt and Thomas Carey. "Rebuttal." *The Plains of San Agustin Controversy, July 1947: Gerald Anderson, Barney Barnett, and the Archaeologists*." Chicago: CUFOS and Washington, D.C.: FUFOR, June 1992: 31.

[788] Randle, Kevin D., Donald R. Schmitt and Thomas Carey. "Gerald Anderson and the Plains of San Agustin." *The Plains of San Agustin Controversy, July 1947: Gerald*

Anderson talked of a "battalion-sized operation" with twin engine aircraft landing on the road. Herbert Dick, who arrived at the Bat Cave on the eastern side of the Plains had a panoramic view of that area all the way to Datil and Horse Springs. He saw nothing to indicate such a military deployment. According to a letter found by Art Campbell in the archives at Harvard, Dick arrived at Bat Cave on July 1. He was in the perfect position to see all that activity claimed by Anderson.[789]

Dick, in fact, said that he knew thirty people in the area from that time and not one of them ever hinted about such a thing. There were no rumors, not gossip about any sort of crash. This included Will Hubble, who had been there on the Plains and never said a thing.[790]

He also made one other comment that was interesting. He said that he didn't care much for the government and he wished he could help. He said, "If I knew anything, I'd tell you."[791]

When it was pointed out that this 1998 statement was different than what had been published before, Friedman wrote:

> *I had forgotten the MUFON JOURNAL letter in 1993. Yes I have definitely not fully accepted that position. And yes, I do have problems with some of John Carpenter's activities. The phone bill was the fraudulent document. I was probably at fault for keeping Gerald appraised [sic] of the nefarious activities of Randle... which would have made anybody ready to trick Randle.*

Anderson, Barney Barnett, and the Archaeologists." Chicago: CUFOS and Washington, D.C.: FUFOR, June 1992: 14

789 Herbert Dick, telephone interview by Kevin Randle, June 23, 1991.

790 Ibid.

791 Ibid.

> *I perhaps should add that I like Gerald and that Berliner wrote the letter and almost all of Crash at Corona. There were well over 20 pages that I had written and wanted in. No such luck. I also wanted more in the extra chapter in the 1997 50th anniversary Edition of Crash at Corona. No luck...*[792]

What is interesting here is that Friedman saying that he wrote very little of the book and blames Berliner for it. Friedman had no power to induce the corrections that he wanted which is difficult to believe. That he couldn't add a chapter to a new edition is easier to believe because the publisher wouldn't want to change the printing plates to incorporate the new material.

And it is interesting that he had problems with John Carpenter's activities but doesn't say anything here about Carpenter's defense of Anderson in his "Fact vs. Fiction" article published in the *MUFON UFO Journal*. It is only the revelations that Carpenter made in the later article "Gerald Anderson: Disturbing Revelations," which provided more evidence that Anderson had been less than candid

What this actually means is that there is no first–hand corroboration for the Plains of San Agustin crash. All that is left are the tales told by friends and relatives of Barney Barnett. There are many of them, and many of them have shared their memories with UFO researchers but there are second-hand testimonies at best.

By contrast, a number of people who lived in that area in July 1947 have been contacted and interviewed. None of them saw anything first hand and few of them said that had something happened, they would have heard about it. John Foard, for example, said that he remembered hearing about the UFO

---

[792] Epilogue #2 at http://www.roswellfiles.com/Witnesses/anderson.htm (accessed January 23, 2015).

crash, but when interviewed, he couldn't remember if it was before or after the publication of *The Roswell incident.*[793]

Another example is Dave Farr who owns the ranch where Anderson claimed the UFO crashed. He said that he had lived in the area all his life and in 1947 lived only eleven miles away. He'd heard nothing about the UFO crash until television crews arrived in the middle of 1991.[794]

Frances Martin, who owned the Navaho Lodge in Datil in 1947, said that she'd never heard a word about any crash. Herbert Dick, who knew her, said that everyone eventually made it into her tavern at one time or another.[795]

No one could be located who had seen anything for him or herself. Each of the witnesses either knew nothing about such a crash, or had heard some vague stories about it. Dr. Peter Harrison said that there was an oral tradition of a crash on the Plains, which meant, simply, that someone was talking about it. But the tradition can be traced back to Robert Drake and offers no corroboration of a crash.[796]

Robert J. Drake said that he had been on the Plains in October 1947, and said that he had been approached by a cowboy who told him about the flying saucer crash a couple of months earlier. Later, Drake claimed that he, and three other scientists including Daniel McKnight, Wesley Hurt and Albert Dittert had discussed the crash as they drove back to Albuquerque.

There was some corroboration in an article that appeared in *American Antiquity*. According to the article, written in 1948, Hurt, McKnight and Drake

---

[793] John Foard, personal interview with Kevin Randle, September 10, 1991.

[794] Dave Farr, personal interview with Kevin Randle, September 10, 1991.

[795] Randle, Kevin D., Donald R. Schmitt and Thomas Carey. "Gerald Anderson and the Plains of San Agustin." *The Plains of San Agustin Controversy, July 1947: Gerald Anderson, Barney Barnett, and the Archaeologists*." Chicago: CUFOS and Washington, D.C.: FUFOR, June 1992: 14; Herbert Dick, telephone interview with Kevin Randle, June 23, 1991; Frances Martin, interviewed in 1989. Friedman dismisses her testimony by referring to her as "some old woman."

[796] Peter Harrison, personal interview with Kevin Randle, November 17, 1991.

had all been to the Bat Cave in October 1946, eight months before the crash. It means that the ride back to Albuquerque took place a year earlier than Drake thought. Besides, none of those others in the car remembered any discussion about a UFO crash.[797]

What it boils down to is simply this. In their original statements, none of the witnesses provided a location or time frame for the Barnett tale. Most of them believed the crash to be out to the west, on the Plains of San Agustin because that was where Barnett spent most of his time when out of the Socorro office and they thought it was prior to 1950, based on some assumptions. Fleck Danley eventually said that it was early summer 1947, but he didn't say that in his first interviews, and in the later interviews it was clear that he really didn't know.

The story of a crash on the Plains is single witness with a number of second-hand testimonies that prove Barnett told the story but not that it was true. In all the years of searching, there has not been a single, reliable witness to a crash in July 1947 in western New Mexico. Without some sort of additional evidence, this tale should be little more than a footnote in the overall history. Until and unless something else is provided, all that we have is Barney Barnett telling friends and family and interesting story, and that is all it is.

---

[797] Hurt, Wesley and Dan McKnight. "Archaeology of the San Augustine Planes: A Preliminary Report," American Antiquity 14 (January 1949): 172 – 194. Also Wesley Hurt, personal interview with Kevin Randle July 15, 1991, Robert Drake, personal interview with Tom Carey, July 1991; Dan McKnight personal interview with Tom Carey, September, 1991; See also Randle, Kevin D., Donald R. Schmitt and Thomas Carey. "Gerald Anderson and the Plains of San Agustin." *The Plains of San Agustin Controversy, July 1947: Gerald Anderson, Barney Barnett, and the Archaeologists*." Chicago: CUFOS and Washington, D.C.: FUFOR, June 1992: 16.

Appendix C: The Flight No. 4 Controversy

All flights of the Mogul arrays, or as Charles Moore called them, the New York University balloon project balloon flights, were accounted for in the records with few exceptions.[1] The first few flights, through Flight No. 3, took place on the east coast before they moved their experiments to New Mexico. Flight No. 4 was probably first scheduled to fly on the morning of June 3, 1947, but according to the diary kept by Dr. Albert Crary[2] it was cancelled because of cloudy weather. Another attempt was made the following morning. Crary's entry for this attempt said:

> June 4 Wed
>
> Out to Tularosa Range and fired charges between 00 and 06 this am. No balloon flight again on account of clouds. Flew

---

1. According to the documentation available concerning the New York University Balloon Flights, the first of those flights were made from New England with Flight No. 1 being launched in Bethlehem, Pennsylvania in April 1947. Two earlier flights were labeled as a and b. Two other attempted flights were also made from Bethlehem, No. 2 on April 18, cancelled because of high winds and radio equipment failure, with the next attempt, No. 3, on May 8 likewise cancelled because of high winds. The next *recorded* flight after Flight No. 1 was in Alamogordo, New Mexico and was Flight No. 5. The next gap in the record is Flight No. 9 which is also missing, cancelled when a coordinated V-2 launch was aborted.

2. Moore, C.B. *The New York University Balloon Flights During Early June, 1947*, privately published in Socorro, New Mexico. See also, Weaver, Richard and James McAndrew. *The Roswell Report*, Government Printing Office, 1995. All references to Crary's diary come from the large Air Force report. Moore did reproduce some of it in his report on the early flights.

> regular sonobuoy mike up in cluster of balloons and had good luck on receiver on ground but poor on plane. Out with Thompson pm. Shot charges 1800 to 2400.

Moore, attempting to interpret what this meant, wrote, "Crary's diary entries for June 4 are puzzling because they contradictory."[3]

He then threw up clouds of dust to hide the facts. He suggested that Crary had copied his field notes into the diary later and that the events of early June, including those on June 4, had been copied in one sitting, which might account for the seeming contradictions. Moore wrote:

> One interpretation of the June 4 entry is that the launch scheduled for making airborne measurements on Crary's surface explosions after midnight was canceled because of clouds but, after the sky cleared around dawn, the cluster of already-inflated balloons was released, later than planned. The initial cancellation and later launch were recorded sequentially, as they occurred, in his field notes which he later transcribed into his permanent diary, without elaboration.[4]

Moore goes on, suggesting that they had just arrived in Alamogordo and that they would not, at that time, have "improvised." He said that after they

---

3. Moore, *Balloon Flights*, p. 4

4. Moore, *Balloon Flights*, p. 4. Note that in this entry Moore is suggesting that the flight was then launched about dawn, which he said was later than planned. All other launches in June were scheduled for right after dawn with nothing flying in the pre-dawn hours of darkness, which would have been in violation of the CAA regulations under which they operated. This revelation would become important in Moore's later writings.

had rigged an entire array, “I think… we would have launched the full-scale cluster, complete with the targets for tracking by the Watson Lab radar.”[5]

In other words, without benefit of documentation, much of which was available in Crary’s diary, Moore has introduced radar into that first New Mexico flight. There is no evidence to support this claim, other than Moore’s suggestion which is qualified with his, “I think.”

Although many reject, out of hand, memories that are forty or fifty years old, many accept what Moore wrote next:

> I have a memory of J. R. Smith watching the June 4th cluster through a theodolite on a clear, sunny morning and that Capt. [Larry] Dyvad reported that the Watson Lab radar had lost the targets while Smith had them in view. It is also my recollection that the cluster of balloons was tracked to about 75 miles from Alamogordo by the crew in the B-17. As I remember this flight, the B-17 crew terminated the chase, while the balloons were still airborne (and J. R. was still watching them), in the vicinity of Capitan Peak, Arabela and Bluewater, New Mexico. I, as an Easterner, had never heard of these exotically-named places but their names have forever been stuck in my memory. This flight provided the only connection that I have ever had with these places. From the note in Crary’s diary, the reason for termination of the chase was due to the poor reception of the telemetered information by the receiver aboard the plane. We never recovered this flight and, because the sonobuoy, the flight gear and the balloons were all expendable

---

5. Ibid. p. 4

> equipment, we had no further concern about them but began preparations for the next flight.[6]

Although Moore claims that "this flight provided the only connection that I have ever had with these places," this is simply not true. Flight No. 17 from September 9, 1947, flew along the same trajectory and passed over the exact same exotically named landmarks Moore associated with the alleged June 4 project flight. In addition, they lost tracking of it in the exact same vicinity of Capitan Peak that Moore said happened for "Flight No. 4."[7] It is quite possible that Moore's 50-year-old memory of "No. 4's" flight path is really a badly distorted recollection of the real Flight No. 17 three months later.

Years later, according to Moore, he heard about the debris found by Brazel and thought it was a good description of the debris that would have been produced by one of their balloon trains. He thought that as the train dragged along the ground, kept partially aloft by the few balloons that hadn't yet burst, it would have shed debris, creating the mess that Brazel described. He said, "It is possible that Brazel found some of the wreckage from the NYU Flight #4."[8]

However, the train being dragged by still inflated balloons to create such a debris field implies the rigging holding everything together to be still there. Yet Brazel, when interviewed by the *Roswell Daily Record* the evening of July 8, 1947, indicated he found no balloon rigging of any kind. A real Mogul constant-altitude flight of this period would have left hundreds of yards of rig-

---

6. Moore, *Balloon Flights*, p. 4

7. Flight No. 17 was a late afternoon flight and they lost visual tracking because of distance and nightfall. The project summaries also note the balloon ended up in Kansas (thus did not crash nearby as Moore claimed for No. 4) "long after flying out of Roswell's [radiosonde] reception range." This contrasts with Moore's recollection of No. 4 that they lost radio tracking reception because the batteries went dead.

8. Ibid. p. 5

ging mixed in with the other crash debris it held together in flight. This discrepancy was noted by Lt. James McAndrew in his interview with Moore, but then the issue was dropped when Moore couldn't come up with an explanation.

The other flights, from Flight No. 5 until the first week in July were accounted for in the history of the balloon project. No other flight disappeared in this fashion in the proper time frame. The records are quite clear on that.[9] If Flight No. 4 does not account for the debris found by Brazel, then Mogul is not the answer.

As Moore said, all this seems to be confusing because of the seeming contradictions in the diary, but there is a way to untangle it. The answer lies in the field notes and diary kept by Crary. He explains what happened and all that needs to be done is to understand what he was saying in the context of the entire diary and not to be confused by the memories of Moore.

There were other flights that were not included in the final reports. These were cancelled for a variety of reasons including high winds and cloud cover at the time of the launch. While it doesn't seem that clouds would affect balloons, the reason is CAA (FAA) requirements. As the balloons ascended and descended and given the length of some of the arrays, these could be a hazard to aerial navigation because they would be invisible in the clouds. This is borne out in *Technical Report No. 1, Balloon Group, Constant Level Balloon Project*, dated April 1, 1948 (covering the period from November 1, 1946 to January 1, 1948) which said that one of the requirements was that the weather be relatively clear so that the balloons could be seen.[10]

---

9. For a complete listing of the balloon flights in June and early July 1947, see Weaver and McAndrew, *The Roswell Report*. See also, Karl Pflock, *Roswell: Inconvenient Facts and the Will to Believe*, Prometheus, pp. 236 – 237, and Moore *Balloon Flights*, p. 2a.

10. *Mogul Technical Report No. 1*, April 1, 1948.

That same report also said, "Notices to airmen [NOTAMs] are to be issued if the balloon is descending within designated regions of dense air traffic."

This establishes a requirement for NOTAMs, both for the launch, as the arrays climbed through the civil air space, and then another for the arrays as they descended. Once at altitude, which would have been somewhere above 50,000 feet, or far above the levels where civilian aircraft operated in 1947, they would no longer be a hazard. Flight No. 5, launched on June 5, eventually came down in the vicinity of the Roswell Army Air Field, which suggests the NOTAM would have been required and everyone in flight operations at the base would then have been aware of these long arrays, if they had not encountered information about them earlier.

Another strong indication that a NOTAM *should* have been issued comes from Flight No. 5 tracking data. This shows it passed only 4-5 miles south of the base as it was descending towards its crash site. It addition, it lingered less than a dozen miles south and west of Roswell air space for over an hour during its slow stratospheric backward drift while a B-17 chase plane circled underneath. The B-17 followed it all the way to its crash site, marked as only 16-17 miles due east of the base. It is difficult to believe that air controllers, plane spotters, or security guards could all have failed to notice the 600-foot-long balloon train and chase plane. At the very least, the B-17 should have been in contact with flight control explaining their presence in Roswell air space. Yet Moore insisted Roswell had no knowledge of their flights at that time.

In an email debate with researcher Brad Sparks mediated by Karl Pflock (a Moore and Mogul supporter), when Sparks pointed out just how close No. 5 came to the base, Moore responded that it came no closer than 15 to 20 miles from the base and maybe Roswell failed to see the balloon train because of cloud cover, yet Mogul flight summaries indicate it was tracked optically through theodolite from Alamogordo for 90% of its flight clear to Roswell. It was only lost from sight during its descent phase when it fell below the horizon formed by the Sacramento Mountains east of Alamogordo. It is difficult to understand how it could be optically tracked for almost 100 miles from Alamogordo yet remain invisible only a few miles away from Roswell base. During that email debate, Moore finally did concede that the B-17 would probably have been in contact with the control tower at Roswell.

A second flight, designated as Flight No. 11A, came down even closer to Roswell on July 7. This too would have entered into the airspace controlled by the Roswell Army Air Field, and a NOTAM would have been required by the CAA. Moore seemed to believe that they had ignored those requirements for these earlier flights,[11] which is a very cavalier way to operate. Such a violation, if caught, would have resulted in a suspension of their experiments.[12]

While all this might suggest that those in Roswell would have been aware of the long balloon arrays, Moore made it even clearer in various interviews. He told researcher Kevin Randle that he, and one or two others, had driven to Roswell to ask for assistance in tracking their balloon arrays. According to Moore, the soldiers in Roswell believed they had more important things to do than to chase balloons for a group of "college boys."[13] This means, of course, that Moore and his colleagues would have explained what they were doing in Alamogordo, and what the arrays would have looked like for those in Roswell.

Moore did alter his story somewhat, or provided another, suggesting that those in Roswell didn't know about the Mogul flights. He wrote:

---

11. A brief mention of the NOTAM requirement was made in Moore's August 10, 1995 letter to New Mexico Representative Steven Schiff. Moore said that no NOTAM had been issued because the ascent would have been over restricted airspace, meaning that civil aviation was prohibited from entering it.

12. To elaborate on the NOTAM issue, on April 17, 1947, C.S. Schneider requested a relaxation of the launch restrictions but that resulted in stricter regulation. He had not requested a relaxation in the requirement to issue NOTAMs.

13. Charles Moore, personal interview with Kevin Randle in Socorro, New Mexico.

As far as the claim that "Roswell AAF" knew about MOGUL operations prior to July, 1947, I have this to offer. On June 5, 1947, after chasing, in an Alamogordo weapons carrier, NYU Flight #5 to its landing about 26 miles east of Roswell, my vehicle was low in fuel so I drove to Roswell AAF and requested entry to refuel. I identified myself, displayed the Alamogordo AAF motor trip ticket to no avail; after lengthy telephone conversations between the guard at the gate and headquarters and an interview by the Officer of the Day to whom I *showed the recovered equipment* [emphasis added] from Flight 5, I was turned away and had to go to a commercial gas station to pay for refueling. Admittedly, I did not use the term Project MOGUL to the Roswell OOD because at that time, the term MOGUL, was not known to any of the NYU balloon crew and was never used by anyone in our hearing at Alamogordo. I did tell the OOD about the NYU balloon operations in Alamogordo. I came away with the impression that the Roswell AAF personnel were so impressed with their own operations and security that they had no interest in what else was occurring in their vicinity. [Somehow I doubt this incident made its way up the chain of command to Blanchard or that Haut or anyone in Marcel's intel shop would have heard about it or, if they did, that they would have connected it with the "flying saucer wreckage" that turned up a month later.] [Parenthetical statement added by Karl Pflock.][14]

14. Letter from Charles Moore to Karl Pflock on June 20, 2002.

There are a couple of statements here that should be addressed. First is this idea that Moore, in his weapons carrier, was turned away from the gate at Roswell. On May 20, 1947, according to Crary's diary, "[Crary and Edmondson] Went over to Roswell Army Air Field, filled up with gas."

In other words, Crary had no trouble at the gate just a few weeks before Moore found himself stranded there. Both Crary and Moore were in a vehicle drawn from the Alamogordo AAF motor pool, both would have had a trip ticket for that vehicle, and both would have had the "book" which would contain the maintenance records, mileage, and the like. That would have demonstrated that the vehicle was military and there should have been no reason for Moore to be turned back at the gate unless he was attempting to prove that those service members at Roswell had no contact at all with the NYU people in New Mexico.

Second is the idea that those on the NYU team in Alamogordo didn't know the name of Project Mogul. Karl Pflock, in an interview conducted by members of New Jersey MUFON on August 27, 1994, said, "This [Mogul] was a top secret, very, very sensitive project that was being run by New York University for the Air Force's Watson lab."

Moore carried on this tradition, not only in the paragraph quoted above, but throughout his writings and statements about the project. According to Dave Thomas, "The Mogul project was so classified and compartmentalized that even Moore didn't know the project's name until Robert Todd informed him of it a couple of years ago."[15] In a handwritten note on a copy of the magazine article sent to Randle by Jim Moseley, Thomas added, "Moore told me this when I met him."

The problem is that this is entirely false. Crary, in his diary, mentions the name, Mogul, more than once. On December 11, 1946, Crary wrote, "Equipment from Johns Hopkins Unicersity [sic] transferred to MOGUL plane."

---

15. Dave Thomas. "The Roswell Incident and Project Mogul. *Skeptical Inquirer*, July/August 1995.

On December 12, 1946, he wrote, "C-54 unloaded warhead material first then all MOGUL eqpt with went to North Hangar."

On April 7, 1947, Crary, according to his diary, "Talked to [Major W. D.] Pritchard re 3rd car for tomorrow. Gave him memo of progress report for MOGUL project to date..."[16]

A report from Wright Field on August 25, 1947, classified only "Confidential", concerned a suspected hoax crash disc from Illinois sent to them by the FBI for analysis. The term "Project Mogul" was explicitly used, saying that the object had nothing to do with it. Another FBI memo a month later, referencing the Wright Field report, uses the term "Operation Mogul" four times even though this memo also had a low classification.

In an unclassified letter, dated May 12, 1949, Cmdr. Robert B. McLaughlin, chief of Naval missile operations at White Sands, is describing, for Dr. James A. Van Allen, who C. B. Moore was.[17] He then wrote "In addition to this, he had been head of Project Mogul for the Air Force." What's additionally interesting about this is that we know about the letter because Moore had it in his own files and gave a copy to Dr. James McDonald in 1968.

The documentation then, shows that the name was known in 1946, was used by the NYU scientists and engineers in that time frame with little concern about security, the name was used by Wright Field and given to the FBI in 1947 in documents of low classification, and the name was even used to introduce Moore to Van Allen in 1949. It would seem that Moore did know the name while in New Mexico with the project and that the claimed classification was not about the activities in New Mexico or the balloon flights, but the ultimate purpose which was to spy on the Soviets. That is the one thing that Moore

---

16.- Weaver and McAndrew, *The Roswell Report*, Chapter 32, "Synopsis of Balloon Research Findings." See also Crary's Diary for April 7, 1947.

17. McLaughlin's letter was about UFO sightings at White Sands, including those of his men and himself during missile launches, and was mentioning a now well-known sighting by Moore and his team during a balloon launch only three weeks earlier on April 24, 1949.

probably didn't know. The ultimate purpose of the constant level balloons was to spy on Soviet experimentation with atomic weapons.

Third, Moore himself said that he showed part of the recovered Mogul array to the Officer of the Day at Roswell, who should have noted the confrontation in his log. It would have been part of the debriefing, and should have come to the attention of the Provost Marshal and the Operations Officer. In other words, Moore was confirming that some of the officers at Roswell had seen one of the balloon arrays, this from Flight No. 5 in early June 1947.

Fourth, there is another fact that shows there was nothing unusual about these arrays, or rather anything that would conceal their nature from those not involved in the project. Crary's diary for Sunday, June 8, said, "Rancher, Sid West, found balloon train south of High Rolls in mountains. Contacted him and made arrangements to recover equipment Monday. Got all recordings of balloon flights…"[18]

Finally, it should be noted that there was no Flight No. 4. Crary's diary is not confusing on the issue. It stated quite clearly that the flight had been cancelled because of clouds, as required by the CAA and their instructions to Crary and his team. The second entry said they flew a cluster of balloons with a sonobuoy but said nothing about radar targets or other equipment or that this was the cancelled Flight No. 4. Moore had said when the flights were cancelled; they stripped the equipment, but let the balloons go because there was no way to get the helium back into the bottles.

Moore, in a letter dated August 10, 1995, wrote about the mythical Flight No. 4 and its cancellation. He said:

---

[18] Crary was referencing Flight No. 6, launched only the previous day, the exact crash site unknown to the Mogul team. The fact that Crary became aware of rancher West's finding only a day later suggests that the No. 6 may very well have carried identification and reward tags encouraging any finder to report it. This is in contradiction to the claim by Moore and others that Flight No. 4 would have carried no such tags because of its high classification and rancher Brazel would not have known who was responsible for the wreckage or where to report it.

> The jury-rigged flight #4 of meteorological balloons that we launched as AMC contractors from Alamogordo Army Air field on July [sic] 4, 1947 was no big deal; it was a test flight, the first in a series and there was no announcement of our plans, either on base or to the Army Air Forces authorities. Since we launched from just within the restricted air space associated with the White Sands Proving ground and expected the balloons to rise high above the civil air space, we did not notify the CAA in El Paso. As I remember, we launched before sunrise with only our Watson Laboratories associates and the B-17 crew knowing about the ascent. This flight was not successful due to the failure of the Watson lab radar to track the balloons and the poor transmission of the acoustic data caused by use of out-dated [sic] World War II batteries. The only mention of these flights in 1947 came in the unclassified progress report for June:

The problem that concerned the CAA wasn't the balloon ascent but their descent. It would be expected that the array would rise quickly and reach stability, or relative stability, at a constant level. Once the balloons began to fail and the array began its fall back to the ground, it would be expected to drift for a time in the civil airspace, and this was the hazard to aerial navigation. This was the point at which the danger existed.

Moore was being disingenuous here. He is attempting to explain the lack of a NOTAM, if records for June 4, 1947 could be found. He knew that no NOTAM had been filed because of the nature of this flight. It was not expected to leave the restricted area of the range. And he knew that the NOTAMs were only necessary for the constant level balloon flights, not the "test flights" that would fall back quickly.

The second point is that there is nothing to suggest that radar was a factor in this flight and nothing to suggest that radar reflectors were included on the cluster. This was a cluster of balloons used to lift a sonobuoy to test its capability of detecting the explosions.

Two previously cancelled flights back East, No. 2 and No. 3, also indicate that *all* reusable equipment was stripped off before the filled (and non-reusable) rubber weather balloons were cut loose. For No. 2, it was stated: "Release was attempted on 18 April. Due to the high wind... and due to malfunctioning of the Army receiver in the plane that was to follow, release was not made. *The already-inflated balloons were cut free and equipment was brought back to NYU.* It is expected that this equipment will be flown about 8 May."[19]

For the new attempted constant-altitude flight on May 8 (No. 3), Crary's diary records the following: "Trouble with winds and *instruments did not go up*. ...recording equipment on B-17 following balloons... B-29 started dropping bombs near Atlantic City about 8..."

With the planned constant-altitude flight No. 3 aborted, all equipment was stripped off ("instruments did not go up"), except apparently for a microphone of some type so that the chase B-17 could test reception by the mic of the explosions on the ground.

The canceled No. 3 provides a model for what Crary's entry meant for the equally canceled constant-altitude flight No. 4. Crary indicates that after cancellation they used the existing balloon cluster to loft a sonobuoy microphone to test reception from the air and ground. This Moore refers to as a "jury-rigged" or unplanned test flight of a specific piece of equipment.

With the cancellation of these planned constant-altitude flights, not only would the constant-altitude gear be removed, so would any tracking gear, such as radiosondes and radar targets, since these were used to 1) carefully chart the flight profiles to see if the flight was indeed achieving constant-level control, and 2) carefully track the path of these typically longer flights to see if they were entering civilian airspace. On the other hand, lacking constant-altitude paraphernalia, simple test flights of equipment would not need their altitude profiles documented nor their trajectories, since the flights would not last long and wander off the White Sands range into civilian or other military flight lanes.

---

19. "NYU Special Report No. 1, Constant Level Balloon, May 1947", p. 27

In fact, there is no evidence that rawin radar reflectors were used in those first flights in New Mexico. According to Crary's diary on June 9, "Bill Godbee and Don Reynolds went out to Sid West's ranch south of High rolls and brought back recovered balloons – clock, 2 radiosondes, sonobuoy and microphone and lower part of dribbler." This is also supported by the flight summary tables which show this flight being tracked by radiosonde and theodolite, but not by radar.

Moore supplied an illustration for Flight No. 5, dated June 5, 1947. There are no radar reflectors on this flight. Given that the balloons sent aloft on June 4 were referred to as a cluster carrying a sonobuoy, there is no reason to suspect that the radar wouldn't work tracking Flight No. 4. In other words, Flight No. 4 would have been configured just as was Flight No. 5 which contained no radar targets, and if there were no radar targets, then one aspect of the Mogul theory for the Roswell debris has been eliminated. There is no mention of radar tracking until Flight No. 8, launched on July 3. An illustration for Flight No. 2, which provided no data, did contain radar reflectors, but again, there is no evidence they were again used until later.

Like with the cluster of balloons on June 4, there was no mention of any radar targets with the recovery at Sid West's ranch. Mogul flight summaries also indicate no radar tracking of this balloon only 3 days after the alleged No. 4. There is almost no mention of radar for the tracking of the balloons, though Moore suggests that the Flight No. 4 proved that the radar wouldn't work so they changed the array. This does not seem to be accurate, based on the records that are available. The only suggestion of radar in these first flights was based on Moore's memory of the targets being included but not from the documentation available now or in the records of the recovered flights.

Moore himself provides some answers to the questions. In the final report on NYU's balloon activities there is a tabulation of all the flights. Both Flight No. 4 and Flight No. 9 are missing. This tabulation also notes about Flight No. 5, "First successful flight carrying a heavy load." Multiple official Air Force and other histories also state that a June 5 flight (i.e. No. 5) was the *first* AAF research balloon in New Mexico. None mention a balloon flight the previous day.[20]

This would seem to suggest that the cluster of balloons was not a full Mogul array. Moore, however, with no documentation to support the conclusion, wrote, "I think that Flight #4 used our best equipment and probably performed about as well as or better than Flight No. 5."[21]

The logical question to be asked is if Flight No. 4 performed as well as or better than Flight No. 5, then why was it not listed in the tabulation or in official histories? It would have been the first successful flight, unless, of course, it wasn't a full Mogul array.

Given the time it took to build the full array and prepare it for launch, it would not have been possible to build a new array for Flight No. 5. Crary's diary is clear on the point. Flight No. 4 was delayed by weather. Flight No. 5 was, in fact, Flight No. 4, using equipment stripped off of the cancelled No. 4 and reattached to a new balloon cluster (much like is indicated happened for No. 3, using equipment stripped from the canceled No. 2), redesignated and launched on June 5. That flight was recovered, as Moore noted and records

---

20. Examples include: 1) *Chronology: From the Cambridge Field Station to the Air Force Geophysics Laboratory 1945-1985* (AFGL, Hanscom AFB, Special Reports, No. 262, 6 Sept 1985): "1947, 5 Jun, The first Army Air Forces research balloon launch was conducted at Holloman Air Force Base, New Mexico, by a New York University team working under contract for the Air Material Command. It featured a cluster of rubber balloons." 2) *U.S. Air Force: A Complete History* (The Air Force Historical Foundation, 2006), p. 300, "1947, 5 June, A New York University team under contract with the Air Materiel Command launches the Army Air Forces' first research balloon. The cluster of rubber spheres is released at Holloman, New Mexico." 3) *Aeronautics and Astronautics: An American Chronology of Science and Technology in the Exploration of Space, 1915-1960* (NASA, 1961), 1945-1949, pp. 49-63: "1947, June 5: First AAF research balloon launch (a cluster of rubber balloons) at Holloman, by New York University team under contract with the Air Materiel Command."

21. Saler, Zeigler and Moore. p. 105

document. In contrast, there is no documented record of a Mogul balloon recovery of a flight from the previous day.

Finally, if the June 4 flight was just a cluster of balloons launched only with a sonobuoy, then it would not have been a constant level balloon and speculation about its flight path is just that… speculation. If it didn't reach the altitude that Moore claimed and/or stayed aloft for the long length of time Moore claimed, then its flight dynamics would have been very different. It would be impossible to provide any flight path for it simply because the data don't exist.

However, a test flight lacking altitude control is unlikely to have stayed up long nor traveled far because it would rapidly have climbed to high altitude, suffered catastrophic balloon failure, then come straight back down. Moore himself suggested this when he claimed no NOTAM was issued because they didn't expect it to leave the White Sands range. In fact, this is exactly what happened to Flight No. 6 only 3 days later (the one found by rancher West), resulting from the constant-altitude equipment being damaged on launch.

In his book, Moore attempted to create a theoretical trajectory. He used June 4, 1947, winds aloft data from the White Sands Missile Range weather station at Orogrande, 32 miles south of Alamogordo, AAF. This high altitude wind data was collected before V-2 rocket launches at White Sands going back to 1946, and employed the same rawin targets said to be exclusive to Project Mogul in 1947. Wind data extended up to about 50,000 feet and Moore spliced in model wind data from the real flights No. 5 and No. 6 to create stratospheric winds up to about 60,000. He then claimed that a model Flight No. 4 constant-altitude balloon, that worked even better than the successful real No. 5, had winds that were "exactly right" to take such a flight "exactly" to the Foster (Brazel) ranch crash site.[22]

Later, independent analysis of Moore's model revealed numerous serious mathematical errors, as well as many hidden assumptions that contradicted his stated ones, such as treating the balloon as faulty in some regards instead of

---

22. Moore quoted in a 1997 Sci Fi channel special on Roswell

one that worked as well or better than No. 5.[23] When these errors and assumptions were removed, the model trajectory would instead have created a 70 to 100 mile overshoot of the area. It turned out that the only way to get such a constant altitude balloon flight to the crash site in a correct model would require drastically different wind speeds and directions.

Other winds aloft data from the U.S. weather service, as measured in 1947, only reached from the surface to 20,000 feet. Anything above that was not measured and certainly not recorded. In addition, the reporting of the winds aloft data was erratic. Some stations missed many reports over the few days in early June that were examined. These data were incomplete.

Moore himself indicated that if he had changed one number in his assumptions, the balloons could have landed as much as 150 miles away. On page 93 of his book, he wrote, "If the balloons had not entered the stratosphere but had continued in the upper troposphere, they would have passed 17 miles south of the actual landing site and would have landed more than 150 miles to the east at the end of the [assumed] 343 minute flight."

But, of course, there were no data for this flight so the height, distance and performance were all speculation built around Moore's memory of the event.[24] That memory is in conflict with the written record about the flight, a flight designed for one thing and that was to test the sonobuoy.

According to Moore, when flights were cancelled, they sometimes conducted experiments using some of the balloons and equipment. Flight No. 4 was not launched in the dark, which would have violated the CAA regulations under which they operated, and was cancelled because of clouds at dawn,

---

23. Moore covertly removed altitude-control, used by Flight No. 5, on the rise and fall segments for No. 4, which greatly shortened the rise and fall times, thus greatly shortened the flight path for No. 4. This was one way he prevented serious overshoot of the crash site, but this was at odds with his stated assumption of equipment that worked as well or better than No. 5.

24. As noted, Moore may have confused the real Flight No. 17 three months later, which followed the same initial trajectory and was lost track of in the same place that Moore attributed to No. 4.

which was demanded by the CAA though Moore said the clouds were gone by then. In his 1995 paper, Moore calculated the launch time as about 5:00 but in the Benson, Saler, Zieler, and Moore book, ironically called *UFO Crash at Roswell: Genesis of a Modern Myth*, on page 102, he changed the time, suggesting that it was actually launched at 3:00 a.m. It is clear that Moore changed the times, not based on newer and better evidence, but on his memory to prove his own theory.

The evidence suggests that Flight No. 4 was cancelled. The evidence suggests that a cluster of balloons lifted a sonobuoy up for testing. Moore himself referred to this as "jury-rigged" which in and of itself suggests it was something thrown together for a specific test. There are no indications that it left the restricted areas around Alamogordo, no evidence that it carried the materials necessary to create the debris field, no evidence that it was what Mack Brazel found, and no evidence that Mogul was so secret that very few knew the name. Since there was no such flight, it could not have explained what Brazel found or what triggered the Roswell incident. Mogul, as an answered, has been disproved.

Appendix D: Deciphering the Ramey Memo

On July 8, 1947, Brigadier Roger Ramey was photographed in his office holding a document that is almost legible. The paper is not directly facing the camera but enlargements revealed that some of the words could be read. It has remained a subject of controversy for over thirty-five years as researchers attempted, using a variety of techniques, computer hardware and software to decipher the message.

In 1980, Brad Sparks obtained a blowup of the photograph in an attempt to read the memo. He saw the word "BALLOONS."[798] In 1985, Sparks was able to pick out another few words that are "unanimously or almost unanimously agreed-upon as being there (weather balloons, Fort Worth, Tex., disc)."[799] In the mid-1980s Barry Greenwood made an attempt to read anything on the memo. In 1991, Don Schmitt, who had been working with Kevin Randle, sent a copy of the photograph to Dr. Richard Haines, a former NASA scientist, asking if he could read anything on the paper. Using a microscope, he scanned the photograph. He reported that he could see vague words but that he couldn't make out individual letters. Haines thought that a better quality enlargement might reveal more of the message.[800]

The next serious attempt to read the message seems to have been made by the Air Force during their investigation into the alleged Roswell UFO crash.

---

[798] Pflock, Karl. Roswell: *Inconvenient Facts and the Will to Believe*. Amherst, NY: Prometheus Books, 2001, pp. 201 – 210; See also Sparks, Brad. (2004). CUFON response to Barry Greenwood.
(See also http://www.cufon.org/contributors/Sparks/Sparks_Rebut_Ramey_Message.pdf Accessed 12 April 2015.)

[799] Ibid. Also, Rudiak, D. Internet letter, 3 March, 2015. (To Martin Dreyer, Kevin Randle)

[800] Richard Haines in a February 13, 1991 to Donald Schmitt.

According to the Air Force:

> Additionally, the researchers obtained from the Archives of the University of Texas – Arlington (UTA), a set of original (i.e., first generation) prints of the photographs taken at the time by the *Fort Worth Star – Telegram*, that depicted Ramey and Marcel with the wreckage. A close review of these photos (and a set of first – generation negatives also subsequently obtained from UTA) revealed several interesting observations... It was also noted that in the two photos of Ramey he had a piece of paper in his hand [though the critical picture is one of Ramey and Dubose]. In one it was folded over so nothing could be seen. In the second, however, there appears to be text printed on the paper. In an attempt to read this text to determine if it could shed any further light on locating documents relating to this matter, the photo was sent to a national – level organization for digitizing and subsequent photo interpretation and analysis.[801] This organization was also asked to scrutinize the digitized photos for any indication of flowered tape (or "hieroglyphics," depending on the point of view) that were reputed to be visible to some of the persons who observed the wreckage prior to it getting to fort Worth. This organization reported on July 20, 1994, that even after digitizing, the photos were of insufficient quality to visualize either of the details sought for analysis...[802]

---

[801] The organization that examined the memo was the National Photographic Interpretation Center, which is part of the CIA. The NPIC evolved into the National Geospatial-Intelligence Agency, NGA.

[802] Weaver, Richard and James McAndrew. The Roswell Report: Fact vs Fiction in the New Mexico Desert. Washington, D.C.: Government Printing Office, 1995, Executive Summary, pp. 29 – 30.

That was where the matter rested until 1998 when J. Bond Johnson, who had taken six of the seven photographs in General Ramey's office,[803] including the one in which the message might be seen, decided to investigate further.[804] Johnson put together a team to inspect the photographs that included Ron Regehr, a space and satellite engineer.[805] Using a huge enlargement of the photograph, a computer and a variety of software and camera equipment, they were able to read more of the message that Ramey held. Or rather, they claimed that they could read it with some degree of certainty.

In the upper left-hand corner, they saw what they believed to be the image of a telephone and concluded that Ramey was holding a "telephone message sheet" because of this "telephone logo." They then claimed to have "positively identified a number of words in the message. There were, quite naturally, gaps in what they could see, and noted that the message had been typed in all capital letters.

Their interpretation of the message was:

> AS THE... 4 HRS THE VICTIMS OF THE... AT FORT WORTH, TEX... THE "CRASH" STORY... FOR 0984 ACKNOWLEDGES... EMERGENCY POWERS ARE NEEDED SITE TWO SW OF MAGDALENA, NMEX...[806]

---

[803] Analysis of the composition of the photographs seems to corroborate this, but Johnson would claim that he had only taken the pictures of Ramey and Ramey and Dubose. He said that he didn't know who had taken the two pictures of Marcel. This statement by Johnson seems to be in error.

[804] Randle, Kevin D. *The Roswell Encyclopedia.* New York: Quill, A HarperResource Book, 2000, p. 294.

[805] Johnson, J. B. Roswell photos revisited to be aired on FOX-TV network. Internet press release. 24 September. (See UFO Updates at http://www.ufomind.com, accessed on 24 Sept).

[806] This seems to illustrate the power of priming. Given what we know about the San Agustin claim of a UFO crash (See Appendix B) it would seem that this conclusion is based more on the belief structure of those attempting to read the memo than it is on the reality of the situation.

> SAFE TALK... FOR MEANING OF STORY AND MISSION... WEATHER BALLOONS SENT ON THE... AND LAND... rOVER CREWS... [SIGNED]... TEMPLE.[807]

If what they found was accurate, and others could corroborate what they had seen, then it was a breakthrough on the Roswell case. Here was a document with an indisputable provenance. General Ramey was holding it in his hand, and copies of the photograph put out over the INS wire provided a time and a date.[808]

Further Analyses of the Ramey Memo

Others began to request copies of the pictures from the Special Collections held at the University of Texas at Arlington Library. They brought their expertise to bear on the message in Ramey's hand. To the delight of many, they could also see letters and images as suggested first by Sparks and then by Johnson and his team. The problem was that many of those doing the work were not seeing the same things as had been published.

For example, the telephone logo that Johnson's team saw looked more like a smudge on the paper than anything else. One researcher said that the telephone looked more like the Liberty Bell as seen on the back of a Franklin half dollar. It was, as Russ Estes described it, "Faces in the clouds."

Neil Morris, a technician who works for the University of Manchester in England, began to work on the message as part of the team created by Johnson. His interpretation of the symbols did not agree exactly with that made by other

---

[807] Johnson, J. B. Roswell photos revisited to be aired on FOX-TV network. Internet press release. 24 September. (See UFO Updates at http://www.ufomind.com, accessed on 24 Sept).

[808] Randle, Kevin and Donald Schmitt. *UFO Crash at Roswell*. New York: Avon Books, 1991, pp. 71 – 79; Randle, Kevin and Donald Schmitt. *The Truth about the UFO Crash at Roswell*. New York: M. Evans and Company, 1994, pp. 48 – 49, 52, 168; Randle, *Encyclopedia*, pp. 195 – 202, 293 – 306.

components of Johnson's team. He did do one thing that was beneficial and that was breakdown the message line by line so that it would be easy to follow his interpretation. He used capital letters to represent the parts of the message of which he was sure, lower case letters to represent his best guess of some letters, an asterisk to denote a letter he couldn't decipher, and a dash where there was little more than a smudge on the message.

Morris' interpretation of the message was:

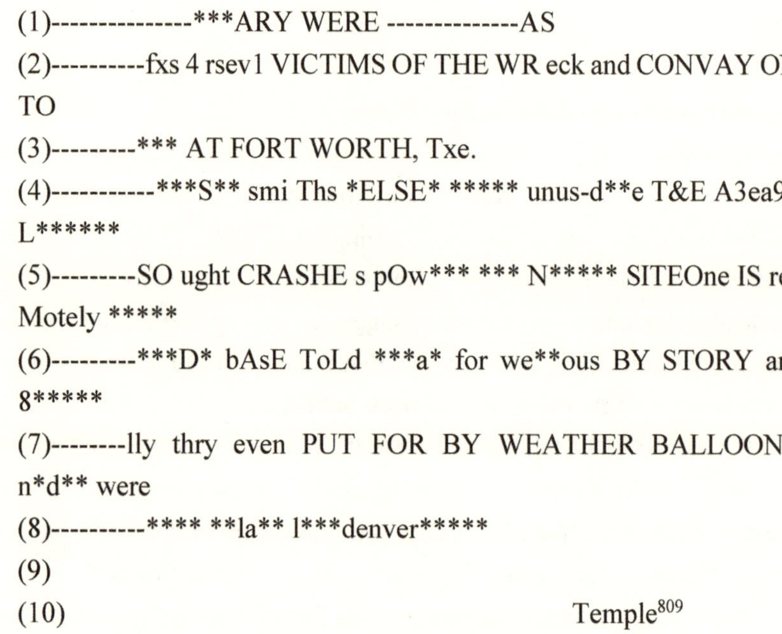

(1)--------------***ARY WERE -------------AS

(2)---------fxs 4 rsev1 VICTIMS OF THE WR eck and CONVAY ON TO

(3)--------*** AT FORT WORTH, Txe.

(4)----------***S** smi Ths *ELSE* ***** unus-d**e T&E A3ea96 L******

(5)--------SO ught CRASHE s pOw*** *** N***** SITEOne IS re-Motely *****

(6)--------***D* bAsE ToLd ***a* for we**ous BY STORY are 8*****

(7)-------lly thry even PUT FOR BY WEATHER BALLOONS n*d** were

(8)---------**** **la** l***denver*****

(9)

(10) Temple[809]

It was not an exact match for what Johnson had released and in fact, went off at a couple of new angles. In the new version, while the word "victims" remains, as does the Fort Worth, Texas, nearly everything else is different. One of the major points in the Johnson version was the wording that suggested,

---

[809] Morris, N. (1998). Internet letter, 2 December. (See UFO Updates at http://www.ufomind.com, accessed on 2 Dec.).

"Emergency Powers are needed Site Two SW of Magdalena, Nmex." It suggests that those interpreting the message were seeing, to some extent, exactly what they wanted to see.

John Kirby, a researcher who is interested in the Roswell case, and who worked for a huge company in the computer field, also looked at the message. Using his expertise and equipment, he was unable to see much of anything. He did agree that the third line were the words, "At Fort Worth, Tex." The second line, which many consider the critical line, said, "…are the remains of the material you commanded we fly."[810]

In still a different version, David Rudiak suggested only a little of what others had seen. According to him, and using the same mix of capitals for what he was sure of and lower case for what he suspected, he reported the message read:

(1) --------------- officer

(2) -----(jul)y 4th the VictIMs of tHE weECK you fOrWArdEd TO The

(3) -------EaM At FORT WORTH, TEX.

(4) -------5 pM THE "DISC" they will ship [swap?] FOR A3 8th Arrived.

(5) ----or 58t(h) bom(be)r sq(?) Assit [Assess] offices? AT ROSwe(ll) AS for

(6) ---54th SAID MIStaken--------[meaning? weather? balloon?] of [is] story And said

(7) news [clip, chat, dirt] out is OF WEATHER BALLOONS which were

(8)----- Add[And, Ask] land d---------[dirt cover?] crews.

(9)

(10) rAMEy

[810] John Kirby, Letter, 7 February 1999 to Kevin Randle.

Those weren't, of course, the only interpretation that was made. Russ Estes, using a 16 x 20 print made by the University of Texas Library, applied his expertise to the examination. Estes, a professional documentarian was able to use a professional quality video camera to capture the image. Then using his huge computer and a variety of software programs, he examined the message every way that he could think of including a jeweler's loupe, magnifying glass and microscope. He said that he could see nothing that he would be willing to swear to in a court.[811] He said there was simply nothing there to see.

Pressed on the point, he did say that he could make a "best guess" about the images on the message. Looking at an 8 x 10 enlargement of the message area, using the same techniques, he could see with a limited amount of confidence, "Fort Work, Tex." On the line below that where one group saw "Disk" and another saw, "ELSE," Estes believed he saw, "ELA*"[812]. He did say that it made no sense to him, just that was what the ambiguous smudges that everyone was attempting to make into words looked like to him.

As for the signature block, he could see nothing that resembled either of the claims. At best, there might have been an "M" in the middle of the word, and the possibility of an "LE" at the end. That gave the nod to "Temple" but Estes said it was more "Faces in the clouds."

Schmitt, Tom Carey and Don Burleson[813] came up with their own interpretations of the memo. Burleson, in the January 7, 2000, issue of *Vision*, a monthly magazine published by the *Roswell Daily Record*, wrote, "A number of attempts have been made to read the Ramey letter. Quite frankly, most of these attempts are amateurish, and even some ufologists have concluded that

---

[811] Russ Estes in unrecorded interview with Kevin Randle, 1998.

[812] Ibid.

[813] Schmitt and Carey were working together on one interpretation while Burleson, although in communication with them, was working on his own.

there is nothing in the Ramey image that advances the case for the Roswell incident. They are MISTAKEN.[814]

Burleson stated that he had spent a year working on deciphering the Ramey memo. He claimed that he had the advantage of being the director of a computer lab with a background in cryptanalysis. According to him, "I'm quite used to reading things that I wasn't meant to read."[815]

Burleson wrote that he had been using several excellent computer image enhancement software packages, "including LUCIS, the most advanced software used today is such fields as microscopy."[816] It should be noted that the Ramey memo is not an encrypted message but a plain text message that is obscured by the distance to and the angle of the camera.

Interestingly, the interpretation of the memo, as given by Burleson, and credited to Schmitt and Carey, does not agree with what Carey himself had written. In a publicly posted email dated March 29, 2000, Carey said that the "take" on the Ramey memo is that of Carey and Schmitt and not "Burlson [sic] or anyone else… All of us continue to work on the memo as best we can, so there will be no doubt be more to say in the future."[817]

Estes pointed out, as had others, that the message was a teletype rather than something from a typewriter. Given that, the message would almost have had to be in all capital letters because many of the teletype machines had no capability for lower case letters.

Stan Friedman contacted Rob Belyea, the owner of ProLab, to examine high resolution scans made of the negative. Friedman paid someone to take the

---

[814] Burleson, Donald. "Looking up." *Roswell Daily Record Vision*, June 7, 2000. p. 22; Randle, Encyclopedia, pp. 299 -302.

[815] Ibid. p. 22.

[816] In should be noted that Burleson wrote those words more than a decade and a half ago. While the programs were, at that time, the most advanced, they have been superseded by other, more powerful programs and means of analysis today.

[817] Carey, Tom, publically posted email, March 29, 2000.

negative from the Special Collections and have a computer lab make the scans. Belyea said that he couldn't spend hours examining the message but that he could rule out or confirm the interpretations made by others by using his software to decide on character count and combinations of letters.[818] It was not at all unlike the work being done by Russ Estes.

While Friedman stood on the sidelines watching and not commenting on the research, Belyea did say that he could not see "Magdalena" in the text. He did say, "They're pulling off all sorts of [readings], but they're making some of it up."[819]

Estes had said much the same thing but much more eloquently when he suggested it was "Faces in the clouds." He added, "Sorry, but I just can't see any of these things."

There is an additional problem, only partially addressed in the search of the message. This was a military message sent to a military installation which means there would have been some military jargon in it. The attempts at reading it have failed to account for any military jargon. The closest is Rudiak's attempt to place military unit designations into the message. Rudiak noted that what he thought as "5 PM" made no sense because the military would have used the twenty-four hour clock and it would have said, "1700 Hrs" rather than "5 PM."[820] That is a valid point.

What it boils down to is that there is no consensus on what the message says, the best way to review it, or what to do next. One researcher said that it had to be assumed that the message had something to do with the Roswell case, but there really is no reason to make that assumption. The message could be

---

[818] Dull, M. (1998). "A New Brunswick Physicist is Fascinated by the Idea of Spaceships in New Mexico." *The Globe and Mail* (Toronto). December 11, 1998.

[819] Ibid.

[820] David Rudiak Internet letter to Kevin Randle, April 2, 2000.

about almost anything and the words and images being seen might be a reflection of what the researcher wants to see rather than what is actually there.[821]

Another researcher suggested that the word “victims” as it appears in the message is the critical word. To him, it “jumped off the page.” The problem is that those looking at the message do not see it as a universal. One man said that he thought it was “remains.” Estes noted that it seemed to be a mix of upper and lower case letters with those doing the viewing seeing what they wanted to see. To Estes the first letter looked more like a “P” than a “V”. He noted that there seemed to be a lower case “I” in the word, and that the last letter looked more like an italic “5" than it did an “S”.

There are those who suggested that the original negative should have been used rather than a photographic print, but the truth is, the negative had been handled so much that it had acquired a bit of dirt and debris.[822] The scan of the negative did not clarify the message to any great extent. In the end, it was an attempt to read letters and words that were sometimes vague to the point of being little more than dark smudges. The message was read in the light of interpretation of the person doing the reading and his or her belief of what it should say.

There is a final complication with the Ramey memo. Johnson, according to some, claimed that he had handed the message to Ramey.[823] That confuses the source of the document that Ramey is holding suggesting that Johnson brought the document into the office. Johnson said that he had received it at

---

[821] Though if some of the interpretations are correct and the words, “weather balloon, Fort Wort, Texas, and disk” are in fact in them, then the consensus would be that this refers to the events transpiring in Roswell.

[822] Those who made the new, high resolution scans in 2015 had requested permission to clean the negative, assuring those at the University of Texas – Arlington that no damage would be caused. That permission was denied, but upon examination of the negative, it was decided that a cleaning would not have improved the overall quality of the scans.

[823] Dennis Balthascr. Internet letter to Kevin Randle, June 9, 2001.

the newspaper office, which suggests that it was one of the teletype messages that had been sent to the newspaper over the news wire that said debris was being sent to Fort Worth from Roswell.[824] If the document Ramey is holding was provided as a prop as Johnson said, then it could relate to the Roswell crash but would be from a civilian source. It would do nothing to establish the extraterrestrial nature of the event. However, when Johnson was challenged on this point, he then changed his mind and said that he had not brought the teletype message into Ramey's office. The events have undergone an evolution from the first interviews conducted in.[825]

Greenwood made additional attempts to read part of the memo. He argued that it more closely matched that transmitted over news wires that it did military teletype communications.[826] He did note that Johnson had said that he had brought the document into Ramey's office and handed it to him, which would mean that it was a news wire teletype rather than a military one. Johnson, however, soon retracted the claim.[827]

Greenwood noted that in some of the phrases in which there was general agreement, were also comment to news reports published at the same time. The phrase, "AT FORT WORTH, TEX," appeared in newspapers just that way. The *Nevada State Journal* on July 9, 1947, said, "…the commanding

---

[824] Ibid.

[825] Houran, James and Kevin Randle. "A message in a bottle: Confounds in deciphering the Ramey memo from the Roswell UFO case." *Journal of Scientific Exploration*, 16, (2002) 45 – 66; Randle, *Encyclopedia*, 293 – 306; Randle, Kevin and Donald Schmitt. *UFO Crash at Roswell*. New York: Avon Books, 1991, pp. 71 – 79.

[826] Greenwood, B. "An observation on the Ramey memo." *U.F.O. Historical Revue*, 2004, 1 – 8

[827] Sparks, B. CUFON response to Barry Greenwood. (See http://www.cufon.org/contributors/Sparks/Sparks_Rebut_Ramey_Message.pdf Accessed 12 April 2015.); See also Balthaser, D. www.truthseekeratroswell.com/interview_James_Bond_Johnson.html. 2001.

general of the 8th air force at Fort Worth, Tex."[828] To Greenwood this seemed to be additional evidence that Ramey was holding a copy of the newspaper teletype handed him by Johnson rather than a classified message that had been delivered to his office.

There was another aspect to this. Greenwood argued that the phrasing in the memo was important. Nearly all military teletype messages of the era did not use punctuation marks but rather wrote out them as "CMA" (comma) and PD (period). He wrote, "The most significant difference is that while newspapers used civilian time formats (AM, PM), the military used "Zulu," or universal 24-hour time for their endings."[829]

In 2009 Greenwood began another examination of the Ramey Memo. Once again he was able to see "AT FORT WORTH, TEX." In the next line, he saw the term, "The 'DISC'" which also agreed with the consensus. It was in the next line down that he made the important change. He noticed that the letters, "GHT" seemed to stand out. Most of those attempting to read the memo interpreted this to be the end of "SOUGHT."[830] Greenwood wrote:

> Having previously read clips in between pondering the photo [Ramey memo], I went back and flipped through it again. There was a press clip from the San Mateo CA *Times* of July 8th. Late edition papers for the 8th had carried the breaking Roswell debris news. Reading down the clip I saw this: "Lt. Warren Haught, public information officer at Roswell said…" And the quote continued to his press release. "HAUGHT"

---

[828] Greenwood, B. "An observation on the Ramey memo." *U.F.O. Historical Revue*, 2004.

[829] Greenwood, B. "An observation on the Ramey memo." *U.F.O. Historical Revue*, 2004, 1 – 8; Randle, *Encyclopedia*, pp. 293 – 306.

[830] Morris, N. Internet letter, December 2, 1998. (See UFO Updates at http://www.ufomind.com, accessed on 2 Dec.).

> stood out like a sore thumb. It was a six-letter word with a "GHT" ending in an article related to Roswell....
>
> In the Ramey document [Greenwood's name for the memo], we don't see the word "Warren" clearly in the text. But... I've determined that the area before "HAUGHT" is a six-letter word and, based on the use of the word, "HAUGHT" in the press coverage, "WARREN" is the most likely fit in that area.[831]

Greenwood's interpretation was not well received by others who attempted to read the memo. His suggestion that it was a newspaper teletype was rejected by other researchers such as Brad Sparks. He noted that in a review of military messages from the era, the showed that contrary to Greenwood, the use of periods and commas rather than abbreviations for them were sometimes found in military teletype messages.[832]

Sparks suggested that the memo might be a "general to general" message which is sometimes referred to as a "back channel." These would be more informal than official communications between commands and were often signed with the originating officer's name rather than the normal date/time group.[833]

The argument made by Sparks was that the memo was not a civilian teletype message brought to Ramey by *Star-Telegram* reporter Bond Johnson. It

---

[831]Greenwood, Barry. "Ramey Memo Redux – Line 5," *U.F.O. Historical Revue*, September, 2009, pp. 5 – 19.

[832] Sparks, Brad. CUFON response to Barry Greenwood. (See http://www.cufon.org/contributors/Sparks/Sparks_Rebut_Ramey_Message.pdf Accessed 12 April 2015.)

[833] Ibid.

was, in fact, a military memo and that it referred to the events that had transpired outside of Roswell. All this demonstrates just how convoluted the attempts to read the memo have become.

Although it had been argued that the memo might not be relevant, Carey has said that the message must refer to the Roswell case because Ramey is holding it while Johnson was in his office to interview and photograph him about the find near Roswell.[834] Sparks reinforced this idea by examining the words in the message and suggesting that the "the disc," "Roswell, NMEX," and "weather balloons," are evidence that the message does refer to the events there.[835] It could be suggested, however, that the words and images being reported by various researchers might be a reflection of what that researcher wanted to see rather than what was actually seen on the document.[836]

Testing for Priming in Reading the Ramey Memo

Given that observation James Houran was interested in researching the variables that guided those interpretations of what was an obviously ambiguous stimuli.[837] The Ramey memo is ambiguous and it seems clear that the bias of the researchers has crept into their analyses. If the document could be more

---

[834] Tom Carey email to Kevin Randle, 1998.

[835] Sparks, Brad. CUFON response to Barry Greenwood. (See http://www.cufon.org/contributors/Sparks/Sparks_Rebut_Ramey_Message.pdf (Accessed 12 April 2015.)

[836] Horan and Randle. "Message in a Bottle." pp. 45 -66.

[837] Houran, J. "Tolerance of ambiguity and the perception of UFOs." *Perceptual and Motor Skills,* (1997), 85, 973 – 974; Houran, J. Toward a psychology of 'entity encounter experiences.' *Journal of the Society for Psychical Research,* (2000) *64*, 141 – 158: Houran, J. and Williams, C. Relation of tolerance of ambiguity to global and specific paranormal experience. *Psychological Reports*, 83, (1998) 807 – 818; Randle, *Encyclopedia*, 293 – 306.

easily interpreted, then this would be a simple task with a consensus of its contents but as demonstrated, even those who have spent months and years in their research do not agree in their interpretations.[838]

They constructed an experiment to test this hypothesis. They performed three related studies in which three groups of self-selected participants were asked to decipher the Ramey memo. The participants were given one of three possible scenarios: that the memo dealt with the Roswell UFO crash; that it dealt with the testing of an atomic bomb or they were told nothing about the contents. The expectation was that each condition would elicit significant differences in the participants' interpretations.[839] They were also interested to see if there was significant agreement in the identity of words in the same locations, regardless of the suggested condition.[840]

The participants studied the memo believing that it had something to do with the Roswell crash for an average of twenty minutes. Those who had been told it was about an atomic bomb averaged sixteen minutes and those who had been told nothing spent fourteen minutes. There were some words seen across all three test conditions. These included "Fort Worth TEX," "Story" and "Balloons". Interestingly, those told the memo was about atomic testing reported seeing "Glasses, Morning, Flash, Atomic, Laboratory, and Land". Those who were given no information only saw "Fort Worth, TEX, Flew, Story, and Balloons".[841]

In the discussion section of the paper prepared for the *Journal of Scientific Exploration*, Houran and Randle wrote:

> The surprisingly high agreement between our participants and previous investigators on specific words in identical locations

---

[838] Houran and Randle. "A message in a bottle." pp. 45 – 66

[839] Ibid.

[840] Ibid.

[841] Ibid.

> in the Ramey memo suggests that some of the document is indeed legible, even without computer enhancement. However, the meaning or context of those words remains ambiguous because the degree of interpretation of the document is strongly influenced by suggestion effects and the interpreter's cognitive style. We are inclined to believe that such effects have also tainted the previous studies on the memo using sophisticated software because there appears to be weak interpreter reliability among the earlier analysts.

There were suggestions for a new study of the negative and attempts to read the memo. According to Houran and Randle:

> First, to be methodologically consistent we recommend that standardized computer enhancement be used on the best raw data that we have using comparable software programs. Analysis should be conducted by at least three independent and blind laboratories that specialized in the area of reading and transcribing archival documents. Their only motivation should be payment for providing professional and objective reports. The laboratories could be provided all available scans of the document... With this triangulation approach, we can reasonably estimate the inter-rater reliability (and hence validity) of the resulting interpretations (i.e., do the laboratories show statistically significant agreement on specific words in precise locations in the text)...

There are problems with these ideas. First, there is not adequate funding for such a project. These laboratories are expensive and the analyses would be expensive. Given the situation, it is nearly impossible to find the money to pay the labs or supply the various documents needed.

Secondly, and more importantly, it is nearly impossible to ensure that the labs would not know what was being asked. The Internet contains the information and a search would not require much time or effort. As the analysis was being arranged for this study, the participants were provided with no information about the source of the document, only that it was a photograph that required scrutiny. They discovered the source of the photograph and nearly severed relations with members of the team.

The Newest Analysis

The research on the Ramey stagnated for a number of years. The scans used for attempted readings had not been upgraded as the technology improved, and no one had examined the negative in that time. One man interested in the memo approached a number of experts in photographic enhancement to ask for advice. It was recommended to:

> ...inspect and re-image the original film negative using a mix of modern analog & digital recording techniques using a digital biological microscope; high-resolution recording film and micro & marco lenses onto a modern digital camera sensor. I sought advice from Mr. J. Morelock in Memphis [TN] USA for his earlier pioneering research work & experience in the development of color micro-film.
>
> There [University of Texas at Arlington], with the assistance of Library Staff and under strict conditions of access and handling of the original film & print materials, work as described commenced on the 21st of April 2015.[842]

---

[842] Schollum, Simon. "Covering Report: Roswell Negatives." The Author, May 8, 2015.

Methodology:[843]

The aim of a direct inspection & re-recording of the negative was to:

- Establish physical condition of the negative/s
- Establish definition, resolution and clarity of target
- Provide a viewing environment for direct reading of text
- Distinguish film base + Fog versus image density/s
- Define silver particles forming individual character-forms
- Identify silver particles (bleed) not forming individual character forms (font letters) - (to be sculpted away from character forms to enhance readability)
- Identify recurring characters among lines of text (aid to readability)
- Identify any 'recurring flaws' or mechanical 'signatures' among fonts (aid to integrity & readability)
- Determine which details are candidates for enhancement

The dual purpose for re-inspecting the negative was to estimate the extent, or whether at all, sufficient information exists in the original to warrant further analysis and if so, to develop a methodology seeking to apply proven imaging practices to render better images of the text.

Visual inspection of film negatives disclosed signs of normal and robust handling in the form of (minor) chemical stains, dust particles and scratches consistent with the age & handling these negatives have been subjected to.

---

[843] The Methodology was created by Simon Schollum for the investigation, examination and scanning of the negative of the memo taken in Ramey's office and housed at the University of Texas at Arlington. This section of this Appendix is taken from portions of Schollum's analysis and explains exactly what was done and how it was accomplished.

Observations and Conditions of the Negative

The densities of the emulsion layers appeared well 'fixed' and readable with no significant damage or degradation of the area of text (memo) forming the purpose of the examination.

These materials are in professional curatorship at the University of Texas at Arlington Special Collections and their longevity assured in their current location. Observable damage to the negatives is consistent both with their age and use prior to being preserved by the Special Collections Library. In particular, the time pressures and techniques of newspaper photographers often required less than optimal processing & drying times before being printed to meet short publication deadlines. Damage consistent with this practice is present.

Exposure levels of the film recorded by flash were adequate and no subject or camera shake evident. The camera was well focused on critical parts of the scene and the 'memo' within the focus zone set by the photographer and diaphragm.

With these negatives in relatively good condition, well exposed and processed and professionally preserved - the problem of whether the text can be read is one mainly of scale. The height & width of any font relative to the size and distribution of the silver halides on the film is the main determinant of whether individual letter forms can be identified and contribute to a full or partial reading of the memo.

For purposes of illustration, the digital file dimensions for the full frame 4x5 negative printed above is 3663.05 by 2743.05 pixels. Whereas the message length is a mere 148.5 pixels wide.

To image the memo in isolation, a Nikon SMZ1500 biological microscope ably operated by a talented Graduate Student at Arlington's School of Engineering was used to view and digitally record aspects of the 'memo' negative.

The magnification factor of the microscope gave better insights into the granular distribution of halides comprising letter forms. Unfortunately the resolution of the images was less than ideal and full copies of original images and enhanced versions are attached.

Negatives were then examined and recorded using the Special Collections digital microfiche system. David Rudiak with assistance of Library staff took a series of image recordings with bracketed exposures and raw & enhanced copies of these files are attached.

Original negatives were imaged using a Canon digital camera with both a macro and micro lens in Canon's proprietary format. I then recorded negatives using a Canon film camera with both a macro and micro lens onto ultra-high resolution Kodak recording film.

Exposures (film & digital) were bracketed and copies of raw and enhanced versions of these files are attached. Films have been sent to Wellington, New Zealand for processing using Kodak proprietary.

Method & Results:

Images from microscope; microfiche and digital camera have been processed into groups of RAW & enhanced files. High Dynamic Range photography has been used to harness the range of tones present with negatives and in particular the 'memo'.

The products of HDR imaging have been processed into working files in the form of image stacks where the interaction among pixels among layers has been influenced variously to:

a) Reduce the visual interference of film grain within the emulsion impinging on the character forms (fonts)

b) Separate out the tones of the paper base from the fonts used in the memo to suppress background interference

c) To isolate and ('lift') tonal values of the fonts away from the background in order to render character forms more clearly.

Resulting files provide a range of image states ranging from low contrast grey tones to contrasty separated tones for interpretive evaluation.

(Please note: pixel destructive approaches using curves or levels has not been used).

Direct examination of the negative rather than viewing positive generations has allowed a clearer picture of the grain structure down to a focused molecular level. No further increase in visual readability can be achieved. Any additional interpretation of the target message will more than likely be left to the application of adequate search algorithms to differentiate between the type fonts and message background.[844]

Conclusions

Although the process is on-going, and new technology was applied in an attempt to clarify the text, the results were less than satisfying. While it seemed that the image was slightly clearer, the difference wasn't enough to make any sort of definitive statement what the memo said, the source of it meaning whether it was military or civilian, or if it was documentation of an alien spaceship crash in New Mexico.

It is quite clear that the material in the photographs are the remains of a neoprene balloon, and the very degraded pieces of a rawin radar reflector made of aluminum foil and balsa wood sticks. There is nothing in the photographs to suggest that the material had laid out in the high desert for more than a month, nor was there any dirt clinging to it. There was no evidence of the strings or the material used to tie a Mogul array together, nor did there seem to be more material than a single balloon and a single target.

---

[844] This is the end of the report prepared by Simon Schollum.

While those points are interesting, they are enough to prove they are not part of a Project Mogul launch, though it certainly suggests that. The pictures, however, do not provide enough clear information to identify a Mogul balloon, and there is testimony that this debris was not part of a Mogul array.

The memo itself is not clear enough to be read with any degree of certainty so all that is left are the various interpretations of the memo suggesting something extraordinary but without the proof that it was. At this point in early 2016, all that can be said is the testing will continue with the expectation that improvements to the software used to analyze the scans will improve to the point where the memo can be read. Today that ability does not exist.

Recommendations for Further Study

An interesting parallel developed in the first months of 2015. Two Kodachrome slides taken in the same time frame as the photograph of General Ramey were discovered years earlier. The slides were color and showed what many thought might be the body of an alien creature. In that photograph there was some sort of placard that was angled away from the camera and that held a legend that was almost legible.

During the three years of the investigation into the Roswell Slides, there were several attempts to read that placard. In the Mexico City presentation held on May 5, 2015, Donald Schmitt reinforced the claim that the slides had been subjected to rigorous testing by experts in the field of photography. According to the newspaper accounts from Mexico City, "Exhaustive investigations by other photographic and medical experts have concluded that the photos are genuine. The experts list presented at the Mexico City event include Dr. David Rudiak, an expert in photographic analysis, Dr. Donald Burleson, a specialist in computer enhancement; Ray Downing, materials expert from the Studio MacBeth, New York; Col Jeffrey Thau associated with the Pentagon's Photo Interpretation Department, and Prof Rod Slemmons, a former Director of the Chicago Museum of Contemporary Photography."

Parts of the statement, however, were not exactly accurate. David Rudiak is not an expert in photographic analysis, but has experience in attempting to read the Ramey Memo as noted. Because of that, he was asked to look at the placard near the body but was unable to unscramble or deblur the image.

Colonel Jeffrey Thau is a retired Air Force officer who once had offices at both Wright-Patterson Air Force Base and the Pentagon. The Photo Interpretation Department had been moved from the Pentagon to Fort Meade, Maryland. Their expertise was not in attempting to read messages that were obscured but in interpreting photo intelligence of various kinds including ground based military facilities and movements. It seems that this failed attempt to read the placard wasn't actually an attempt by the experts at the Pentagon or Fort Meade, but friends seeing if they could make out anything on the placard as a favor to Colonel Thau.

Adam Dew posted to his web site, in the hours after the presentation in Mexico City, a higher quality scan of one of the slides and a number of different people around the world downloaded it, beginning to work to deblur the placard. Within forty-eight hours of the presentation, these independent researchers had been able to read the placard. The first line said, "Mummified body of a two year old boy."

Although some of the words were still obscured, the rest of the placard added details that confirmed the first line. According to Curt Collins at Blue Blurry Lines, the placard said:

> MUMMIFIED BODY OF TWO YEAR OLD BOY
> At the time of burial the body was clothed in a xxx-xxx cotton
> shirt. Burial wrappings consisted of these small cotton blankets.
> Loaned by the Mr. Xxxxxx, San Francisco, California

Tony Bragalia confirmed the words on the placard, and identified the location in which the photograph had been taken. He wrote:

> Working with a colleague from Europe and with the text of the de-blurred placard, I discovered… that this interpretation of the text was correct. Found in the September 1938 Volume VIII, Number 1 *Mesa Verde Notes* that was published by the

National Park Service was an article that definitively solves the mystery of the "Roswell Slides." In paragraph four of the section of the publication entitled *Around The Mesa* was found this:

> A splendid mummy was received by the Park Museum recently when Mr. S.L. Palmer Jr. of San Francisco returned one that his father had taken from the ruins in 1894. The mummy is that of a two year old boy and is in an excellent state of preservation. At the time of burial the body was clad in a slip-over cotton shirt and three small cotton blankets. Fragments of these are still on the mummy.

The deblurring process was not available when most of the work had been done on the memo. With these programs in wide spread use, the high resolution scans of the Ramey Memo made from the original negative should be made available to those who can duplicate the process. Given that, it will be possible for the work done to be enhanced, replicated and confirmed.

Bibliography

"AAF Finds 'Saucer'." *Boston Herald*, July 9, 1947, p.2

Air Intelligence Report No. 100-201-79, "Analysis of Flying Object Incidents in the U.S., 10 December 1948.

Aldrich, Jan. "Investigating the Ghost Rockets." *International UFO Reporter* 23,4 (Winter 1998): 9 – 14.

------. "Project 1947: An Inquiry into the Beginning of the UFO Era." *International UFO Reporter* 21,2 (Summer 1996): 18 – 20.

Alexander, John B. *UFOs: Myths, Conspiracies, and Realities*. New York: St. Martin's Press, 2011.

Allan, Christopher D. "Dubious Truth about the Roswell Crash," *International UFO Reporter* 19, 3 (May/June 1994), 12 – 14.

Andrus, Walt. "Air Intelligence Report No. 100-203-79." *MUFON UFO Journal* 207 (July 1985): 3 – 19.

"AP Wires Burn with 'Captured Disk' Story." *Daily Illini*, July 9, 1947, p. 5.

"Army, Navy Move on 'Flying Disc' Rumors." *El Paso Herald Post*, July 9, 1947: pp. 1, 11.

"Army Says New Mexico 'Disc' Wind Balloon." *Oregon Journal*, July 9, 1947.

ATIC UFO Briefing, April 1952, Project Blue Book Files.

Barret, William P. “Now Where was it Those Alien Crashed.” *Crosswinds*, August 1996: pp. 14-16, 34.

Barker, Gray. "America's Captured Flying Saucers - The Cover-up of the Century," *UFO Report* 4, 1 (May 1977). 32 – 35, 64, 66 – 73.

------. "Archives Reveal More Crashed Saucers." *Gray Barker's Newsletter* (14 March 1982). 5—6.

------. “Chasing Flying Saucers.” Gray Barker’s Newsletter 17 (December 1960), 22 – 28.

------. "Von Poppen Update." *Gray Barker's Newsletter* (December 1982): 8.

Berliner, Don. “The Ghost Rockets of Sweden.” *Official UFO* 1,11 (October 1976): 30 – 31, 60 – 64.

Berliner, Don with Marie Galbraith and Antonio Huneeus. *Unidentified Flying Objects Briefing Document: The Best Evidence Available*. Washington, D.C. 1995. 33 – 35.

Berlitz, Charles and Moore, William L. *The Roswell Incident*. New York: Berkley, 1988.

"Big Fire in the Sky: A Burning Meteor," *New York Herald Tribune* (10 December 1965).

Binder, Otto. *What We Really Know About Flying Saucers*. Greenwich, Conn.: Fawcett Gold Medal, 1967.

------. *Flying Saucers Are Watching Us*. New York: Tower, 1968.

------. "The Secret Warehouse of UFO Proof." *UFO Report*, 2,2 (Winter 1974),16 –19, 50, 52.

Birnes, William J. with Mark Magruder, Merrit Magruder and Natalie Magruder. "Squiggly." *UFO Magazine* 21,4 (June 2006): pp. 32-39.

Bloecher, Ted. *Report on the UFO Wave of 1947*. Washington, D.C.: Author, 1967.

Bloecher, Ted and Cerny, Paul. "The Cisco Grove Bow and Arrow Case of 1964." *International UFO Reporter* 20,5 (Winter 1995): 16 – 22, 32.

Blum, Howard. *Out There: The Government's Secret Quest for Extraterrestrials*. New York: Simon and Schuster, 1991.

Blum, Ralph, with Blum, Judy. *Beyond Earth: Man's Contact with UFOs*. New York: Bantam Books, 1974.

Bowen, Charles (ed). *The Humanoids*. Chicago: Henry Regency, 1969.

Bourdais, Gildas. *Roswell.* Agnieres, France: JMG Editions, 2004.

Braenne, Ole Jonny. "Legend of the Spitzbergen Saucer." *International UFO Reporter* 17,6 (November/December 1992): pp. 14 – 20.

Brew, John Otis and Danson, Edward B. "The 1947 Reconnaissance and the Proposed Upper Gila Expedition of the Peabody Museum of Harvard University." *El Palacio* (July 1948): 211-222.

Briefing Document: Operation Majestic 12, November 18, 1952.

"Brilliant Red Explosion Flares in Las Vegas Sky," *Las Vegas Sun* April 19, 1962, 1.

Britton, Jack, and Washington, George, Jr. *Military Shoulder Patches of the United States Armed Forces*. Tulsa, Okla.: MCN Press, 1985.

Brown, Eunice H. *White Sands History*. White Sands, N.M.: Public Affairs Office, 1959.

Bullard, Thomas E. *The Myth and Mystery of UFOs*. Lawrence, KS: University of Kansas Press, 2010.

Burleson, Donald R. "Deciphering the Ramey Memo," *International UFO Reporter* 25,2 (Summer 2000): 3 –6, 32.

Cahn, J.P. "The Flying Saucers and the Mysterious Little Men," *True* (September 1952): 17 – 19, 102 --12.

------. "Flying Saucer Swindlers." *True*. (August 1956): 36 – 37, 69 –72.

Carey, Tom. "Will the Real Sheridan Cavitt Please Stand Up." *International UFO Reporter*, Fall 1998: pp. 14-21.

Carey, Thomas J. and Schmitt, Donald R. *Witness to Roswell Revised and Expanded*. Pompton Plains NJ: New Page Books, 2009.

Catoe, Lynn E. *UFOs and Related Subjects: An Annotated Bibliography*. Washington, D.C.: Government Printing Office, 1969.

Cerny, Pau and Neville, Robert. "U.S. Navy 1942 Sighting." *MUFON UFO Journal* 185 (July 1983): 14 – 15.

Chester, Keith. *Strange Company: Military Encounters with UFOs in WW II*. San Antonio, TX: Anomalist Books, 2007.

Clark, Jerome. "The Great Unidentified Airship Scare." *Official UFO* (November, 1976).

------. "The Great Crashed Saucer Debate." *UFO Report* (October 1980): 16-19, 74,76.

------. "Crashed Saucers - Another View." *Saga's UFO Annual 1981* (1981). 44 – 47, 66.

------. *UFO's in the 1980s*. Detroit: Apogee, 1990.

------. “The Great Crashed Saucer Debate,” *UFO Report* 8,5 (February 1980), 16 – 19, 74, 76.

------. "Crash Landings." *Omni* (December 1990): 92-91.

------. "UFO Reporters. (MJ-12)". *Fate* (December 1990).

------. “A Catalog of Early Crash Claims,” *International UFO Reporter*, (July/August 1993), 7 – 14.

------. *The UFO Encyclopedia*. Detroit: Omnigraphics, 1998.

------. *Hidden Realms, Lost Civilizations and Beings from Other Worlds*, Detroit: Visible Ink Press 2010.

Committee on Science and Astronautics, report, 1961.

Cohen, Daniel. *Encyclopedia of the Strange*. New York: Avon, 1987.

------. *UFOs - The Third Wave*. New York: Evans, 1988.

Cooper, Milton William. *Behold a Pale Horse*. Sedona, AZ: Light Technology, 1991.

Corley, Linda. "For the Sake of My Country." *In MUFON 2000 International UFO Symposium Proceedings,* Sequin, TX: MUFON, 2000: 111 – 129.

"Could the Scully Story Be True?" *The Saucerian Bulletin* 1,2 (May 1956), 1.

Creighton, Gordon. "Close Encounters of an Unthinkable and Inadmissible Kind." *Flying Saucer Review*. (July/August 1979).

------. "Further Evidence of 'Retrievals." *Flying Saucer Review*. (Jan 1980).

------. "Continuing Evidence of Retrievals of the Third Kind." *Flying Saucer Review*. January/February 1982).

------. "Top U.S. Scientist Admits Crashed UFOs." *Flying Saucer Review*. (October 1985).

Davison, Leon, ed. *Flying Saucers: An Analysis of Air Force Project Blue Book Special Report No. 14*. Clarksburg, Va.: Saucerian Press, 1971.

Davies, John K. *Cosmic Impact*. New York: St. Martin's, 1986.

Davis, Richard. "Results of a Search for Records Concerning the 1947 Crash Near Roswell, New Mexico." Washington, D.C.: GAO, 1995

Dennett, Preston. "Project Redlight: Are We Flying The Saucers Too?" *UFO Universe,* May 1990: 39.

"Disc Solution Collapses." *San Francisco Chronicle*, July 9, 1947, p. 1.

"'Disc-overy' Near Roswell Identified as Weather Balloon by FWAAF Officer,"*Fort Worth Star-Telegram*, July 9, 1947: pp. 1 – 2.

"'Disk' Near Bomb Test Site Is Just a Weather Balloon," New York Times, July 9, 1947: pp. 1 – 2.

Dobbs, D.L. "Crashed Saucers - The Mystery Continues." *UFO Report* September 1979, 28 – 31, 60 – 61.

"DoD News Releases and Fact Sheets," 1952 – 1968.

Dolan, Richard M. *UFOs and the National Security State*. Charlottesville, VA.: Hampton Roads Publishing Company, 2000.

------. *UFOs and the National Security State: The Cover-Up Exposed, 1973 – 1991*. Rochester, NY: Keyhole Publishing, 2009.

Douglas, J.V. and Henry Lee. "The Fireball of December 9, 1965 - Part II." *Royal Astronomical Society of Canada Journal* 62, no. 41.

Earley, George W. "Crashed Saucers and Pickled Aliens, Part I.) *Fate* 34, 3 (March 1981), 42 – 48.

------. "Crashed Saucers and Pickled Aliens, Part II.) *Fate* 34, 4 (April 1981), 84 – 89.

Eberhart, George, *The Roswell Report: A Historical Perspective*. Chicago: CUFOS, 1991.

Ecker, Don. "The New Tale of a Roswell Witness." UFO Magazine, 12,9 (1994): 12 – 13.

Edwards, Frank. *Flying Saucers - Here and Now*! New York: Bantam,1968.

------. *Flying Saucers - Serious Business*. New York: Bantam, 1966.

------. *Strange World*. New York: Bantam, 1964.

"Effect of the Tunguska Meteorite Explosion on the Geomagnetic Field," Office of Technical Services U.S. Department of Commerce, 21 December 1961.

Eighth Air Force Staff Directory, Texas: June 1947.

Fact Sheet, "Office of Naval Research 1952 Greenland Cosmic Ray Scientific Expedition," October 16, 1952.

Farish, Lucius and Clark, Jerome. "The Mysterious 'Foo Fighters' of World War II." *Saga's UFO Report*, 2,3 (Spring 1974) 44 – 47, 64 -66.

------. "The 'Ghost Rockets' of 1946." *Saga's UFO Report*, 2,1 (Fall 1974) 24 – 27, 62 – 64.

Fawcett, Lawrence and Barry J. Greenwood. *Clear Intent: The Government Cover-up of the UFO Experience*. Englewood Cliffs, N.J.: Prentice-Hall, 1984.

"Gen. Ramey Empties Roswell 'Saucer.'' *Roswell Daily Record*, July 9, 1947: p. 1

Gillmor, Daniel S., ed. *Scientific Study of Unidentified Flying Objects*. New York: Bantam Books, 1969.

Good, Timothy. *Above Top Secret*. New York: Morrow, 1988.

------. *The UFO Report*. New York: Avon Books, 1989.

------. *Alien Contact*. New York: Morrow, 1993.

Gordon, Stan and Vicki Cooper, Vicki. "The Kecksburg Incident." *UFO*, 6, 1 (1991): 16-19.

Gordon, Stan. "After 25 Years, New Facts on the Kecksburg, Pa. *UFO* Retrieval are Revealed." *PASU Data Exchange #15* Dec 1990): 1.

------. "Kecksburg Crash Update." *MUFON UFO Journal* (September 1989).

------. "Kecksburg Crash Update." *MUFON UFO Journal* (October,1989) 3-5,9.

------. "The Military UFO Retrieval at Kecksburg, Pennsylvania." *Pursuit*, 20, no. 4 (1987): 174-179.

Graeber, Matt "The Reality, the Hoaxes and the Legend." The Author, 2009.

Greenwell, J. Richard. "UFO Crash/Retrievals: A Critque." *MUFON UFO Journal* 153 (November 1980), 16 – 19.

Gribben, John. "Cosmic Disaster Shock." *New Scientist* (Mar 6, 1980):750-52.

"Guidance for Dealing with Space Objects Which Have Returned to Earth, Department of State Airgram, July 26, 1973.

Hall, Michael. "Was There a Second Estimate of the Situation," *International UFO Reporter*, 27,1 (Spring 2002), 10 – 14, 32.

Hall, Michael and Connors, Wendy. "Alfred Loedding: New Insight on the Man Behind Project Sign." *International UFO Reporter* 23,4 (Winter 1999): 3 – 8, 24 – 28.

Hall, Richard. "Crashed Discs - Maybe," *International UFO Reporter*, 10, 4 (July/August 1985).

------. *Uninvited Guests*. Santa Fe, NM: Aurora Press, 1988.

------. ed. *The UFO Evidence*. Washington, D.C.: NICAP, 1964.

------. "Pentagon Pantry: Is the Cupboard Bare?" *MUFON UFO Journal* 108 (November 1976): 15 – 18.

------. *Contributions of Balloon Operations to Research and Development at the Air Force Missile Development Center 1947 - 1958*. Alamogordo, NM: Office of Information Services, 1959.

"Harassed Rancher Who Located 'Sorry' He Told About It." *Roswell Daily Record*, July 9, 1947: p. 1.

Hastings, Robert. *UFOs and Nukes*. Bloomington, Ind.: Author House, 2008.

Haugland, Vern. "AF Denies Recovering Portions of 'Saucers.'" *Albuquerque New Mexican*, 23 March 1954.

Hazard, Catherine. "Did the Air Force Hush Up a Flying Saucer Crash?" *Woman's World*, (February 27, 1990): 10.

Henry, James P. and John D. Mosely "Results of the Project Mercury Ballistic and Orbital Chimpanzee Flights," *NASA SP-39*, NASA, 1963.

Hessmann, Michael and Philip Mantle. *Beyond Roswell: The Alien Autopsy Film, Area 51 and the U.S. Government Cover-up of UFOs*. New York, N.Y.: Marlowe and Company, 1991.

Hippler, Robert H. "Letter to Edward U. Condon," January 16, 1967.

"History of the Eighth Air Force, Fort Worth, Texas," (Microfilm) Air Force Archives, Maxwell Air Force Base, AL.

"History of the 509th Bomb Group, Roswell, New Mexico," (Microfilm)Air Force Archives, Maxwell Air Force Base, AL.

Hogg, Ivan U. and J. B. King. *German and Allied Secret Weapons of World War II.* London: Chartwell, 1974.

Houran, James and Randle, Kevin. "Interpreting the Ramey Memo," *International UFO Reporter*, 27, 2 (Summer 2002) 10 – 14, 26 – 27.

Humble, Ronald D. "The German Secret Weapon/UFO Connection." UFO 10,4 (July/August 1995): 21 – 25.

Huneeus, J. Antonio. "Roswell UFO Crash Update." *UFO Universe,* Winter1991): 8-13, 52, 57.

"Hunting Old and New UFOs in New Mexico." *International UFO Reporter* 7,2 (March 1982) 12 – 14.

Hurt, Wesley R. and Daniel McKnight. "Archaeology of the San Augustine Plains: A Preliminary Report." *American Antiquity* (January 1949): 172-194.

Hynek, J. Allen. *The UFO Experience: A Scientific Inquiry*. Chicago: Henry Regency, 1975.

------. *The Hynek UFO Report*. New York: Dell 1977.

Hynek, J. Allen and Jacques Vallee. *The Edge of Reality*. Chicago: Henry Regency, 1972.

"Ike and Aliens? A Few Facts about a Persistent Rumor." *Focus* 1, 2 (April 30, 1985), 1, 3 – 4.

"International Reports: Tale of Captured UFO." *UFO*, 8, 3 (1993):10-11.

Jacobs, David M. *The UFO Controversy in America.* New York: Signet, 1975.

Jeffrey, Kent. “Roswell – Anatomy of a Myth.” *The MUFON UFO Journal* 350 (June 1997): pp. 3 – 17.

Johnson, J. Bond. "'Disk-overy' Near Roswell Identified As Weather Balloon by FWAAF Officer," *Fort Worth Star-Telegram*, 9 July 1947.

Jones, William E. and Rebecca D. Minshall "Aztec, New Mexico – A Crash Story Reexamined." *International UFO Reporter* 16, 5 (September/October 1991): 11.

Keel, John. "Now It's No Secret: The Japanese 'Fugo Balloon.'" *UFO* (January/February 1991): 33 - 35.

------. *UFOs: Operation Trojan Horse*. New York: G.P. Putnam's Sons, 1970.

------. *Strange Creatures from Space and Time*. New York: Fawcett, 1970.

Kellahin, Jason. “NM Rancher Sorry He Said Anything about ‘Disc Find.’” *Albuquerque Journal*, July 9, 1947” p. 2.

Kennedy, George P. "Mercury Primates," American Institute of *Aeronautics and Astronautics* (1989).

Keyhoe, Donald E. *Flying Saucers from Outer Space*. New York: Henry Holt and Company, 1953.

------. *Aliens From Space*. New York: Signet, 1974.

Klass, Philip J. *UFOs Explained.* New York: Random House, 1974.

------. "Crash of the Crashed Saucer Claim," *Skeptical Inquirer* 10, 3 (1986).

------. *The Real Crashed-Saucer Coverup*. Amherst, NY: Prometheus Books, 1997.

Knaack, Marcelle. *Encyclopedia of U.S. Air Force Aircraft and Missile*

Systems. Washington, D.C.: Office of Air Force History, 1988.

Kent, Jeffrey. "Roswell - Anatomy of a Myth." *The MUFON UFO Journal* 350 (June 1997): pp. 3 -17,

LaPaz, Lincoln and Albert Rosenfeld. "Japan's Balloon Invasion of America," *Collier's*, January 17, 1953, 9.

Lasco, Jack. "Has the US Air Force Captured a Flying Saucer?" *Saga* (April 1967), 18 – 19, 67 – 68, 70 – 74.

Library of Congress Legislative Reference Service, "Facts about UFOs," May 1966.

"Little Frozen Aliens." *The APRO Bulletin* (January/February 1975), 5 – 6.

Lore, Gordon, and Harold H. Deneault. *Mysteries of the Skies: UFOs in Perspective*. Englewood Cliff, N.J.: Prentice-Hall, 1968.

Low, Robert J. "Letter to Lt. Col. Robert Hippler," January 27, 1967.

Maccabee, Bruce. "Hiding the Hardware." *International UFO Reporter*. (September/October 1991): 4.

-----. "Did Sheridan Cavitt Visit the Same Crash Site?" found at http://brumac.8k.com/Roswell/CavittEmptor.html.

------. "What the Admiral Knew." *International UFO Reporter.* (November/December 1986).

Mantle, Phillip. *Roswell Alien Autopsy.* Edinburg, TX: RoswellBooks, 2012.

------. “Alien Autopsy Film, R.I.P.,” *International UFO Reporter*, 32,1 (August 2008), 15 – 19.

Marcel, Jesse and Linda Marcel. *The Roswell Legacy.* Franklin Lakes, N.J.: New Page Books, 2009.

McAndrews, James. *The Roswell Report: Case Closed.* Washington, D.C.: Government Printing Office, 1997.

McCall, G. J. H. *Meteorites and their Origins.* New York: Wiley & Sons, 1973.

McClellan, Mike. "The Flying Saucer Crash of 1948 is a Hoax," *Official UFO* 1, 3 (October 1975): 36-37, 60, 62-64.

"McClellan Sub-Committee Hearings," March 1958.

"McCormack Sub-Committee Briefing," August 1958.

McDonald, Bill. "Comparing Descriptions, An Illustrated Roswell." *UFO* 8, 3 (1993): 31-36.

Moore, Charles B. "The New York University Balloon Flights During Early June, 1947," The Author, 1995.

Moore, Charles B., Benson Saler and Charles A. Ziegler. *UFO Crash at Roswell: Genesis of a Modern Myth.* Washington, D.C.: Smithsonian Institute Press, 1997.

Mueller, Robert. *Air Force Bases: Volume 1, Active Air Force Bases within the United States of American on 17 September 1982*. Washington, D.C.: Office of Air Force History, 1989.

National Security Agency. Presidential Documents. Washington, D.C.: Executive Order 12356, 1982.

Neilson, James. "'Secret U.S./UFO Structure." *UFO*, 4,1, (1989): 4-6.

"New Mexico Rancher's 'Flying Disk' Proves to Be Weather-Balloon-Kite." *Fort Worth Star-Telegram*, July 9, 1947, p. 1.

Nickell, Joe. "The Hangar 18 Tales" *Common Ground* (June 1984).

"No Reputable Dope On Disks," *Midland (Texas) Reporter Telegram,* 1 July 1947.

Norry, George. "Coast to Coast AM: Once a Hero." UFO Magazine, 21,4 (June 2006) pp. 26-27, 75.

Northrup, Stuart A. *Minerals of New Mexico*. Albuquerque: University of New Mexico, 1959.

"No Sign of 'UFO' – *NSRI*. News24, 5 May 2006

Oberg, James. "UFO Update: UFO Buffs May Be Unwitting Pawns in an Elaborate Government Charade," *Omni* 15, no. 11 September 1993: 75.

"Officers Say Disc." *Roswell Morning Dispatch*, July 9, 1947: p. 8

Packard, Pat and Terry Endres. "Riding the Roswell-go-round." *A.S.K. UFO Report* 2, (1992): 1, 1-8

Peebles, Curtis. *The Moby Dick Project.* Washington, D.C.: Smithsonian Institution Press, 1991.

------. *Watch the Skies!* New York, N.Y.: Berkley Books, 1995.

Pflock, Karl. *Roswell in Perspective*. Mt. Rainier, MD: FUFOR, 1994

------. "In Defense of Roswell Reality." *HUFON Report* (Feb. 1995) 5-7

------. "Roswell, A Cautionary Tale: Facts and Fantasies, Lessons and Legacies." In Walter H. Andrus, Jr., ed. *MUFON 1995 International UFO Symposium Proceedings* Seguin, TX: MUFON, 1990: 154-68.

------. "Roswell, The Air Force, and Us." *International UFO Reporter* (November/December 1994): 3-5, 24

------. *Roswell: Inconvenient Facts and the Will to Believe*. Amherst: N.Y.: Prometheus Books. 2001

"Press Release - Monkeynaut Baker is Memorialized," Space and Rocket Center, Huntsville, AL (December 4, 1984).

"Project Blue Book" (microfilm). RG [Record Group] 341, T-1206 National Archives, Washington, D.C.

Prytz, John M. "UFO Crashes" *Flying Saucers* (October 1969): 24-25.

Randle, Kevin D. "Mysterious Clues Left Behind by UFOs," *Saga's UFO Annual* (Summer 1972).

------. *The UFO Casebook.* New York: Warner, 1989.

------. *A History of UFO Crashes*. New York: Avon, 1995

------. *Conspiracy of Silence*. New York: Avon, 1997.

------. *Project Moon Dust*. New York: Avon, 1998.

------. *Scientific Ufology*. New York: Avon, 1999.

------. *Roswell Encyclopedia*. New York: Avon, 2000.

------. *Roswell Revisited*, Lakeville, MN: Galde Press, 2007

------. *Crash: When UFOs Fall from the Sky*, Franklin Lakes: NJ, 2010

------. "MJ-12's Fatal Flaw and Robert Willingham," *International UFO Reporter*, 33,4 (May 2011) 3 – 7.

------. *Reflections of a UFO Investigator*, San Antonio, TX: Anomalist Books, 2012.

------. *Roswell, UFOs and the Unusual*. Kindle eBooks, 2012.

Randle, Kevin D. and Schmitt, Donald R. *UFO Crash at Roswell*. New York, N.Y.: Avon, 1991.

------. *The Truth about the UFO Crash at Roswell*. New York, N.Y.: M. Evans and Company, 1994.

Randles, Jenny. *The UFO Conspiracy*. New York: Javelin, 1987.

"Report of Air Force Research Regarding the 'Roswell Incident'," July 1994.

"Rocket and Missile Firings," White Sands Proving Grounds, Jan - Jul 1947.

"Rocket Craft Encounter Revealed by World War 2 Pilot." *The UFO Investigator* 1,2 (August/September 1957): 15.

Rodeghier, Mark. "Roswell, 1989." *International UFO Reporter.* (September/October 1989): 4.

Rodeghier, Mark and Chesney, Mark. "The Air Force Report on Roswell: An Absence of Evidence." *International UFO Reporter,* September/October 1994).

Rosignoli, Guido. *The Illustrated Encyclopedia of Military Insignia of the 20th Century*. Secaucus, N.J.: Chartwell, 1986.

Russell, Eric. "Phantom Balloons Over North America," *Modern Aviation* (February 1953).

Schmitt, Donald R. "New Revelations from Roswell." In Walter H. Andrus, Jr., ed. *MUFON 1990 International UFO Symposium Proceedings* Seguin, TX: MUFON, 1990: 154-68.

------. *UFO Crash at Roswell II.* The Author, 1997: p. 60.

Schmitt, Donald R. and Randle, Kevin D. "Second Thoughts on the Barney Barnett Story." *International UFO Reporter* (May/June 1992): 4-5, 22.

------. "What Really Happened in Ramey's Office." *The MUFON UFO Journal* 276 (April 1991): pp. 3 – 9.

------. "Fort Worth, July 8, 1947: The Cover Up Begins," *International UFO Reporter* 15,2 (March/April 1990): pp. 21 – 23.

-----. "The Fort Worth Press Conference: The J. Bond Johnson Connection," *International UFO Reporter*, 15,6 (November/December 1990): pp. 5 – 16.

Shandera, Jaime. "New Revelations about Roswell Wreckage." *The MUFON UFO Journal* 273 (January 1991).

Shandera, Jaime and William Moore. "Three Hours that Shook the Press." *The MUFON UFO Journal* 269 (September 1990): pp. 3 – 9.

Slate, B. Ann "The Case of the Crippled Flying Saucer." *Saga* (April1972): 22-25, 64, 66-68, 71, 72.

Smith, Scott. "Q & A: Len Stringfield." *UFO* 6,1, (1991): 20-24.

"The Space Men at Wright-Patterson." UFO Update.

Spencer, John. *The UFO Encyclopedia.* New York: Avon, 1993.

Spencer, John and Evans, Hilary. *Phenomenon.* New York: Avon, 1988.

Steiger, Brad. *Strangers from the Skies*. New York: Award, 1966.

------. *Project Blue Book.* New York: Ballantine, 1976.

------. *The Fellowship.* New York: Dolphin Books, 1988.

Steiger, Brad and Steiger, Sherry Hanson. *The Rainbow Conspiracy.* New York: Pinnacle, 1994.

------. *Conspiracies and Secret Societies*. Canton, MI: Visible Ink Press, 2006.

------. *Real Aliens, Space Beings, and Creatures from Other Worlds.* Canton, MI: Visible Ink Press, 2011.

Stone, Clifford E. *UFO's: Let the Evidence Speak for Itself.* Calif: The Author, 1991.

------. "The U.S. Air Force's Real, Official Investigation of UFO's." private report: The Author, 1993.

Stonehill, Paul. “Former Pilot Tells of Captured UFO.” *UFO* 8,2 (March/April 1993): 10 —11.

Story, Ronald D. *The Encyclopedia of UFOs*. Garden City, New York: Doubleday, 1980.

------. *The Encyclopedia of Extraterrestrial Encounters*. New York: New American Library, 2001.

Stringfield, Leonard H. *Situation Red: The UFO Siege*! Garden City, New York: Doubleday, 1977.

------. *Inside Saucer Post... 3-0 Blue*. Cincinnati, OH: Civilian Research, Interplanetary Flying Objects, 1975.

------. *UFO Crash/Retrieval Syndrome: Status Report* II. Seguin, TX: MUFON, 1980.

------. *UFO Crash/Retrieval: Amassing the Evidence: Status Report III* Cincinnati, Ohio: The Author, 1982.

------. *UFO Crash/Retrievals: The Inner Sanctum Status Report VI*, Cincinnati, Ohio: The Author, 1991.

------. “Retrievals of the Third Kind.” In *MUFON Symposium Proceedings* (1978): 77 – 105.

------. "Roswell & the X-15: UFO Basics," *MUFON UFO Journal* No. 259 (November 1989): 3-7.

Sturrock, P.A. "UFOs - A Scientific Debate," *Science* 180 (1973): 593.

Swords, Michael, et.al. *UFOS and the Government: A Historical Inquiry*. San Antonio, TX: Anomalist Books, 2012.

-----. "Too Close for Condon: Close Encounters of the 4th Kind," *International UFO Reporter*, 28,3 (Fall 2003) 3 – 6.

Tech Bulletin, "Army Ordnance Department Guided Missile Program," Jan 1948.

Technical Report, "Unidentified Aerial Objects, Project SIGN," Feb. 1949.

Technical Report, "Unidentified Flying Objects, Project GRUDGE," August 1949.

Templeton, David., "The Uninvited," *Pittsburgh Press* (May 19, 1991): 10-15.

"The Search for Hidden Reports." *The U.F.O. Investigator* 4,5 (March 1968) 7-8.

Thomas, Dick. "'Flying Saucers' in New Mexico." *Denver Post* (May 3, 1964).

Thompson, Tina D., ed. *TRW Space Log*. Redondo Beach, Calif.: TRW 1991.

Torres, Noe and Ruben Uriarte. *The Other Roswell*. Edinburg, TX: Roswell Books, 2008.

"UFOs and Lights: 12 Aliens on Ice in Ohio?" *The News* 10 (June 1975), 14 – 15.

Unit Histories of the 509th Bomb Group for June, July and August 1947 available on Microfilm Roll No. BO679, Index 1898 AP-509-Hi 7/47.

U.S. Congress, House Committee on Armed Forces. Unidentified Flying Objects. Hearings, 89th Congress, 2nd Session, April 5, 1966. Washington D.C.: U.S. Government Printing Office, 1968.

U.S. Congress Committee on Science and Astronautics. Symposium on Unidentified Flying Objects. July 29, 1968, Hearings, Washington, D.C.: U.S. Government Printing Office, 1968.

Vallee, Jacques. *Anatomy of a Phenomenon*. New York: Ace, 1966.

------. *Challenge to Science*. New York: Ace, 1966.

------. *Dimensions*. New York: Ballantine, 1989.

------. *Revelations*. New York: Ballantine, 1991.

War Department. Meteorological Balloons (Army Technical Manual) Washington, D.C.: Government Printing Office, 1944.

Webber, Bert. Retaliation: *Japanese Attacks and Allied Countermeasures on the Pacific Coast in World War II*. Corvallis: Oregon State University Press, 1975.

Weaver, Richard L. and James McAndrew. *The Roswell Report: Fact vs Fiction in the New Mexico Desert*. Washington, D.C.: Government Printing Office, 1995.

Wilcox, Inez. "Four Years in the County Jail." Unpublished manuscript. The Author.

Wilkins, Harold T. *Flying Saucers on the Attack.* New York: Citadel, 1954.

------. *Flying Saucers Uncensored.* New York: Pyramid, 1967.

Wise, David and Ross, Thomas B. *The Invisible Government.* New York: 1964.

Wood, Robert M. "Forensic Linguistics and the Majestic Documents." In *6th Annual UFO Crash Retrieval Conference.* Broomfield, Co.: Wood and Wood Enterprises, 2008, 98 – 116.

------. "Validating the New Majestic Documents. In *MUFON Symposium Proceedings* (2000): 163 – 192.

Wood, Ryan. *Majic Eyes Only.* Broomfield, Co: Wood Enterprises, 2005

"World Round-up: South Africa: Search for Crashed UFO," Flying Saucer Review 8, 2 (March/April 1962), 24.

Zeidman, Jennie. "I Remember Blue Book." *International UFO Reporter*(March/April 1991): 7.

Coming Soon!

Case MJ-12

by

Kevin D. Randle
Lt. Col. USAR (Ret)

To fully understand the Majestic Twelve committee, the twelve men originally assigned to investigate the crash of an extraterrestrial craft outside of Roswell, New Mexico, it is necessary to understand some of the history that surrounds the UFO field that led up to the creation of the Majestic Twelve documents. We can say, without fear of contradiction that the documents *do* exist. The problem for researchers, historians, scientists, journalists and Ufologists is their authenticity. Are the MJ-12 documents, as they have come to be called, authentic, meaning, were they created by the government agency charged with the examination and the exploitation of the craft found outside Roswell, or were they created by Ufologists, or hoaxers, as a way of establishing the reality of the crash?

FOR MORE EXCITING BOOKS, E-BOOKS, AUDIOBOOKS AND MORE
J.R. ROBERTS
Bleeders
THOM REESE
TAC
CRISIS!
SEALS
MAAN MEYERS
ROBERT J. RANDISI
Henry and the IDIOTS
WHITE MALE INFANT
BARBARA D'AMATO
BOOBY TRAP
BILL PRONZINI
DON BENDELL
BARBARA D'AMATO
HARD EVIDENCE
SURVIVALIST
BENEATH THE ASHES
SUE HENRY
GHOST
COLORLAND
ANNETTE MEYERS
GROANING BOARD
10:07
THE SLEEPING
L. SYDNEY ABEL
DEMONS
BILL PRONZINI
visit us at
www.speakingvolumes.us

www.ingramcontent.com/pod-product-compliance
Lightning Source LLC
Chambersburg PA
CBHW031229090426
42742CB00007B/130

* 9 7 8 1 6 2 8 1 5 5 1 3 6 *